Contactless Vital Signs Monitoring

Contactless Vital Signs Monitoring

Edited by

Wenjin Wang
Philips Research and
Eindhoven University of Technology (TU/e)
Eindhoven, The Netherlands

Xuyu Wang
Department of Computer Science
California State University
Sacramento, CA, United States

ACADEMIC PRESS
An imprint of Elsevier

Academic Press is an imprint of Elsevier
125 London Wall, London EC2Y 5AS, United Kingdom
525 B Street, Suite 1650, San Diego, CA 92101, United States
50 Hampshire Street, 5th Floor, Cambridge, MA 02139, United States
The Boulevard, Langford Lane, Kidlington, Oxford OX5 1GB, United Kingdom

Library of Congress Cataloging-in-Publication Data
A catalog record for this book is available from the Library of Congress

British Library Cataloguing-in-Publication Data
A catalogue record for this book is available from the British Library

ISBN: 978-0-12-822281-2

For information on all Academic Press publications
visit our website at https://www.elsevier.com/books-and-journals

Publisher: Mara Conner
Acquisitions Editor: Sonnini R. Yura
Editorial Project Manager: Chiara Giglio
Production Project Manager: Nirmala Arumugam
Cover Designer: Miles Hitchen
Cover Artist: Xingyu Wang

Typeset by VTeX

Working together
to grow libraries in
developing countries

www.elsevier.com • www.bookaid.org

Contents

3. Model-based camera-PPG: pulse rate monitoring in fitness

Albertus C. den Brinker and Wenjin Wang

4. Camera-based respiration monitoring: motion and PPG-based measurement

Wenjin Wang and Albertus C. den Brinker

5. Camera-based blood oxygen measurement

Izumi Nishidate

6. Camera-based blood pressure monitoring

*Keerthana Natarajan, Mohammad Yavarimanesh, Wenjin Wang,
and Ramakrishna Mukkamala*

7. Clinical applications for imaging photoplethysmography

Sebastian Zaunseder and Stefan Rasche

8. Applications of camera-based physiological measurement beyond healthcare

Daniel McDuff

Part II
Wireless sensor-based vital signs monitoring

9. Radar-based vital signs monitoring

Jingtao Liu, Yuchen Li, and Changzhan Gu

10. Received power-based vital signs monitoring

Jie Wang, Alemayehu Solomon Abrar, and Neal Patwari

11. WiFi CSI-based vital signs monitoring

Daqing Zhang, Youwei Zeng, Fusang Zhang, and Jie Xiong

12. RFID-based vital signs monitoring

Yuanqing Zheng and Yanwen Wang

List of contributors

Alemayehu Solomon Abrar, Microsoft Azure, Redmond, WA, United States

Albertus C. den Brinker, AI Data Science & Digital Twin Department, Philips Research, Eindhoven, the Netherlands

Changhong Fu, Nanjing University of Science and Technology, Nanjing, China

Changzhan Gu, MoE Key Lab of Artificial Intelligence, AI Institute, Shanghai Jiao Tong University, Shanghai, China
MoE Key Lab of Design and EMC of High-Speed Electronic Systems, Department of Electronic Engineering, Shanghai Jiao Tong University, Shanghai, China

Hong Hong, Nanjing University of Science and Technology, Nanjing, China

Alexei A. Kamshilin, Institute of Automation and Control Processes of Far Eastern Branch of the Russian Academy of Sciences, Vladivostok, Russia
Department of Circulation Physiology, Almazov National Medical Research Center, St. Petersburg, Russia

Changzhi Li, Texas Tech University, Lubbock, TX, United States

Yuchen Li, MoE Key Lab of Artificial Intelligence, AI Institute, Shanghai Jiao Tong University, Shanghai, China
MoE Key Lab of Design and EMC of High-Speed Electronic Systems, Department of Electronic Engineering, Shanghai Jiao Tong University, Shanghai, China

Jingtao Liu, MoE Key Lab of Artificial Intelligence, AI Institute, Shanghai Jiao Tong University, Shanghai, China
MoE Key Lab of Design and EMC of High-Speed Electronic Systems, Department of Electronic Engineering, Shanghai Jiao Tong University, Shanghai, China

Oleg V. Mamontov, Department of Circulation Physiology, Almazov National Medical Research Center, St. Petersburg, Russia

Shiwen Mao, Department of Electrical and Computer Engineering, Auburn University, Auburn, AL, United States

Daniel McDuff, Microsoft Research, Redmond, WA, United States

Ramakrishna Mukkamala, Department of Bioengineering and Department of Anesthesiology and Perioperative Medicine, University of Pittsburgh, Pittsburgh, PA, United States

Keerthana Natarajan, Department of Electrical and Computer Engineering, Michigan State University, East Lansing, MI, United States

Izumi Nishidate, Graduate School of Bio-applications and Systems Engineering, Tokyo University of Agriculture and Technology, Tokyo, Japan

Neal Patwari, Washington University in St. Louis, St. Louis, MO, United States

Stefan Rasche, Faculty of Medicine Carl Gustav Carus, TU Dresden, Dortmund, Germany

Dangdang Shao, Biodesign Institue, Arizona State University, Tempe, AZ, United States

Jie Wang, Washington University in St. Louis, St. Louis, MO, United States

Wenjin Wang, Patient Care and Monitoring Department, Philips Research, Eindhoven, the Netherlands
Electrical Engineering Department, Eindhoven University of Technology, Eindhoven, the Netherlands

Xuyu Wang, Department of Computer Science, California State University, Scaramento, CA, United States

Yanwen Wang, The Hong Kong Polytechnic University, Hong Kong, China

Jie Xiong, College of Information and Computer Sciences, University of Massachusetts, Amherst, Amherst, MA, United States

Mohammad Yavarimanesh, Department of Electrical and Computer Engineering, Michigan State University, East Lansing, MI, United States

Sebastian Zaunseder, Faculty of Information Engineering, FH Dortmund, Dortmund, Germany

Youwei Zeng, School of Electronics Engineering and Computer Science, Peking University, Beijing, China

Daqing Zhang, School of Electronics Engineering and Computer Science, Peking University, Beijing, China

Fusang Zhang, School of Electronics Engineering and Computer Science, Peking University, Beijing, China

Li Zhang, Nanjing University of Science and Technology, Nanjing, China

Yuanqing Zheng, The Hong Kong Polytechnic University, Hong Kong, China

Foreword

In the Mesopotamian epic Gilgamesh written in 2600 BC, the hero-king Gilgamesh uttered the following lament at the death of his best friend Enkidu: "*I touch his heart, but it does not beat at all.*" This is the earliest reference acknowledged by historians that shows humankind understood that heartbeat is an important indicator for health and well-being and that it is measurable by (contact) sensing. Almost in the same period, the Egyptians and ancient Chinese developed various non-invasive methods to measure the vital signs of the human body, such as heart-rate assessment by pulse-taking (i.e., pressing a finger on the wrist to feel the blood pulsation, as is still common in today's Chinese Traditional Medicine), respiration measurement by breath counting, and fever detection by feeling the temperature of the forehead. Ancient civilizations already established the relationships between vital signs and health issues like deterioration or sickness, based on the knowledge built up by "observations and experiences".

With the rapid development of science and technology since the 20th century, the monitoring of vital signs no longer relies on manual inspections, but rather on biomedical sensors and wearables featured for automatic sensing and processing of physiological data, such as the extraction of heart rate from electrocardiograms, respiration rate using accelerometers, blood oxygen saturation from photoplethysmography (PPG) and body temperature using thermometers. These new sensing technologies have standardized the procedures for physiological measurement and made it reproducible in various scenarios (e.g., hospitals and homes), and, most importantly, enabled a tremendous amount of health-monitoring solutions that significantly improve the life and care of a vast population worldwide, spanning aspects of disease diagnosis, emergency detection, well-being management, etc.

The fast progress made in sensing and computing fuels the development of healthcare technology. In recent decades, pioneer researchers found that traditional monitoring approaches proposed by our ancestors can actually be achieved by modern techniques: "observation" by non-contact cameras and wireless sensors and "experience" learned by artificial intelligence (AI). This opens up the new era of vital signs monitoring based on contactless sensors, where physiological information can be collected from the human body without mechanical contact between sensors and body, such as using optical cameras to measure skin blood perfusion (from contact-PPG to camera-PPG), or radio-frequency sensors (e.g., radar and WiFi) to acquire respiratory or ballistocardiographic motion signals at a remote distance. The new trend of contactless vital signs monitoring has led to numerous solutions addressing issues associated with contact-based monitoring by: eliminating the discomfort and inconvenience caused by contact probes or electrodes; improving user experience and personnel workflow efficiency; giving rise to more scalable and versatile health-monitoring solutions for pervasive healthcare (e.g. by the use of regular cameras and WiFi). Contactless monitoring not only justifies the ancient wisdom in healthcare, but also sparks new opportunities for the future.

This is the first book in the emerging field of contactless vital signs monitoring that covers various topics and aspects from multiple interdisciplinary fields. It has 14 chapters organized into two parts, camera-based monitoring (Part I) and wireless sensor-based monitoring (Part II). These are two different communities evolved from different domains (e.g., biomedical optics, wireless sensing) but with very similar goals in healthcare applications and complementary features in two functionalities (non-contact camera and wireless sensors). The book attempts to unite the pioneer researchers from both sides to review the progress, share thoughts, and discuss the potential of fusion because it would provide new insights and impetus for health informatics. The book presents a structured introduction of various contactless sensing modalities (e.g., camera, radar, WiFi, acoustic, and RFID) and a clear roadmap for how to create novel health-monitoring systems based on these technologies, as well as a rich set of health applications that demonstrate the current achievements and great potential in this direction. This book will be helpful for readers looking for a comprehensive and educational introduction to this new direction, and those who seek inspiration for novel concepts and techniques in health monitoring.

Wenjin Wang   Xuyu Wang

Preface

Contactless vital signs monitoring is an emerging healthcare technology that has grown rapidly in both the academic and industrial domains in the past decade. Different from conventional biomedical sensors that require mechanical contact with the skin for physiological measurement, contactless and wireless sensors (e.g., cameras and radio-frequency and acoustic sensors) can be used to measure physiological signals remotely with electromagnetic and acoustic waves reflected from the human face and body, such as using cameras to measure blood hemodynamics from optical signals reflected from skin tissues, or RF sensors to measure respiratory motion from RF signals reflected from body surface. The typically measured vital signs include heart rate, respiration rate, blood oxygen saturation (SpO2), skin temperature, and blood pressure, and these can be converted into physiological markers that provide medical insights or diagnostic support, i.e., for patient monitoring or screening, cardiovascular health assessment, sleep staging, deterioration alarming, etc.

The health-monitoring systems enabled by contactless and wireless sensing technology reflect the current needs in clinical practice and present future directions for pervasive healthcare, from hospital care units to assisted-living homes and from precision care to personal care. Given that the contactless sensors (e.g., cameras, radar, WiFi, and acoustics) are ubiquitous and cost-effective, contactless monitoring will lead to numerous healthcare applications and opportunities that: directly improve the care experience of patients; increase the clinical workflow efficiency; reduce the cost of care; and eliminate the risk of infection (e.g., COVID-19) caused by contact sensing. It can either be shaped into stand-alone monitoring solutions to serve the end-users directly or integrated with existing medical devices to solve specific challenges, as evidenced in (neonatal) intensive-care units, general wards and triage, sleep/senior centers, baby/elderly care at home, telemedicine and e-health, fitness and sports, automotive, contactless cardiac/respiratory gating for magnetic resonance (MR), etc. It is unquestionable that contactless vital signs monitoring will evolve into a key technology in healthcare, advancing medical diagnosis and prognosis, patient care and treatment, and management of chronic diseases and personal well-being.

Contactless Vital Signs Monitoring gives a systematic overview of the progress made in this emerging field and attempts to create a synergy between the communities of camera-based and RF-based sensing for health monitoring. The book covers expertise and insights of sensing and processing for healthcare, dealing not only with technical issues but also with issues involving compliance with ethical standards and privacy in health applications. It presents: (i) an in-depth overview of principles and fundamentals for vital signs monitoring using contactless sensors; (ii) a detailed introduction of state-of-the-art methodologies developed for this purpose, including software algorithms and hardware setups; (iii) thorough benchmarks and validations in the context of healthcare; and (iv) a balanced and comprehensive discussion of opportunities, challenges, and future directions of contactless health monitoring. This book has 14 topical chapters organized into two parts: Part I is on camera-based vital signs monitoring (Chapters 2–8), and Part II is on wireless sensor-based vital signs monitoring (Chapters 9–14).

- Chapter 1 is an introduction that broadly reviews human physiology and general approaches used for contactless vital signs monitoring in an electromagnetic spectrum (from optical wavelengths to radio wavelengths).
- Chapter 2 delves into the physiological origins and fundamental mechanisms of camera-based photoplethysmographic (PPG) imaging, providing a solid basis for camera-PPG.
- Chapter 3 is focused on camera-based heart-rate monitoring that exploits the camera-PPG technology, with a detailed introduction to optical-physiological model-based PPG-extraction approaches with an illustration of performance in fitness applications.
- Chapter 4 is dedicated to camera-based respiration monitoring, with a particular focus on motion-based and PPG-based respiratory signal extraction in the application of respiratory gating for MR.
- Chapter 5 presents camera-based SpO_2 monitoring and highlights a simple and affordable solution (RGB camera based) for measuring SpO_2 and hemoglobin concentration in biological tissues.
- Chapter 6 introduces camera-based blood-pressure monitoring, specifically the transition from contact-PPG to camera-PPG based measurement, and its potential advantages and limitations.
- Chapter 7 summarizes clinical applications and trials deploying camera-based PPG imaging and their medical merits.
- Chapter 8 envisions the applications of vital signs cameras beyond healthcare and discusses the issues involving compliance with privacy and ethical standards in customized applications.
- Chapter 9 focuses on radar-based vital signs monitoring and potential healthcare applications using continuous-waves (CW) radar and frequency-modulated continuous-waves (FMCW) radar.
- Chapter 10 presents a received power-based vital signs monitoring system that can detect respiration and pulse rates using a single pair of low-cost radio transceivers.

- Chapter 11 reviews contactless monitoring techniques for human-respiration based on WiFi channel state information (CSI) including pattern-based and model-based methods.
- Chapter 12 describes vital signs monitoring with RFID systems and introduces the key characteristics of RFID sensing systems and implementation of breathing monitoring systems.
- Chapter 13 discusses acoustic-based healthcare-monitoring techniques and the implementation of acoustic-based respiration-rate monitoring with smartphones.
- Chapter 14 considers RF and camera-based vital signs monitoring and provides two contactless vital signs monitoring applications (i.e., body-movement cancellation and emotion recognition) with CW radar and RGB cameras.

This book is written for a broad audience from a wide range of interdisciplinary fields, such as AI healthcare, biomedical engineering, contactless and wireless sensing, radar monitoring, computer vision, ubiquitous computing, and specifically for those interested in novel health-monitoring technologies, including academics, scientists, engineers, industrial manufacturers, and bachlelor/master/doctoral students, et al. This book will raise awareness in both the scientific and industrial community that contactless sensors is creating a new value stream in healthcare by vital signs monitoring. A multidisciplinary research path with endeavors from various communities is needed to create momentum to meet the promises currently available in this field. We hope that this book will spark further research in this promising direction, leading towards an integrated and comprehensive knowledge base for contactless healthcare.

Chapter 1

Human physiology and contactless vital signs monitoring using camera and wireless signals

Xuyu Wang[a] and Dangdang Shao[b]

[a]*Department of Computer Science, California State University, Scaramento, CA, United States,* [b]*Biodesign Institue, Arizona State University, Tempe, AZ, United States*

Contents

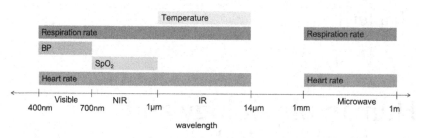

FIGURE 1.1   Vital signs monitoring in various wavelengths.

1.1   Contactless vital signs monitoring with cameras and wireless

Currently, camera and wireless signals have been used for vital signs monitoring, which offers an effective solution for long-term, contactless, low-cost healthcare monitoring. In Fig. 1.1, we can see that visible light, near-infrared (NIR), and infrared IR in wavelength range (400 nm–14 μm), as well as microwave or radio frequency (RF) in wavelength range (1 mm–1 m), can be used for monitoring respiration rate and heart rates. In addition, visible light can be used for blood-pressure measurement, while NIR can be also exploited for blood-oxygen saturation (SpO_2). Infrared (1–14 μm) can be used to measure body temperature. In this chapter, we will discuss and summarize the fundamental features and related works in camera-based and wireless-based vital signs monitoring.

1.2   Camera-based vital signs monitoring

Monitoring vital signs, such as heart rates and breathing patterns, are basic requirements in the diagnosis and management of various diseases. Traditionally, these signals are measured only in hospital and clinical settings. An important recent trend is the development of portable devices for tracking the physiological signals based on optical methods using cameras in a contactless manner. These portable devices, when combined with cell phones, tablets, or other mobile devices, provide a new opportunity for everyone to monitor vital signs anytime and anywhere.

1.3   Current techniques of camera-based vital signs monitoring

1.3.1   Camera-based pulse monitoring

Different types of cardiac waveforms have been obtained using cameras in the past decade.

Photoplethysmogram (PPG) is one of the cardiac waveforms that have attracted a lot of research interest and been widely studied. PPG is an optical signal that is proportional to the pulsatile blood-volume change in the microvascular tissue bed beneath the skin [9]. PPG carries useful information related to the cardiovascular system and can be obtained with low-cost optical techniques. Traditional PPG measurement uses a light source and photodetector to measure the volumetric variations of blood circulation at the skin surface. Pulse oximeter is one of the most widely used devices to monitor PPG in such a non-invasive manner. There are two types of pulse oximeters based on the mode of operation. In transmission mode, the LED and photodiode are placed on the opposite sides of the tissue allowing light going through the tissue. In reflection mode, the LED and photodiode are placed on the same side of the tissue, and the photodiode will detect the light reflected by the tissue.

In contactless PPG monitoring, the photodiode is replaced by camera as the light detector. The light source and camera are placed on the same side of tissue. The optical path is similar to the one in reflectance pulse oximeter, but the measured pulsatile energy can be much weaker. It is probably because the "banana-shape" optical path in the contact device ensures the measurement of arterial blood from deeper vessels, and the contactless mode suffers from mirror-like specular reflections directly from the skin surface. The selection of light source includes quite a few options in the visible and NIR ranges. Sufficient light and stable illumination condition are required to obtain a reliable PPG signal. After the light travels from the light source and reaches skin, it will be partially absorbed by the tissue and partially get scattered, and the rest of it will be reflected from the skin surface and finally be detected by the light-sensitive component inside the camera, like s charge-coupled device (CCD) or complementary metal oxide semiconductor (CMOS).

The first contactless PPG monitoring work was reported by Takano et al. [81]. In their work, the heart rate was obtained from the video-based PPG. Other early work includes the ones reported by Verkruysse et al. [87], Poh et al. [65, 66], Sun et al. [80], and others.

The applicability of using PPG to obtain different vital signs has been discussed based on the light properties and human dermatology. Haan et al. [13] analyzed the light reflection and diffusion at the skin surface. Based on that, they proposed a method to obtain robust PPG by combining RGB channels from video to compensate for the influence of motion. Moco et al. [49] studied the skin reflectance with visible and infrared light and further drew the conclusion that green light probes dermal arterioles and red/NIR reach subcutaneous blood-volume variations.

PPG can be technically monitored from the forehead, earlobe, wrist, fingertip, and ankle with contact sensors. When considering the feasibility of using contactless methods to detect PPG that require both the convenience of taking videos with exposed skin surfaces of the users and richly perfused skin layers containing strong pulsatile component of the cardiac cycle, the facial region including forehead, cheek, and lips, may be preferable and have been used in most of the published work.

Ballistocardiogram (BCG) is another cardiac waveform that has been monitored using contactless methods in recent years. The concept of BCG was originally brought up in 1800s by Gordon [20] as a rhythmic movement that is synchronous at each occurrence of systole in cardiac cycle. The early BCG apparatuses were bulky and hard to implement compared with other medical procedures.

In the past few years, researchers have explored the possibilities of monitoring BCG using cameras. Heart rate has been extracted from head motions caused by the cyclical ejection of blood from the heart to the head. The head motion can be mixed with other involuntary and voluntary head movements [51]. Balakrishnan et al. [5] detected cardiac-related waveform contained BCG using camera by tracking the vertical movement of the head. Krug et al. [32] used a camera to track the motion of a marker worn by subjects on the nasal bridge and obtained BCG waveforms. Shao et al., [74] detected various different outlooks of BCG waveform with supine and sitting positions using camera with ambient light. Pereira et al., [6] used thermal cameras to estimate heart rate from head motion.

The most-measured cardiac waveform in clinic settings is the electrocardiogram (ECG), which is a recording of the electrical activities of the heart. Compared with ECG, both PPG and BCG have the advantages that the measurements can be performed using easy and low-cost techniques without electrodes, and thus the irritation and skin rash during the ECG procedure could be avoided. Moreover, the measurements can be potentially taken in a more flexible and portable way by the user in normal living conditions with a camera. However, ECG is still the most accurate signal to monitor cardio health and diagnose cardiovascular diseases because it can provide more cardiovascular health metrics based on the unique features that are only carried by ECG, for instance, the QRS complex.

1.3.2 Cardiac-related physiological signals using camera-based methods

1.3.2.1 Heart rate

Heart rate, or pulse, is one of the earliest vital signs that has been obtained using a camera. In an ideal lab setting, where illumination is well-controlled and subject remains still, heart rate can be easily obtained by finding the largest peak amplitude in the FFT spectrum within certain frequency range, which corresponds to the strongest heartbeat signal. However, in real application scenarios, compared with traditional contact-based methods, camera-based methods are less tolerant of motion. While the contact-free interface between sensor and subject makes measurement convenient, the lack of mechanical coupling between sensor and subject [9] adds unwanted noise due to the existence of motion artifact and may contaminate the desired physiological signal, for example, PPG.

Researchers have proposed multiple algorithms to improve heart rate accuracy from PPG. One of the simplest and mostly implemented methods is

TABLE 1.1 Comparison of various camera systems for vital signs monitoring (o means Yes).

Wavelength		Pulse					Respiration	
		HR	HRV	SpO$_2$	PTT	BP	BR	Tidal volume
Visible	[400 nm, 700 nm]	o	o		o	o	o	o
Near Infrared	[700 nm, 1 µm]	o	o	o			o	
Infrared	[1 µm, 14µm]	o					o	o

temporal filtering. Other methods include blind-source separation, independent component analysis [66], alternative reflectance models, spatial pruning, autoregressive model [82], face tracking [39], chrominance combination [13], and spatial redundancy utilization [97].

1.3.2.2 Blood oxygen saturation

Blood-oxygen saturation (SpO$_2$) is another important physiological parameter that has been monitored based on PPG using camera-based techniques. SpO$_2$ is the percentage of oxygenated hemoglobin in a peripheral capillary. Traditional pulse oximetry measures SpO$_2$ based on the differential absorption of light by the concentration of oxygenated hemoglobin (HbO$_2$) and the concentration of deoxygenated hemoglobin (Hb) at two wavelengths using contact methods.

Researchers have explored the possibility of noncontact SpO$_2$ monitoring using camera [19,21,26,27,29,36,85,86,88,111,112]. The measurement sites include the face, finger, forearm, and palm. The selection of light source in the most reported work covers visible and NIR ranges from 450 to 1,000 nm. When NIR light sources is involved, the camera sensor needs to cover the NIR spectral range. However, if the method only requires ambient light, it is feasible to use just RGB camera [22].

One of the major challenges of monitoring SpO$_2$ using cameras is the selection of a light source. SpO$_2$ detection requires employing at least two wavelengths to obtain two PPG signals. The choice of the wavelength combination preferably meets these conditions: (1) The PPG signals obtained from both wavelengths should have high signal-to-noise (SNR); (2) optical absorption associated with HbO$_2$ is opposite to that associated with Hb, and the differences between them are large at the two wavelengths; (3) the two wavelengths should have sufficient penetration depth into the skin to reach arterial blood with a similar optical path; (4) the selected wavelengths need to be within the spectral sensitivity range of the camera's CCD or CMOS sensor.

1.3.2.3 Blood pressure

Blood pressure (BP) is a key physiological parameter, and its timely measurement helps to prevent and manage various diseases. A traditional BP monitor uses the oscillometric method, which requires the user to wear a cuff during the measurement. While valuable and popular, it provides only sporadic BP readings.

Researchers have tried to extract BP-related features from the PPG signal [15,113]. Recent efforts have been devoted to developing cuffless BP measurement methods. One of the methods is based on pulse-wave velocity (PWV) and pulse transit time (PTT), which both can be potentially obtained using contactless sensors, for example, cameras. PWV is an independent predictor of the longitudinal increase in SBP and of incident hypertension [54]. PTT is the time interval for a pulse signal to travel from one body location to another [51]. By using cameras, PTT can be measured from the time difference between one type of cardiac signal (e.g., two PPGs) obtained from the different body sites [73] or from two types of cardiac signals obtained from the same or different body sites, for example, BCG and PPG [12]. One of the potential advantages of using PWV/PTT-based methods to estimate BP, when compared with contact cuff-based method, is that the measurement could be continuous with high temporal resolution. Another cuffless BP measurement method is based on the morphology analysis of PPG waveform. The shape and phase information of systolic and diastolic peaks of PPG may be correlated with BP [48,114].

The main challenges for camera-based BP monitoring include: (1) Calibration may be needed with each individual. It is challenging to have a single calibration curve to provide BP estimation for a population; (2) Calibration may have posture dependency. If the calibration curve is obtained from sitting, the BP results may not be accurate when the measurement is taken during standing or reclining; (3) Absolute BP values may be hard to obtain, and the methods may only indicate relative BP changes.

1.3.3 Camera-based respiration monitoring

Respiratory signals carry parameters associated with respiratory activities. Measuring respiration functions can help screen various types of disease, for example, pneumonia, lung cancer, and sleeping disorders. Significant reduction in respiratory rate is an indication of patient deterioration. A few types of psychological disorders, such as anxiety and depression, can also be characterized by abnormal respiratory activities.

Traditional contact-based methods for respiration monitoring include chest-movement tracking and nasal/oral airflow assessment. For measurements taken at the chest, the user may wear a flexible belt strap around the chest, and the strap is integrated with a motion or force sensor to track the chest motions associated with breathing. For measurements taken at the nose and mouth, the user may need to wear a facial mask integrated with a valve to collect the breathing flow during inhalation and exhalation.

There have been increasing efforts to develop contactless respiratory monitoring technologies using cameras. The existing work can be mostly divided into three categories:

1. Extracting respiratory signals embedded in PPG signals that are based on the principle that respiration activity will modify PPG signals in the low-frequency range [65]. The movement of the thoracic cavity affects the blood flow during breathing, which leads to amplitude and frequency modulations in the PPG signal during respiration. Since the extraction of respiratory signal from PPG is indirect, the obtained respiratory parameters are limited, mostly only the respiratory rate.

2. Detecting subtle body motions induced by respiration [67]. The targeted body parts could be chest, trunk, shoulders, or other places where the respiratory signal is sensitive enough to be captured by a regular camera or RGB-D depth camera in the visible and near-infrared ranges [41,62,63,73,79]. Besides respiratory rate, the obtained respiration parameters include forced expiratory volume in the first second (FEV1), forced vital capacity (FVC), and peak expiratory flow rate (PEF). These parameters are critical for the diagnosis and management of asthma and chronic obstructive pulmonary disease (COPD).

3. Using infrared thermal imaging system to measure the air temperature change associated with exhaled breath near the mouth and nose regions of a subject [52,53], which is based on the fact that the air exhaled by human during respiration normally has a higher temperature than the typical background indoor environments [53]. Respiration parameters, such as respiratory rate and tidal volumes, have been captured by thermography [35,53].

1.3.4   Camera-based body temperature monitoring

Human body temperature is controlled by the central nervous system. It carries a wealth of information about health status and reflects body metabolic rate. Body—temperature values can vary by multiple factors—such as measurement site, sex, age, time of the day, and activity level. A normal body temperature is about 36.5–37.5 °C (97.7–99.5 °F) in most normal cases. A measured temperature of 38 °C (100 °F) or above could suggest a fever, which can be an indication of many medical conditions ranging from trivial to fatal (e.g., viral infection, bacterial infection, medication side effect). A measured temperature of 35 °C (95 °F) or below is a warning sign of hypothermia, which can happen either when the body is exposed to extreme cold temperatures or experiencing certain medical conditions (e.g., low blood sugar and alcohol intoxication). A thermometer is the most used device to measure body temperature. With contact-based mercury or electronic thermometers, the measurements can be taken in the mouth, anus, or under the arm. With a noncontact infrared thermometer or thermal camera, the temperature can be measured at the forehead from a distance between the device and subject. The application of a thermal camera is within the scope of this book. It captures the infrared radiation emitted from the object and creates an image representing the object's temperature distribution.

Temperature tracking provides early warnings of the infection of many diseases. Measuring body temperature has been used in the early screening solution in the previous Severe Acute Respiratory Syndrome (SARS) outbreak in 2003 and the ongoing pandemic of coronavirus disease 2019 (COVID-19). As one of the body's first reactions to the virus, one of the commonest symptoms of people with infection is fever. Compared with other symptoms (e.g., fatigue, muscle pain), fever can be detected in a more straightforward, objective, and efficient way by measuring body temperature. Infrared thermal camera screening systems have been implemented in many public places for this purpose. It only takes a few seconds to make the temperature measurement, which can have a significant impact to prevent the spread of the infectious diseases in hospitals, railway stations, airports, marketplaces, warehouses and other environments with a high population density.

1.4 Applications of camera-based vital signs monitoring

Camera-based vital signs monitoring has been extended from the research lab to real life in various applications.

1.4.1 Clinical applications

In clinic settings, conventional patient monitoring systems mostly require probes to be attached to the patients. Since the measurements usually need to be performed over long time periods instead of short time periods, it is likely the patients may become discomforted with the burden of the devices or skin irritation caused by direct contact with the sensors. It may also increase the risk of infection. Using camera-based methods to monitor vital signs may help to avoid some of the disadvantages imposed by contact-based methods. A camera-based vital signs monitoring system may also be useful for unconscious patients since they do not need to wear device or provide cooperation to the medical staff during measurements.

Camera-based techniques have already been applied in a few hospital scenarios (e.g., the ICU, MRI testing, and surgery) to recover a broad range of vital signs without interfering with regular patient care [7,32,68,84,90]. The obtained signals include heart rate, respiratory rate, SpO_2 and blood volume. Meanwhile, the camera-based vital signs can also be used in respiratory- or cardiac-triggered medical imaging acquisitions (e.g., CT and MRI). The synchronization of vital sign recording and image acquisition can help to avoid the motion artifacts caused by the corresponding physiological activities, and therefore the quality of the medical images can be improved.

Moreover, researchers have also reported using non-contact methods in neonatal intensive care unit (NICU) for pre-mature infants during their medical treatments [1,17,30,33,89]. The developed systems can measure vital signs for infants continuously. They can also avoid inflicting stress and pain from potential damage to infants' fragile skin [30]. Contact-free methods are friendly for long-time monitoring since the measurements will have no extra burden and require no attention from the users, which might be even more useful for pediatric patients compared with adult patients, since the latter can more easily cooperate with if their health conditions guarantee their consciousness.

1.4.2 Free-living applications

In free-living conditions, researchers have also explored using camera-based techniques to monitor vital signs for various applications.

Researchers have reported using camera in sleep monitoring to detect respiration disorders, obstructive sleep apnea (OSA) [16,78], sleeping posture change [34,55], and cardiopulmonary conditions (heart rate and respiratory rate) [37]. An NIR camera is preferable for sleeping monitoring since a visible light source could be avoided so as not to disturb the subject's sleeping.

Another application is fitness monitoring. Professional fitness tracking for physiological monitoring may require the user to wear a face mask to collect and analyze breathe components. Other widely used personal trackers include smart-phone-based or wrist-band-based ones with an integrated accelerometer and photodiode. Wang et al., [99] measured heart rates from PPG in fitness activities using cameras. The proposed algorithm was able to measure heart rate during vigorous body motions when the subjects were running on a treadmill.

Camera-based vital signs monitoring has also been used for biometrics and forensics, for example, anti-spoofing. Many personal devices currently include a face metric as one of the identity methods. A corresponding problem is that the attacker can use face spoofs (e.g., static photo, mask) to pretend to be the person who is authorized by the system. Adding vital sign detection can be an enhancement to the original face detection. It may improve the robustness of the security system by avoid spoofing happening since a live signal (e.g., existence of heart rate) will be much harder to synthesize. Researchers have been using PPG-based methods to identify heart rate, living skin-tissue, and the liveness of the user's face [18,44,46,61,98]. Based on the aforementioned information, they could find more robust solutions to avoid spoofing attack.

Stress analysis is another application of camera-based vital signs monitoring in recent years. Heart-rate variability (HRV) is a quantification of variations in the intervals between heartbeats, which can be used to evaluate the stress level of a human. HRV can be obtained from cardiac waveforms, either PPG or BCG [70,77]. In addition to HRV, the contour change of PPG may also indicate the stress level of the subject [126].

Camera-based health monitoring system is also useful for elderly people. Besides measuring physiological parameters, the recorded video could contain other useful information to reflect the elderly people's health and wellness. It provides an interface for users, family members, and doctors to manage and communicate the health conditions of elderly people. It could also help to detect medical emergencies (e.g., shortness of breath, a sudden drop of SpO_2) and accidents (e.g., fainting, occurrence of a sudden fall).

Another useful application scenario is driver monitoring. The camera-based monitoring system can be run in the background to retrieve vital signs without attention from the driver. The related physiological information can be further processed to indirectly estimate the driver's stress level. It is also possible to have additional computer–vision algorithms to detect the drowsiness of the driver, which can be useful to prevent driving accidents [24,76,125].

In the entertainment business, a virtual reality (VR) and augmented reality (AR) headset can enhance user's experience by adding extra features using the physiological parameters obtained from the camera sensing system on-the-fly. For example, the user's real-time heart rate and blood pressure values could be displayed on the user's view to create a more immersive feeling in the game.

In recent years, camera-based approaches have also been applied to animal husbandry and zoological research to measure the vital signs of the animals [4, 31]. The obtained physiological measurements provide useful information that reflect the health status of the animals on farms and in zoos. The non-contact methods can also relieve the workload of the workers since they don't need to exert effort to control animals for the vital sign measurements.

1.5 Wireless-based vital signs monitoring

With the rapid development of mobile techniques and improvements in the living standard, healthcare problems become more important. It is reported that three-fourths of the total US healthcare cost address chronic health conditions such as heart diseases, lung disorders, and diabetes [8]. Breathing signals are useful for physical health monitoring because such vital signs can offer important information on health problems, such as sudden-infant-death syndrome (SIDS) of sleeping infants [28]. Traditional systems require a person to wear special devices, such as a pulse oximeter [75] or a capnometer [50], which are inconvenient and uncomfortable, especially for the elders and infants. There is a compelling need for technologies that can enable contact-free, easy deployment, and long-term vital signs monitoring for healthcare. Currently, wireless-based sensing techniques can provide effective solutions for vital signs monitoring.

1.6 Current techniques of wireless-based vital signs monitoring

Existing wireless vital signal monitoring systems mainly focus on radio-frequency (RF) based techniques, which use radio frequency (RF) signals to capture breathing and heart activities. Existing RF-based schemes include radar-based, RSS-based, WiFi CSI-based, and RFID-based vital signs monitoring. In addition, we will also discuss acoustic-based breathing monitoring techniques.

1.6.1   Radar-based vital signs monitoring

Radar-based techniques can be used for several applications including vital signs monitoring, activity recognition, movement detection, localization, sleep apnea monitoring, and driver behavior monitoring [83]. Currently, there are three radar-modulation techniques: (1) continuous wave (CW) or Doppler radar; (2) frequency-modulated continuous-wave (FMCW); (3) impulse radio ultra-wideband (IR-UWB).

In CW or Doppler radar, the phase or Doppler shift can be exploited for vital signs monitoring [14,59]. In particular, the transmitter of CW radar can generate a CW signal (i.e., a periodic signal with the fixed frequency that is generated by an oscillator). By using a power amplifier (PA), this CW signal can be directed at the human body (e.g., the chest of the body). Then, the receiver can capture the reflected signal that is modulated by the body movements caused by breathing. The received signal is also amplified by a linear amplifier, which is then down-converted into two orthogonal signals. By using an arctangent demodulator, analog-to-digital converter (ADC), and digital signal processing, the useful phase information from two orthogonal signals can be extracted, which can measure distance change, thus capturing breathing and heart signals. Generally, CW radar is simple and can be used for monitoring movements like vital signs. However, CW radar cannot measure the absolute distance or distinguish different persons for vital signs monitoring.

FMCW radar can use a synthesizer to generates a chirp that is a linear frequency signal in a fixed transmitting bandwidth and a pulse repetition period [3]. Then, the chirp is transmitted by the transmitter antennas, which is reflected by the chest of a person. In addition, the receiver antenna can capture the reflected chirp, which is mixed by the transmitting signal to obtain ian intermediate frequency (IF) signal. Then, using a low-pass filter, ADC, and Fast Fourier transform (FFT), the measured distance can be estimated. In summary, FMCW can measure distance and distance changes that are caused by physiological activities. The wireless bandwidth can determine the distance resolution. To monitor small changes like vibrations of breathing and heart signals, FMCW radar uses a large bandwidth, thus obtaining good distance resolution.

IR-UWB radar can estimate the time-of-arrival (TOA) of the reflected signals from the body, which can be leveraged for vital signs monitoring [71]. Generally, IR-UWB radar includes two operating modes (i.e., emitting and silence). In the emitting mode, a short pulse can be sent, while the echo signal can be obtained in IR-UWB in silent mode. By comparing the transmitted short pulse and the received signal, the distance and velocity of the object can be estimated. In particular, IR-UWB radar can utilize a pulse generator to create a desired pulse where a local oscillator can determine the pulse repeat frequency.

Before being amplified by PA and sent by the transmitting antenna, the short pulse is modulated. Then, the reflected pulse from the chest can be amplified and down-converted to the base band. Thus, the round-trip time or TOA can be estimated to measure the distance to the target. Similar to FMCW, IR-UWB radar also measures the distance to the target with a large bandwidth. However, the multi-user interference and resolution are still limitation for IR-UWB radar.

In summary (see Table 1.2), radar-based vital signs monitoring mainly uses CW, FMCW, and IR-UWB radars to measure breathing and heart rates with special hardware. Currently, some commercial radar devices have also been used (e.g., XeThru IR-UWB radar with 3.1 to 10.6 GHz, TI 60-GHz mmWave radar, and Infineon's 60-GHz BGT60TR13C radar). From Table 1.2, radar-based methods use a large bandwidth (e.g., FMCW and IR-UWB radars). Also, they generally use directional antennas like horn antennas to improve the received signal power. However, radar-based vital signs monitoring has a high cost and a large received signal path loss that limits the monitoring distance.

1.6.2   RSS-based vital signs monitoring

Received signal strength (RSS) or received power has been used for vital signs monitoring or other wireless sensing applications in WiFi, ZigBee, and narrowband wireless networks [2,25,64]. For example, UbiBreathe can detect a human respiration signal with RSS values received by a commercial WiFi device [2], where a cell phone serves as the receiver and a laptop as the transmitter. To extract the respiration signal, raw RSS values are first filtered by a bandpass filter with a cutoff frequency from 0.1 Hz to 0.5 Hz. Then, a Discrete Wavelet Transform (DWT) is exploited to remove the noise and outlier. Breathing rate estimation is implemented by FFT. Besides, a threshold method is exploited to detect abnormal breathing (i.e., apnea), when a user is sleeping. The experimental results show the estimation error of breathing rate in the UbiBreathe system is less than 1 beats per minute (bpm). In fact, the system can only detect human breathing signal when the receiver is on the user's chest or the user is in the LOS path of the transmitter and the receiver.

Generally, RSS values collected from 2.4 GHz and 5 GHz WiFi devices can hardly monitor human heart rate because of the low resolution of RSS values. Typically, the RSS is quantized with a 1 dB step size or higher. In fact, breathing and heart signals have amplitude on the order of 0.1 dB and 0.01 dB, respectively. It is challenging using the raw RSS values for vital signs monitoring. To address this issue, some systems use link diversity, frequency diversity, time diversity, and even increasing the noise, to improve the vital signs monitoring accuracy [25,64].

Moreover, the received RSS values can be easily influenced by other nearby persons, because of using omnidirectional antennas. Compared with 2.4/5 GHz WiFi devices, mmVital system also uses RSS values to estimate the human both breathing rate and heart rate with 60 GHz WiFi devices [122]. Because mmVital

TABLE 1.2 Features of wireless sensing techniques.

Wireless	Sensing techniques	Sensingr features	Commodity device	Pros. and Cons.
Radar	CW FMCW IR-UWB	Doppler Shift Phase Distance	XeThru IR-UWB radar with 3.1–10.6 GHz TI 60-GHz mmWave radar Infineon's 60-GHz BGT60TR13C radar	Pros: Large bandwidth; Directional performance Cons. High cost;
RSS	Narrowband Zigbee WiFi	sub-dB RSS Zigbee RSS WiFi RSS	CC1200 900-MHz radio CC2530 2.4-GHz Zigbee WiFi 2.4-GHz router	Pros: Low cost Cons: Low RSS resolution; Limited coverage
CSI	WiFi OFDM	CSI amplitude CSI phase	Intel WiFi 5300 NIC at 2.4 GHz and 5 GHz Atheros WiFi NIC at 2.4 GHz and 5 GHz	Pros: High CSI resolution; Ubiquitousness; Low cost Cons: Susceptible to environment influence
RFID	CW	RFID Phase	Impinj Speedway R420 RFID with 902.5–927.5 MHz	Pros: Directional performance; Cheap tag Cons: Channel hopping;
Acoustic	CW FMCW OFDM	Acoustic Phase Acoustic Distance	Smartphones with 18–20 KHz (e.g., Samsung Galaxy S6 and S7)	Pros: High resolution Cons: Susceptible to environment; small coverage

uses a directional beam to transmit the 60 GHz WiFi signal, the monitoring user can be separated from environments noise. The experimental results demonstrate that the mean absolute errors of breathing rate and heart rate estimation are 0.43 bpm and 2.15 bpm, respectively. Although the accuracy of the system is high, the system still includes some issues. For example, the user is required to be stay in the LOS or the reflection path of the transmitter and the receiver. Also, this technique uses a special mmWave hardware.

In summary (see Table 1.2), RSS-based vital signs monitoring can employ sub-dB, ZigBee, and WiFi RSS to monitor breathing rate and heart rates with special hardware (WiFi 60 GHz device) or a commercial device (e.g., CC1200 900 MHz radio, CC2530 2.4 GHz ZigBee, and WiFi 2.4 GHz router). From Table 1.2, low-frequency RSS-based methods have a low-cost, low RSS resolution, and small coverage. mmWave (i.e., 60 GHz WiFi) RSS-based methods can have high accuracies for estimating breathing and heart rates. However, they require a large bandwidth and a special device.

1.6.3 CSI-based vital signs monitoring

Current WiFi protocols (e.g., IEEE 802.11a/g/n/ac) mainly use Orthogonal Frequency Division Multiplexing (OFDM) technique in the Physical Layer (PHY) [23,103,109]. OFDM can use multiple orthogonal subcarriers for data transmission, thus addressing frequency selective fading in indoor environments. Some off-the-shelf WiFi network interface cards (NIC), e.g., Atheros AR9390 chipset and Intel 5300 NIC, can be leveraged to report CSI of subcarriers with modified open-source device drivers. CSI values can represent fine-grained PHY channel measurements that reflect channel features (e.g., power distortion, shadowing, multipath, and reflections). Compared with RSS, CSI is a complex value with a higher resolution including CSI amplitude and phase values [121]. In addition, each packet can extract multiple CSI values over subcarriers (30 subcarriers with Intel 5300 NIC) in WiFi networks. Thus, CSI can provide much more channel information in PHY than a RSS signal in the link layer. Currently, several CSI-based systems are proposed for higher accuracy vital signs monitoring using amplitude and phase difference data.

WiFi-sleep is the first device able to detect human respiration and different sleeping postures by analyzing the amplitudes of CSI [45]. Beside estimating human breathing, human heart rate is also estimated from CSI amplitude [42]. Compared with the amplitude values, CSI phase information is more robust for monitoring human vital signs at various distances and orientations. However, due to the unsynchronized issues between the transmitter and the receiver, the measured phase information cannot be directly used to detect vital signs. Thus, the Phasebeat system uses the phase difference between two receiving antennas to track human vital signs instead of using phase information directly [100,106]. Although the distance has less effect on phase-difference information, the sensitivities of CSI phase difference and amplitude can be affected by the reflection

TABLE 1.3 Comparison of various systems for vital signs monitoring (o is Yes, x is No).

	Breathing rate	Heart rate	Multiple users	Special hardware
Wi-sleep [45]	o	x	x	x
Amplitude based [42]	o	o	o	x
Phasebeat [106]	o	o	o	x
Resbeat [107]	o	x	x	x
Tensorbeat [108]	o	x	o	x
TR-BREATH [11]	o	x	o	x

from surroundings or by the posture of the user [92]. To improve the robustness of CSI-based systems, the ResBeat leverages CSI bi-modal values (i.e., phase difference and amplitude of CSI) to track human vital signs [101,107]. The Resbeat system implements an adaptive signal selection algorithm to ensure that only the most sensitive amplitude and the phase differences can be used for extracting a respiration signal.

To monitor breathing rates for multiple persons, the TensorBeat system is available, which can separate multiple breathing signals using tensor decomposition [108]. Experimental results show that TensorBeat can achieve high accuracy under various different environments for multi-person breathing-rate monitoring, even for more than five people simultaneously. In addition, TR-BREATH system is also suggested for monitoring breathing rates for multiple persons [11]. CSI values is used for obtaining time-reversal resonating strength (TRRS) features, which can be thus analyzed by root-MUSIC and affinity propagation algorithms to estimate multiple persons' breathing rates. The difference between TensorBeat and TR-BREATH is that the TensorBeat system can obtain different breathing signals (i.e., breathing waveforms) for different persons, while TR-BREATH system only estimates the breathing rates (without obtaining breathing waveforms) for multiple persons. In fact, both systems cannot know which breathing rate belongs to a particular monitored person. Table 1.3 provides the comparison of various systems for vital signs monitoring in detail.

In summary (see Table 1.2), CSI-based vital signs monitoring mainly uses CSI amplitude and phase information to measure breathing rates and heart rates with commercialy WiFi NICs (e.g., Atheros chipset and Intel 5300 NIC) using 2.4 GHz and 5 GHz. From Table 1.2, CSI-based methods have high resolution of CSI amplitude or phase information, as well as low cost. However, CSI-based vital signs monitoring is susceptible to environment change.

1.6.4 RFID-based vital signs monitoring

Current commercial RFID readers (e.g., Impinj Speedway R420) can report the low-level data such as timestamps, RSS, information, and Doppler-shift [60]. In particulary, RFID phase data is mainly used for wireless sensing applications (e.g., vital signs monitoring) [115,117]. The phase information can be used to measure the total distance between the reader's antenna to tag and back to the antenna, which is used to detect vital sign signals.

Recently, RFID low-level data (e.g., phase information) has been used for vital signs monitoring systems. For example, RFID tags that can be attached on the chest of a person have been leveraged for breathing rate estimation, heart rate estimation, and sleeping posture recognition. In particular, the RF-ECG system is used for estimating heart rate variability with an RFID tag array [91]. To address the channel-hopping offset in RFID systems, the AutoTag system is exploited for breathing monitoring ands use a variational autoencoder method for apnea detection [115,117]. However, AutoTag and RF-ECG systems are designed for an indoor, static environment. In fact, the RFID-based system is also designed in driving environments to monitor breathing rates with tensor decomposing methods and driver fatigue with the variational autoencoder [116,118]. The current work also considers RFID-based breathing monitoring in dynamic environments (e.g., when the person moves in the vicinity of the monitored person) by using a matched filter to detect the monitored breathing cycles.

In addition to vital signs monitoring, RFID tags have also been applied for many other applications, such as indoor localization, user authentication, material identification, object orientation estimation, vibration sensing, anomaly detection, and drone localization and navigation [38,40,47,72,93,94,110,119, 124,127]. Compared with RSS values, recent works mainly focus on the phase for wireless sensing, which can be used to derive the distance and direction of arrival (DOA) for indoor localization with a sparse tag array [119]. In addition, synthetic aperture radar (SAR) and the hologram techniques are proposed to address the phase ambiguity problem [120]. The RFID phase is also used for temperature estimation with the median error of $2°C$ [102]. The basic idea is that the impedance of the tag can be affected by the tag temperature, thus leading to the received phase change.

In conclusion (see Table 1.2), RFID-based vital signs monitoring mainly uses RFID phase information to measure breathing rates, heart rates, and human poses with commercial RFID device. From Table 1.2, RFID uses directional antennas to improve the coverage range and reduce the interference, as well as inexpensive RFID tags. However, RFID systems need to address channel hopping for vital signs monitoring.

1.6.5 Acoustic-based vital signs monitoring

Similar to RF-based vital signs monitoring, wireless sensing systems based on acoustic signal with smartphones also have attracted great attention in industrial applications and among researchers [10]. These systems offer great convenience for healthcare monitoring where acoustic sensing systems only require software with smartphones to design new sensing application (e.g., vital signs monitoring). Currently, acoustic sensing systems can be classified into two types (i.e., passive and active sensing systems). First, passive acoustic sensing systems mainly focus on how to exploit microphones on smartphones to detect and process the surrounding audible signal for wireless sensing applications. Fine-grained sleeping monitors use the microphone in an earphone to record human breathing sound and further extract the respiration signal when the users are sleeping [69]. Moreover, passive acoustic sensing can be exploited for other mobile applications. For example, the AAmouse system uses inaudible sound signal at various frequencies to transform a mobile device into a mouse based on the Doppler shifts and speed and distance estimation [123]. Furthermore, the Keystroke snooping [43] and Ubik systems [95] can receive the sound signal with a single smartphone or dual microphones, and then use time-different-of-arrival(TDOA) measurements to monitor finger strokes on a table. Then, those strokes are transformed into related alphabets in the same position and computer keyboard.

On the other hand, active acoustic sensing system [56] can treat a smartphone as an active sonar with the speaker and microphone using inaudible sound wave at 18–22 kHz. Generally, three different sensing methods (i.e., modulations) for active acoustic systems are OFDM, CW, and FMCW. For example, OFDM-based acoustic sensing systems such as FingerIO can track finger movements in the 2-D domain, where tracking echoes of the fingers received by microphone can be used to estimate the positions of fingers [58]. In fact, OFDM-based acoustic sensing methods require high implementation complexity, which is challenging for vital signs monitoring. In addition, the Apnea system uses an active sonar with smartphone to monitor the breathing signal [57]. The system uses FMCW technique for breathing monitoring, which can track the distance between the smartphone and the chest of the person before tracking breathing. However, when the body of the person suddenly moves (e.g., should he/she turn over), the Apnea system needs to estimate the new distance, thus requiring additional time complexity. Moreover, LLAP system uses CW methods on a smartphone with a speaker and two microphones to measure the distance, thus implementing device-free hand tracking [96]. Similar to LLAP, SonarBeat system with active acoustic sensing also uses the phased-based CW signal to monitor breathing rates, which will be discussed in the chapter on acoustic-based vital signs monitoring in detail [104,105]. The SonarBeat system is more robust in different environments by using an adaptive median-filter technique.

In summary, in Table 1.2, acoustic-based vital signs monitoring can use acoustic phase information or distance to monitor breathing and heart rates using three different methods (i.e., OFDM, FMCW, and CW), where smartphones (e.g., Samsung Galaxy S6 and S7) are used in the acoustic frequency range with 18–20 KHz. Generally, OFDM, and FMCW can estimate the distance between the device and a person's chest with high complexity, while CW can only measure the phase change to monitor breathing signals with low complexity. Compared with RF-based vital signs monitoring, acoustic-based sensing has high resolution in acoustic phase or estimated distance. However, the main challenge of acoustic-based sensing is susceptible to environment audio noise and small coverage range with smartphones. Also, compared with radar/WiFi, acoustic-based sensing cannot penetrate multi-layer cloths/bed covers/obstructions because of the low transmitting power of smartphones.

1.7 Conclusions

In this chapter, we discuss human physiology and contactless vital monitoring using cameras, wireless RF, and acoustic-based sensing techniques. The camera methods have advantages such as low-cost and burden-free and easy implementation. The wireless RF methods are suitable for long-term monitoring using radar-based, RSS-based, and CSI based monitoring. The acoustic-based methods can be realized with the microphone and the speaker of a smartphone, so the implementation can also be very accessible.

Although these methods are introduced separately, if they can be combined through sensor fusion methods, the hybrid system could measure a variety of physiological signals simultaneously. There are a few potential benefits for multi-sensor approaches based on cameras and wireless signals.

First, even though there have some similarities of various types of physiological signals, each of them may still have some unique characteristics that are not fully exploited with others. The measurement could be made more comprehensive by integrating data from multiple sources. For example, the camera and wireless RF sensor can be combined to be a multi-modal sensing platform. In particular, the facial PPG signal obtained from a camera and the respiration activities obtained from the chest based on RF sensor can be analyzed together to evaluate the cardiovascular and respiratory systems.

Second, the influence of certain types of noise on different types of measurement may vary. For example, camera-based approaches may be less affected by high-frequency environment noise, while RF-based approaches won't be affected by light fluctuation. A multi-sensor fusion platform can help to enhance the overall quality and reliability of the measurements. In particular, the camera can help the wireless sensor to calibrate the body movement for vital signs monitoring, while wireless sensor can complement the camera-based vital signs monitoring in weak light conditions (e.g., driving at night).

Third, the signal obtained from one sensor could be the input information to assist the measurement of another sensor. A potential fusion example is to use the camera as a beacon reference to help the training of the RF-signals for vital signs monitoring and pose estimation using knowledge distillation (i.e., teacher-student learning). Then, the wireless sensors can be alone used for privacy-preserving health monitoring without cameras.

Acknowledgments

This work is supported in part by the US National Science Foundation (NSF) under Grant CNS-2105416. Any opinions, findings, and conclusions or recommendations expressed in this material are those of the authors and do not necessarily reflect the views of the foundation.

References

[1] Abbas K. Abbas, et al., Neonatal non-contact respiratory monitoring based on real-time infrared thermography, Biomedical Engineering Online 10 (1) (2011) 93.

[2] H. Abdelnasser, K.A. Harras, M. Youssef, Ubibreathe: a ubiquitous non-invasive WiFi-based breathing estimator, in: Proc. IEEE MobiHoc'15, Hangzhou, China, 2015, pp. 277–286.

[3] F. Adib, et al., Smart homes that monitor breathing and heart rate, in: Proc. ACM CHI'15, Seoul, Korea, 2015, pp. 837–846.

[4] Ali Al-Naji, et al., A pilot study for estimating the cardiopulmonary signals of diverse exotic animals using a digital camera, Sensors 19 (24) (2019) 5445.

[5] Guha Balakrishnan, Fredo Durand, John Guttag, Detecting pulse from head motions in video, in: Proceedings of the IEEE Conference on Computer Vision and Pattern Recognition, 2013, pp. 3430–3437.

[6] Carina Barbosa Pereira, et al., Monitoring of cardiorespiratory signals using thermal imaging: a pilot study on healthy human subjects, Sensors 18 (5) (2018) 1541.

[7] Nikolai Blanik, et al., Hybrid optical imaging technology for long-term remote monitoring of skin perfusion and temperature behavior, Journal of Biomedical Optics 19 (1) (2014) 016012.

[8] O. Boric-Lubeke, V.M. Lubecke, Wireless house calls: using communications technology for health care and monitoring, IEEE Microwave Magazine 3 (2002) 43–48.

[9] M.J. Butler, et al., Motion limitations of non-contact photoplethysmography due to the optical and topological properties of skin, Physiological Measurement 37 (5) (2016) N27.

[10] Chao Cai, Rong Zheng, Menglan Hu, A survey on acoustic sensing, arXiv:1901.03450, 2019.

[11] Chen Chen, et al., TR-BREATH: time-reversal breathing rate estimation and detection, IEEE Transactions on Biomedical Engineering 65 (3) (2018) 489–501.

[12] Zhihao Chen, et al., Noninvasive monitoring of blood pressure using optical ballistocardiography and photoplethysmograph approaches, in: 2013 35th Annual International Conference of the IEEE Engineering in Medicine and Biology Society (EMBC), IEEE, 2013, pp. 2425–2428.

[13] Gerard De Haan, Jeanne Vincent, Robust pulse rate from chrominance-based rPPG, IEEE Transactions on Biomedical Engineering 60 (10) (2013) 2878–2886.

[14] A. Droitcour, O. Boric Lubecke, G. Kovacs, Signal-to-noise ratio in Doppler radar system for heart and respiratory rate measurements, IEEE Transactions on Microwave Theory and Techniques 57 (10) (2009) 2498–2507.

[15] Mohamed Elgendi, et al., The use of photoplethysmography for assessing hypertension, npj Digital Medicine 2 (1) (2019) 1–11.

[16] Jong Yong, A. Foo, Pulse transit time in paediatric respiratory sleep studies, Medical Engineering & Physics 29 (1) (2007) 17–25.

[17] Mark van Gastel, Sander Stuijk, Gerard de Haan, Robust respiration detection from remote photoplethysmography, Biomedical Optics Express 7 (12) (2016) 4941–4957.

[18] Guillaume Gibert, David D'Alessandro, Florian Lance, Face detection method based on photoplethysmography, in: 2013 10th IEEE International Conference on Advanced Video and Signal Based Surveillance, IEEE, 2013, pp. 449–453.

[19] Sylvain Gioux, et al., First-in-human pilot study of a spatial frequency domain oxygenation imaging system, Journal of Biomedical Optics 16 (8) (2011) 086015.

[20] J.W. Gordon, Certain molar movements of the human body produced by the circulation of the blood, Journal of Anatomy and Physiology 11. Pt 3 (1877) 533.

[21] Alessandro R. Guazzi, et al., Non-contact measurement of oxygen saturation with an RGB camera, Biomedical Optics Express 6 (9) (2015) 3320–3338.

[22] Alessandro R. Guazzi, et al., Non-contact measurement of oxygen saturation with an RGB camera, Biomedical Optics Express 6 (9) (2015) 3320–3338.

[23] D. Halperin, et al., Predictable 802.11 packet delivery from wireless channel measurements, in: Proc. ACM SIGCOMM'10, New Delhi, India, 2010, pp. 159–170.

[24] Jennifer A. Healey, Rosalind W. Picard, Detecting stress during real-world driving tasks using physiological sensors, IEEE Transactions on Intelligent Transportation Systems 6 (2) (2005) 156–166.

[25] Peter Hillyard, et al., Experience: cross-technology radio respiratory monitoring performance study, in: Proceedings of the 24th Annual International Conference on Mobile Computing and Networking, 2018, pp. 487–496.

[26] K. Humphreys, T. Ward, Charles Markham, A CMOS camera-based pulse oximetry imaging system, in: 2005 IEEE Engineering in Medicine and Biology 27th Annual Conference, IEEE, 2006, pp. 3494–3497.

[27] Kenneth Humphreys, Tomas Ward, Charles Markham, Noncontact simultaneous dual wavelength photoplethysmography: a further step toward noncontact pulse oximetry, Review of Scientific Instruments 78 (4) (2007) 044304.

[28] C. Hunt, F. Hauck, Sudden infant death syndrome, CMAJ. Canadian Medical Association Journal 174 (13) (2006) 1309–1310.

[29] Ryan Imms, et al., A high performance biometric signal and image processing method to reveal blood perfusion towards 3D oxygen saturation mapping, in: Imaging, Manipulation, and Analysis of Biomolecules, Cells, and Tissues XII, vol. 8947, International Society for Optics and Photonics, 2014, 89470X.

[30] Joao Jorge, et al., Non-contact monitoring of respiration in the neonatal intensive care unit, in: 2017 12th IEEE International Conference on Automatic Face & Gesture Recognition (FG 2017), IEEE, 2017, pp. 286–293.

[31] Maria Jorquera-Chavez, et al., Modelling and validation of computer vision techniques to assess heart rate, eye temperature, ear-base temperature and respiration rate in cattle, Animals 9 (12) (2019) 1089.

[32] Johannes W. Krug, et al., Optical ballistocardiography for gating and patient monitoring during MRI: an initial study, in: Computing in Cardiology 2014, IEEE, 2014, pp. 953–956.

[33] Mayank Kumar, Ashok Veeraraghavan, Ashutosh Sabharwal, DistancePPG: robust non-contact vital signs monitoring using a camera, Biomedical Optics Express 6 (5) (2015) 1565–1588.

[34] Jaehoon Lee, Min Hong, Sungyong Ryu, Sleep monitoring system using kinect sensor, International Journal of Distributed Sensor Networks 11 (10) (2015) 875371.

[35] Gregory F. Lewis, Rodolfo G. Gatto, Stephen W. Porges, A novel method for extracting respiration rate and relative tidal volume from infrared thermography, Psychophysiology 48 (7) (2011) 877–887.

[36] Jun Li, et al., A reflectance model for non-contact mapping of venous oxygen saturation using a CCD camera, Optics Communications 308 (2013) 78–84.

[37] Michael H. Li, Azadeh Yadollahi, Babak Taati, Noncontact vision-based cardiopulmonary monitoring in different sleeping positions, IEEE Journal of Biomedicalandhealth Informatics 21 (5) (2016) 1367–1375.

[38] Li Ping, et al., Towards physical-layer vibration sensing with rfids, in: Proc. IEEE INFOCOM 2019, Paris, France, 2019, pp. 892–900.

[39] Xiaobai Li, et al., Remote heart rate measurement from face videos under realistic situations, in: Proceedings of the IEEE Conference on Computer Vision and Pattern Recognition, 2014, pp. 4264–4271.

[40] Qiongzheng Lin, et al., Beyond one-dollar mouse: a battery-free device for 3d human-computer interaction via rfid tags, in: Proc. IEEE INFOCOM 2015, Kowloon, Hong Kong, 2015, pp. 1661–1669.

[41] Chenbin Liu, et al., Noncontact spirometry with a webcam, Journal of Biomedical Optics 22 (5) (2017) 057002.

[42] J. Liu, et al., Tracking vital signs during sleep leveraging off-the-shelf wifi, in: Proc. ACM Mobihoc'15, Hangzhou, China, 2015, pp. 267–276.

[43] Jian Liu, et al., Snooping keystrokes with mm-level audio ranging on a single phone, in: Proc. ACMMobicom'15, ACM, Paris, France, 2015, pp. 142–154.

[44] Siqi Liu, et al., 3D mask face anti-spoofing with remote photoplethysmography, in: European Conference on Computer Vision, Springer, 2016, pp. 85–100.

[45] Xuefeng Liu, et al., Wi-sleep: contactless sleep monitoring via WiFi signals, in: Real-Time Systems Symposium (RTSS), IEEE, 2014, pp. 346–355.

[46] Yaojie Liu, Amin Jourabloo, Xiaoming Liu, Learning deep models for face anti-spoofing: binary or auxiliary supervision, in: Proceedings of the IEEE Conference on Computer Vision and Pattern Recognition, 2018, pp. 389–398.

[47] Yunfei Ma, Nicholas Selby, Fadel Adib, Drone relays for battery–free networks, in: Proc. ACMSIGCOMM2017, Los Angeles, CA, 2017, pp. 335–347.

[48] Daniel McDuff, Sarah Gontarek, Rosalind W. Picard, Remote detection of photoplethysmographic systolic and diastolic peaks using a digital camera, IEEE Transactions on Biomedical Engineering 61 (12) (2014) 2948–2954.

[49] Andreia V. Mogo, Sander Stuijk, Gerard de Haan, New insights into the origin of remote PPG signals in visible light and infrared, Scientific Reports 8 (1) (2018) 1–15.

[50] M.L.R. Mogue, B. Rantala, Capnometers, Journal of Clinical Monitoring (1988).

[51] Ramakrishna Mukkamala, et al., Toward ubiquitous blood pressure monitoring via pulse transit time: theory and practice, IEEE Transactions on Biomedical Engineering 62 (8) (2015) 1879–1901.

[52] Jayasimha N. Murthy, et al., Thermal infrared imaging: a novel method to monitor airflow during polysomnography, Sleep 32 (11) (2009) 1521–1527.

[53] Ramya Murthy, Ioannis Pavlidis, Noncontact measurement of breathing function, IEEE Engineering in Medicine and Biology Magazine (2006) 57–67.

[54] Samer S. Najjar, et al., Pulse wave velocity is an independent predictor of the longitudinal increase in systolic blood pressure and of incident hypertension in the Baltimore Longitudinal Study of Aging, Journal of the American College of Cardiology 51 (14) (2008) 1377–1383.

[55] Kazuki Nakajima, Yoshiaki Matsumoto, Toshiyo Tamura, Development of real-time image sequence analysis for evaluating posture change and respiratory rate of a subject in bed, Physiological Measurement (2001) N21.

[56] Rajalakshmi Nandakumar, Shyamnath Gollakota, Unleashing the power of active sonar, IEEE Pervasive Computing 16 (1) (2017) 11–15.

[57] Rajalakshmi Nandakumar, Shyamnath Gollakota, Nathaniel Watson, Contactless sleep apnea detection on smartphones, in: Proceedings of the 13th Annual International Conference on Mobile Systems, Applications, and Services, ACM, 2015, pp. 45–57.

[58] Rajalakshmi Nandakumar, et al., Fingerio: using active sonar for finegrained finger tracking, in: Proc. ACM CHI'16, ACM, Santa Clara, CA, 2016, pp. 1515–1525.

[59] Phuc Nguyen, et al., Continuous and fine-grained breathing volume monitoring from afar using wireless signals, in: Proc. IEEE INFO-COM'16, San Francisco, CA, 2016.

[60] Impinj Speedway Revolution Reader Application Note, Low level user data support, in: Impinj, Seattle, Washington, USA, 2013.

[61] Ewa Magdalena Nowara, Ashutosh Sabharwal, Ashok van Veeraragha, Ppgsecure: biometric presentation attack detection using photo- pletysmograms, in: 2017 12th IEEE International Conference on Automatic Face & Gesture Recognition (FG 2017), IEEE, 2017, pp. 56–62.

[62] Sarah Ostadabbas, et al., A passive quantitative measurement of airway resistance using depth data, in: 2014 36th Annual International Conference of the IEEE Engineering in Medicine and Biology Society, IEEE, 2014, pp. 5743–5747.

[63] Sarah Ostadabbas, et al., A vision-based respiration monitoring system for passive airway resistance estimation, IEEE Transactions on Biomedical Engineering 63 (9) (2015) 1904–1913.

[64] Neal Patwari, et al., Monitoring breathing via signal strength in wireless networks, IEEE Transactions on Mobile Computing 13 (8) (2013) 1774–1786.

[65] Ming-Zher Poh, Daniel J. McDuff, Rosalind W. Picard, Advancements in noncontact, multiparameter physiological measurements using a webcam, IEEE Transactions on Biomedical Engineering 58 (1) (2010) 7–11.

[66] Ming-Zher Poh, Daniel J. McDuff, Rosalind W. Picard, Non-contact, automated cardiac pulse measurements using video imaging and blind source separation, Optics Express 18 (10) (2010) 10762–10774.

[67] A.P. Prathosh, et al., Estimation of respiratory pattern from video using selective ensemble aggregation, IEEE Transactions on Signal Processing 65 (11) (2017) 2902–2916.

[68] S. Rasche, et al., Camera-based photoplethysmography in critical care patients, Clinical Hemorheology and Microcirculation 64 (1) (2016) 77–90.

[69] Y. Ren, et al., Fine-grained sleep monitoring: hearing your breathing with smartphones, in: Proc. IEEE INFOCOM'15, Hong Kong, China, 2015, pp. 1194–1202.

[70] Angel Melchor Rodriguez, J. Ramos-Castro, Video pulse rate variability analysis in stationary and motion conditions, Biomedical Engineering Online 17 (1) (2018) 11.

[71] J. Salmi, A.F. Molisch, Propagation parameter estimation, modeling and measurements for ultrawideband mimo radar, IEEE Transactions on Microwave Theory and Techniques 59 (11) (2011) 4257–4267.

[72] Longfei Shangguan, Zimu Zhou, Kyle Jamieson, Enabling gesture-based interactions with objects, in: Proc. ACMMobiSys 2016, Niagara Falls, NY, 2017, pp. 239–251.

[73] Dangdang Shao, et al., Noncontact monitoring breathing pattern, exhalation flow rate and pulse transit time, IEEE Transactions on Biomedical Engineering 61 (11) (2014) 2760–2767.

[74] Dangdang Shao, et al., Simultaneous monitoring of ballistocardiogram and photoplethysmogram using a camera, IEEE Transactions on Biomedical Engineering 64 (5) (2016) 1003–1010.

[75] N.H. Shariati, E. Zahedi, Comparison of selected parametric models for analysis of the photoplethysmographic signal, in: Proc. 1st IEEE Conf. Comput., Commun. Signal Process, Kuala Lumpur, Malaysia, 2005, pp. 169–172.

[76] Heung-Sub Shin, et al., Real time car driver's condition monitoring system, in: SENSORS, 2010 IEEE, IEEE, 2010, pp. 951–954.

[77] Jae Hyuk Shin, Kwang Suk Park, HRV analysis and blood pressure monitoring on weighing scale using BCG, in: 2012 Annual International Conference of the IEEE Engineering in Medicine and Biology Society, IEEE, 2012, pp. 3789–3792.

[78] Robin P. Smith, et al., Pulse transit time: an appraisal of potential clinical applications, Thorax 54 (5) (1999) 452–457.

[79] Vahid Soleimani, et al., Depth-based whole body photoplethysmography in remote pulmonary function testing, IEEE Transactions on Biomedical Engineering 65 (6) (2018) 1421–1431.

[80] Yu Sun, et al., Motion-compensated noncontact imaging photoplethysmography to monitor cardiorespiratory status during exercise, Journal of Biomedical Optics 16 (7) (2011) 077010.

[81] Chihiro Takano, Yuji Ohta, Heart rate measurement based on a timelapse image, Medical Engineering & Physics 29 (8) (2007) 853–857.

[82] L. Tarassenko, et al., Non-contact video-based vital sign monitoring using ambient light and auto-regressive models, Physiological Measurement 35 (5) (2014) 807.

[83] Van Nguyen Thi Phuoc, et al., Microwave radar sensing systems for search and rescue purposes, Sensors 19 (13) (2019) 2879.

[84] Alexander Trumpp, et al., Camera-based photoplethysmography in an intraoperative setting, Biomedical Engineering Online 17 (1) (2018) 1–19.

[85] Hsin-Yi Tsai, et al., A noncontact skin oxygen-saturation imaging system for measuring human tissue oxygen saturation, IEEE Transactions on Instrumentation and Measurement 63 (11) (2014) 2620–2631.

[86] Hsin-Yi Tsai, et al., A study on oxygen saturation images constructed from the skin tissue of human hand, in: 2013 IEEE International Instrumentation and Measurement Technology Conference (I2MTC), IEEE, 2013, pp. 58–62.

[87] Wim Verkruysse, Lars O. Svaasand, J. Stuart Nelson, Remote plethys-mographic imaging using ambient light, Optics Express 16 (26) (2008) 21434–21445.

[88] Wim Verkruysse, et al., Calibration of contactless pulse oximetry, Anesthesia and Analgesia 124 (1) (2017) 136.

[89] Mauricio Villarroel, et al., Continuous non-contact vital sign monitoring in neonatal intensive care unit, Healthcare Technology Letters 1 (3) (2014) 87–91.

[90] Mauricio Villarroel, et al., Non-contact vital sign monitoring in the clinic, in: 201712th IEEE International Conference on Automatic Face & Gesture Recognition (FG 2017), IEEE, 2017, pp. 278–285.

[91] Chuyu Wang, et al., Rf-ecg: heart rate variability assessment based on cots rfid tag array, Proceedings of the ACM on Interactive, Mobile, Wearable and Ubiquitous Technologies 2 (2) (2018) 1–26.

[92] Hao Wang, et al., Human respiration detection with commodity wifi devices: do user location and body orientation matter?, in: Proceedings of the 2016 ACM International Joint Conference on Pervasive and Ubiquitous Computing, ACM, 2016, pp. 25–36.

[93] Ju Wang, et al., Are RFID sensing systems ready for the real world?, in: Proc. ACMMobiSys 2019, New York, NY, 2019, pp. 366–377.

[94] Ju Wang, et al., TagScan: simultaneous target imaging and material identification with commodity RFID devices, in: Proc. ACMMobiCom, Snowbird, Utah, 2017, pp. 288–300.

[95] Junjue Wang, et al., Ubiquitous keyboard for small mobile devices: harnessing multipath fading for fine-grained keystroke localization, in: Proc. ACM Mobisys'14, ACM. Bretton, Woods, NH, 2014, pp. 14–27.

[96] WeiWang, Alex X. Liu, Ke Sun, Device-free gesture tracking using acoustic signals, in: Proc. ACM MobiCom'16, ACM, New York City, NY, 2016, pp. 82–94.

[97] Wenjin Wang, Sander Stuijk, Gerard De Haan, Exploiting spatial redundancy of image sensor for motion robust rPPG, IEEE Transactions on Biomedical Engineering 62 (2) (2014) 415–425.

[98] Wenjin Wang, Sander Stuijk, Gerard De Haan, Unsupervised subject detection via remote PPG, IEEE Transactions on Biomedical Engineering 62 (11) (2015) 2629–2637.

[99] Wenjin Wang, et al., Robust heart rate from fitness videos, Physiological Measurement 38 (6) (2017) 1023.

[100] X. Wang, C. Yang, S. Mao, On CSI-based vital sign monitoring using commodity WiFi, ACM Transactions on Computing for Healthcare 1 (3) (2020) 12:1–12:27, https://doi.org/10.1145/3377165.

[101] X. Wang, C. Yang, S. Mao, Resilient respiration rate monitoring with realtime bimodal CSI data, IEEE Sensors Journal 20 (17) (2020) 10187–10198, https://doi.org/10.1109/JSEN.2020.2989780.

[102] Xiangyu Wang, et al., On remote temperature sensing using commercial UHF RFID tags, IEEE Internet of Things Journal 6 (6) (2019) 10715–10727.

[103] Xuyu Wang, Lingjun Gao, Shiwen Mao, CSI phase fingerprinting for indoor localization with a deep learning approach, IEEE Internet of Things Journal 3 (6) (2016) 1113–1123.

[104] Xuyu Wang, Runze Huang, Shiwen Mao, Demo abstract: sonarbeat: sonar phase for breathing beat monitoring with smartphones, in: Sensing, Communication, and Networking (SECON), 201714th Annual IEEE International Conference on. IEEE, 2017, pp. 1–2.

[105] Xuyu Wang, Runze Huang, Shiwen Mao, SonarBeat: sonar phase for breathing beat monitoring with smartphones, in: Computer Communication and Networks (ICCCN), 2017 26th International Conference on. IEEE, 2017, pp. 1–8.

[106] Xuyu Wang, Chao Yang, Shiwen Mao, PhaseBeat: exploiting CSI phase data for vital sign monitoring with commodity WiFi devices, in: Distributed Computing Systems (ICDCS), 2017IEEE 37th International Conference on. IEEE, 2017, pp. 1230–1239.

[107] Xuyu Wang, Chao Yang, Shiwen Mao, ResBeat: resilient breathing beats monitoring with realtime bimodal CSI data, in: GLOBECOM 2017-2017 IEEE Global Communications Conference, IEEE, 2017, pp. 1–6.

[108] Xuyu Wang, Chao Yang, Shiwen Mao, Tensorbeat: tensor decomposition for monitoring multiperson breathing beats with commodity WiFi, ACM Transactions on Intelligent Systems and Technology (TIST) 9 (1) (2017) 8.

[109] Xuyu Wang, et al., CSI-based fingerprinting for indoor localization: a deep learning approach, IEEE Transactions on Vehicular Technology 66 (1) (2017) 763–776.

[110] Teng Wei, Xinyu Zhang, Gyro in the air: tracking 3D orientation of batteryless Internet-of-things, in: Proc. ACMMobiCom'16, New York City, NY, 2016, pp. 55–68.

[111] Fokko P. Wieringa, Frits Mastik, Antonius F.W. van der Steen, Contactless multiple wavelength photoplethysmographic imaging: a first step toward "SpO 2 camera" technology, Annals of Biomedical Engineering 33 (8) (2005) 1034–1041.

[112] F.P. Wieringa, et al., in: In Vitro Demonstration of an SpO2-Camera, IEEE, 2007.

[113] Y.S. Yan, Y.T. Zhang, Noninvasive estimation of blood pressure using photoplethysmographic signals in the period domain, in: 2005 IEEE Engineering in Medicine and Biology 27th Annual Conference, IEEE, 2006, pp. 3583–3584.

[114] Y.S. Yan, Y.T. Zhang, Noninvasive estimation of blood pressure using photoplethysmographic signals in the period domain, in: 2005 IEEE Engineering in Medicine and Biology 27th Annual Conference, IEEE, 2006, pp. 3583–3584.

[115] C. Yang, X. Wang, S. Mao, AutoTag: recurrent vibrational autoencoder for unsupervised apnea detection with RFID tags, in: Proc. IEEE GLOBECOM2018, Abu Dhabi, United Arab Emirates, 2018, pp. 1–7.

[116] C. Yang, X. Wang, S. Mao, Respiration monitoring with RFID in driving environments, IEEE Journal on Selected Areas in Communications 39 (2) (2021), https://doi.org/10.1109/JSAC.2020.3020606.

[117] C. Yang, X. Wang, S. Mao, Unsupervised detection of apnea using commodity RFID tags with a recurrent variational autoencoder, IEEE Access Journal 7 (1) (2019) 67526–67538, https://doi.org/10.1109/ACCESS.2019.2918292.

[118] C. Yang, X. Wang, S. Mao, Unsupervised drowsy driving detection with RFID, IEEE Transactions on Vehicular Technology 69 (8) (2020) 8151–8163, https://doi.org/10.1109/TVT.2020.2995835.

[119] C. Yang, et al., SparseTag: high-precision backscatter indoor localization with sparse RFID tag arrays, in: Proc. IEEE SECON2019, Boston, MA, 2019.

[120] Lei Yang, et al., Tagoram: real-time tracking of mobile RFID tags to high precision using COTS devices, in: Proc. ACM MobiCom'14, Maui, HI, 2014, pp. 237–248.

[121] Z. Yang, Z. Zhou, Y. Liu, From RSSI to CSI: indoor localization via channel response, ACM Computing Surveys 46 (2) (2013).

[122] Z. Yang, et al., Monitoring vital signs using millimeter wave, in: Proc. IEEE MobiHoc'16, Paderborn, Germany, 2016.

[123] Sangki Yun, Yi-Chao Chen, Lili Qiu, Turning a mobile device into a mouse in the air, in: Proc. ACM MobiSys'15, ACM, 2015, pp. 15–29.

[124] Jian Zhang, et al., RFHUI: an intuitive and easy-to-operate human- UAV interaction system for controlling a UAV in a 3D space, in: Proc. EAI MobiQuitous 2018, New York City, NY, 2018, pp. 69–76.

[125] Qi Zhang, et al., Webcam based non-contact real-time monitoring for the physiological parameters of drivers, in: The 4th Annual IEEE International Conference on Cyber Technology in Automation, Control and Intelligent, IEEE, 2014, pp. 648–652.

[126] Xiao Zhang, et al., Evaluating photoplethysmogram as a real-time cognitive load assessment during game playing, International Journal of Human-Computer Interaction 34 (8) (2018) 695–706.

[127] Cui Zhao, et al., RF-mehndi: a fingertip profiled RF identifier, in: Proc. IEEE INFOCOM 2019, Paris, France, 2019, pp. 1513–1521.

Part I

Camera-based vital signs monitoring

Chapter 2

Physiological origin of camera-based PPG imaging

Alexei A. Kamshilin[a,b] and Oleg V. Mamontov[b]

[a]*Institute of Automation and Control Processes of Far Eastern Branch of the Russian Academy of Sciences, Vladivostok, Russia,* [b]*Department of Circulation Physiology, Almazov National Medical Research Center, St. Petersburg, Russia*

Contents

2.1 Introduction

The term "photoelectric plethysmography", later transformed to "photoplethysmography" (PPG), was first introduced by Hertzman in 1938 [1]. This technology was proposed to estimate the blood supply of various skin areas. In this pioneering work, the capability of the PPG waveform to follow the time varying changes in blood volume of superficial vessels below the skin during the cardiac cycle was demonstrated. Hertzman called the photoelectrically detectable waveform a pulse volume. Interestingly, over 80 years ago, it was claimed that the amplitude of the pulse volume reflects changes in the blood supply caused by painful and psychic stimuli, cold, amyl nitrite, voluntary apnea, and the Valsalva experiment [1]. Moreover, the parameters affecting the amplitude of PPG waveform were listed and discussed in this pioneering work, among them skin movement relative to the sensing element, variations and spectrum of the illumination, character of the contact of the device with the skin, and the ratio of the reduced to the oxygenated hemoglobin. In 1972, Takuo Aoyaqi proposed to exploit the dependence of the light modulation on the hemoglobin state for noninvasive assessment of arterial-blood-oxygen saturation with an optical device referred to as the pulse oximeter [2,3]. Since 1983, pulse oximeters became commercially available. Today these devices have been adopted to carry out routine measurements in clinics worldwide. The remarkable feature of the PPG waveform to be modulated in time due to the heart contractions is commonly used for estimation and monitoring the heart rate. Currently, PPG modules built into ubiquitous smartphones and watches are used for heart-rate monitoring [4,5]. Moreover, PPG probe serves as a sensitive element in the FINAPRES (an acronym for FINger Arterial PRESsure) system used for continuous measurement of arterial blood pressure [6]. In recent years, there has been increasing interest in expanding the use of PPG beyond these areas. Several studies have been undertaken to prove the ability of the PPG system for assessing pulse-rate variability, pulse-arrival time, vascular tone, tissue perfusion, sympathetic activity, etc. [7–18]. However, despite intensive investigations over a long time and the wide range of applications, the physiological model for the formation of the PPG waveform remains a subject of continuing debate.

In this chapter, we discuss various factors that affect the light modulation of tissue containing vessels with traveling red blood cells (RBC). The most attention will be paid to the reflection mode of the light–tissue interaction, which is typical for a camera-based PPG imaging system. The chapter is organized as following. First, we describe the conventional model of the PPG-waveform appearance, and then the inability of this model to explain the observed largest pulsation amplitude under green illumination is pointed out. Thereafter, an alternative PPG model considering compression of the capillary bed by large arteries is presented and discussed. The rest of the chapter is devoted to a description of experimental observations that support the alternative model.

2.2 Conventional PPG model: blood volume modulation

In any PPG system, a biological tissue is illuminated by an incoherent light source, and the power of light either transmitted through or reflected from the tissue is measured by a photodetector. Correspondingly, there are two modes of operation: transmittance and reflection. In most camera-based PPG experiments, a light source (including ambient illumination) and a light-sensitive camera are located on one side of the biological tissue to be studied, i.e., the PPG imaging system operates in the reflection mode. As is well known, after the light interaction with living tissue containing blood vessels, its intensity acquires modulation in time with the heart rate [19]. The conventional model of the PPG waveform generation was developed on the basis of experimental observations exploiting a pair of light sources and a photodetector in contact with the tissue. It assumes that the main reason for the light-intensity modulation registered by the photodetector is the variation of the blood volume in the tissue [1,12,20–22]. Further in the Chapter, the conventional model is called a blood-volume (BV) model. This modulation mechanism is easy to understand in the transmittance-mode PPG operating at the near infrared: An increase of blood volume absorbs more photons resulting in a decrease of light intensity measured by the photodetector and vice versa. It is less obvious why the inverse relationship between remitted light power and blood volume should hold in the reflectance-mode PPG, although experimental observations seem to support this standpoint. Therefore, it is generally accepted that PPG measures blood-volume variations in the vascular bed [7,23–26]. The essence of this understanding of PPG signal formation is expressed by the fact that these systems are often called pulse-volume monitor [27–30].

An example of the PPG waveform measured by a camera-based system in the subject's cheek is shown in Fig. 2.1. In this particular example, the subject's face was illuminated by six LEDs arranged around the camera lens, emitting green light at 530 nm with the spectral bandwidth of 40 nm [9]. Raw waveform calculated as a time concatenation of average pixel values in the consecutive frames is shown in Fig. 2.1(A). Averaging was carried out within a selected area sized 7×7 pixels (about 2×2 mm^2). This waveform is presented without any filtering. As can be seen, there are two different terms: an alternating component (AC) with modulation at a frequency of about 1 Hz and a slowly varying component (DC). Both components are proportional to the incident light intensity [31]. By calculating the AC/DC ratio, it is possible to compensate for the spatial unevenness of tissue illumination. In addition, subtracting the unity from this ratio during time-dependence analysis allows one to deal with waveforms with an average level of zero. Usually, AC/DC ratio is measured as a percentage.

For a clearer view of the waveform details, the five-second segment of the PPG waveform in Fig. 2.1(A) is shown in Fig. 2.1(B) after transformations and inversion of the sign. The graph in Fig. 2.1(C) shows a respective segment of the electrocardiogram (ECG) that was recorded simultaneously with video. During recording, the digital camera and ECG were synchronized with an accuracy of one ms. One can clearly see in Fig. 2.1 (B and C) that variations of the light intensity follow the cardiac ejections indicated by R-peaks on the ECG. It should be noted that inverted AC-components correlate positively with variations of arterial blood pressure [32,33]. The relationship of PPG waveform with changes of blood volume was supported by the observed similarity between PPG and simultaneously measured volume of a limb by the strain gauge [34], and by correlation between PPG and arterial diameters measured by Doppler ultrasound [35]. In the systole phase, when the arterial blood pressure reaches its maximum, vessels walls are expanded leading to the maximal momentary blood volume, with reversion to the minimum blood volume as blood runs off in diastole [28]. Therefore, the higher the blood pressure, the greater the blood volume in the arteries, leading to increased light absorption and, accordingly, to a decrease in the pixels value of the camera. Consequently, the minima of the AC/DC ratio shown in Fig. 2.1(C) are reached at the end of diastole when

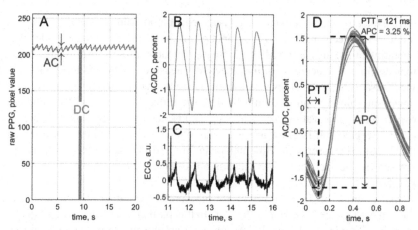

FIGURE 2.1 Typical PPG waveform measured by an imaging PPG system in a subject's cheek under green-light illumination. (A) Raw signal averaged in the region of 7 × 7 pixels (2 × 2 mm²) without any filtration. (B) The ratio of AC and DC components after low-pass filtering (passband frequency of 20 Hz) and sign inversion. (C) ECG signal synchronously recorded with video frames. (D) One-cardiac-cycle pulse wave (thick green line (mid gray in print version)) obtained after averaging of particular pulse waves (thin gray lines) over 30 cycles.

the blood pressure starts to increase. To verify PPG–ECG correlation, we plotted in Fig. 2.1(D) the PPG pulses of 30 subsequent cardiac cycles, so that each respective R-peak of the ECG is at the beginning of the time scale [9]. The thick greenish line in Fig. 2.1(D) shows an average PPG pulse obtained by averaging the filtered one-cardiac-cycle PPG pulses shown by thin gray lines. The average PPG pulse was used to estimate two important parameters of the PPG waveform. The first is the amplitude of the pulsatile component (APC) which is defined as the difference between maximum and minimum values of the average PPG pulse (Fig. 2.1D). This parameter is usually expressed in percentages. The second is the transit time of the pulse wave from the heart to the measuring point, PTT (pulse transit time), which is the time delay between the R-peak (zero of the abscissa axis in Fig. 2.1D) and the minimum of the average PPG pulse (yellow circle (ligth gray in print version) in Fig. 2.1D). Note that the APC is the amplitude of the AC/DC ratio, which is also referred to as a perfusion index [8,12,36].

2.3 How to explain the largest modulation of the green light?

It should be mentioned that the BV-model, according to which the PPG waveform is treated as originating from pulsatile blood-volume modulation, was developed considering experimental data obtained by contact oximeters operating with red and infrared lights. However, several research groups reported that the largest heartbeat-related AC/DC modulation was observed at the green light [31,37–39]. The conventional BV-model fails to explain these observations because of the inability of green light to interact directly with pulsatile blood vessels. Usually green light does not penetrate human skin deeper than 0.5 mm [40,41]. There is no pulsating arteries and arterioles at this depth [42], whereas the pulsatile capillary pressure [43,44] and variance in time velocity profile of RBCs [45] are unlikely to exert true blood volume modulation in a single capillary due to the rigid nature of these vessels [46,47].

Nevertheless, many researchers still believe that the conventional BV-model is able to explain features of the PPG waveform with green light, giving the following argument. While the peak of the depth influence of green light is shown to be less than 0.5 mm, some light returning to the surface still can be seen to penetrate deeper into the tissue, most likely encountering subcutaneous arterioles that are probably pulsating. Consequently, this part of the returned light might carry pulsatile (at the heart rate) information. To verify the correctness of this statement, let us estimate the fraction of photons that can reach the arterioles, assuming their location at a depth of 1.2 mm from the stratum corneum. By using statistical analysis of photon diffusion in tissue [48], the probability of the green-light ($\lambda = 560$ nm) fluence at this depth is estimated as $8.25 \times 10^{-4}\%$. This is 3×10^5 times smaller than the fluence probability in the peak depth of 0.5 mm. Most of the light interacts with non-pulsating capillaries, which returns to the detector as the DC component of the PPG. It means that the capillary bed plays the role of efficient screening for light interaction with deeper situated pulsating arterioles. Consequently, the theoretical estimation of AC/DC gives us 0.0032%. In contrast, experimentally measured AC/DC as shown in Fig. 2.1(D) as three orders of value higher (3.25%). This waveform was recorded in the reflection mode by a digital monochrome camera when the subject's face was illuminated by green light (the wavelength of 530 ± 20 nm). RBCs have a peak of light absorption at this wavelength, which leads to the very shallow penetration depth of green light into tissue with blood vessels. Recent rigorous simulation of multiwavelength light–tissue interaction by using a Monte-Carlo model [49] also revealed that green light does not reach the arterioles, interacting exclusively with the upper capillaries.

Therefore, the question arises: Where does the clearly observed modulation at the heartbeat frequency (Fig. 2.1) come from? The same question arises also when analyzing experimental data obtained in PPG systems exploiting a color digital camera. These studies always emphasize that the largest intensity modulation by heartbeats is observed precisely in the green channel of the camera [31,37,39]. It is worth noting that camera-based PPG systems collect light that has traveled through much shallower tissue depths over much smaller distances than takes place in conventional contact PPG probes of oximeters [50,51]. This difference may be due to different net angles of multiple light scattering, which is smaller in the PPG contact sensor, thus forming the characteristic "banana-shaped" optical path. We believe that the heart-related modulation in camera-based PPG systems originates mainly due to mechanical compression/decompression of the intercapillary tissue caused by pulsatile

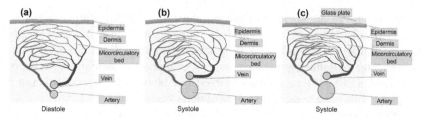

FIGURE 2.2 Simplified diagram of the new concept of the PPG signal formation. (a) A part of the artery that is situated near the dermis in the diastole phase; (b) the artery in the systole phase in the case of free skin surface; (c) the systole phase in the case of skin contact with the glass. Note that the density of vessels in the microcirculatory bed is the lowest in (a) and the highest in (c).

transmural arterial pressure as was suggested by an alternative PPG model in 2015 [33]. Further, this alternative model is called the tissue-compression (TC) model.

2.4 Alternative PPG model: tissue compression modulation

The concept of the TC-model [33] is schematically illustrated in Fig. 2.2. In the end-diastole phase (Fig. 2.2(a)), the arterial pressure is at its minimum, which suggests that tissue located between the artery and epidermis is under minimal compression. In this phase, the reflected light intensity is maximal [37,52], and the inverted AC/DC ratio reaches its minimum, indicated by a yellow circle (light gray in print version) in Fig. 2.1(D). Then the fast increase of the arterial pressure during the systole provides the force affecting the surrounding tissue. The resulting compression of the tissue depends on the boundary conditions, which are defined by the elastic properties of the stratum corneum and epidermis.

Due to the dermis compression, the distance between adjacent capillaries diminishes [53] and leads to an increase of the capillary density that follows the transmural pressure in the location of the measurement. The blood volume interacting with light is certainly larger in more compressed dermis resulting in increased light absorption and a respective growth in the inverted AC/DC ratio after the end-diastole (Fig. 2.1D). It should be noted that mechanical deformations might lead also to changes in the orientation or structure of the connective tissue. Consequently, both the absorption and scattering coefficients of the compressed tissue increase [53], resulting in a decrease in the back-reflected light intensity. However, the relationship between the influence of changes in absorption and scattering coefficients on the formation of the PPG waveform has not yet been established and may vary from one subject to another. The degree of compression depends both on the pressure exerted by the walls of the artery on the adjacent tissue and on the boundary conditions.

2.5 Boundary conditions and influence of skin contact

The effect of the boundary conditions on the degree of dermis compression is illustrated in Fig. 2.2. If the skin motion is not limited by any external contact (Fig. 2.2B), the compression depends primarily on the skin's elasticity. As it was experimentally shown, tighter skin, typical of older people, leads to a greater degree of compression and, accordingly, to a higher APC [54,55]. Note that skin elasticity usually varies spatially across skin areas. In the frames of the TC-model, this variability can explain typically observed heterogeneity of the spatial distribution of APC. Moreover, increased APC is observed in the area of moles or scars, which are also characterized by altered skin elasticity. Dermis compression in the systole phase for the contactless setting is shown in Fig. 2.2(b). One can increase the degree of dermis compression by restricting the skin motion with a glass plate lightly contacting the skin as illustrated in Fig. 2.2(c). A multiple increase in APC (up to seven times) after slight skin contact with the glass plate was experimentally demonstrated by applying imaging PPG with green light to assess blood pulsations in subjects' palms [33,56]. Such a large enhancement of the heart-related modulation amplitude cannot be explained in the framework of the conventional BV-model. Note that the external pressure caused by the glass-plate contact in these experiments was on average of 35-mm Hg [55], thus preventing occlusion of the blood vessels. The frequently observed improvement of the waveform shape by an external force in contact PPG probes [25,57,58] can be also associated with changes in mechanical properties of the tissue. Noise-like PPG waveforms observed without skin-probe contact might be caused by inefficient transmission of the pressure wave from the arteries to the capillary bed, whereas the contact restricts skin motion and provides conditions under which pulsatile arterial pressure efficiently modulates the capillary density.

2.6 Pulsatile dermis compression and modulation of IR light

Even though the reflection-mode oximeters operate at the frequencies of red and near-infrared light, which penetrates much deeper than green light, the influence of the dermis compression should be also significant for the infra-red light because any light interacts twice with upper level of the dermis before reaching the photo-receiver. Such an influence was observed in the experiments with simultaneous measurements of PPG signals at green (525 nm) and NIR (810 nm) illuminations in the palms of 34 healthy subjects [56]. In these experiments, the temporal modulation of both wavelengths was assessed in a quasi-synchronous mode. A single monochrome camera was used to record images of the subjects' palms when they were illuminated with rapidly switching green and NIR light emitted by different LEDs synchronized with the camera. A trigger pulse generated by the camera at the beginning of each frame switched on either green or NIR LEDs. Thus, each even-numbered camera frame was recorded when the palm was illuminated by green LEDs, but each odd-numbered frame by NIR

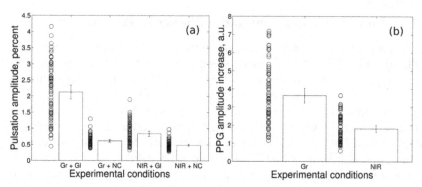

FIGURE 2.3 Influence of the skin-glass contact on the PPG amplitude measured at green and NIR wavelengths. (a) Average PPG amplitude measured in the contact experiment (Gl) and in the same areas in the contactless experiment (NC). Here Gr and NIR stand for green (525 nm) and NIR (810 nm) illumination, respectively. (b) Ratio of the mean amplitude measured in the glass-contact to that in the contactless experiment for both wavelengths. The individual measurements are shown by circles, whereas the bars represent mean values for the whole data sets.

LEDs [56]. A strong positive correlation between the changes caused by the skin contact with a glass plate in both wavelengths was found, suggesting that the same signal was read independently from the depth of penetration. Specifically, changes in APC due to the skin contact were found to be correlated for the two wavelengths both in space and in amplitude. The glass plate increased APC measured at NIR illumination at the same locations as with green-light illumination [56]. It was also found that the contact with glass increased the light modulation up to seven times with green illumination and up to 3.6 times with NIR illumination (see Fig. 2.3b).

Analysis of the data shows that an essential part of remitted NIR light is modulated in time as a result of elastic deformations of dermis caused by variable blood pressure in the arteries [56]. These observations suggest that, in contrast with the classical BV-model, photoplethysmographic waveform originates from the density modulation of the upper capillaries due to the variable pressure applied to the dermis from large blood vessels. Given that camera-based PPG imaging is at the heart of the development of contactless methods for measuring oxygen saturation [50,59], application of the TC model in the development of algorithms for new SpO_2 measurement systems can improve the accuracy of their calibration.

2.7 Light modulation in a single capillary

The alternative TC-model is based on the assumption that blood volume modulation in a single capillary at a heart rate is unlikely. Let us take a closer look at this issue. On the one hand, the average density of capillaries in human tissue is approximately $600/mm^3$, which implies a mean separation of about 40 μm between adjacent capillaries, with their average length 1.1 mm and their average diameter about 8 µm [60]. On the other hand, the average size of human RBCs is of the same order of value [61]. In these conditions, one hardly can assume that the amplitude of blood volume pulsations in the same vessels might grow up to seven times after slight contact with glass. Nevertheless, such a large increase in the pulsation amplitude when the skin motion is restricted by light contact with a glass plate (see Fig. 2.3b) was experimentally demonstrated [56].

Detailed study of the PPG signal formation in the spatially resolved images of nailfold capillaries provided by a microscope [45] revealed that the spatial distribution of APC does not always coincide with the position of visible capillaries (see Fig. 2.4). Note that, unlike the experiments discussed in the previous sections that used a monochrome camera and active illumination by green light, this study used a color camera and white light. However, only the green channel of the color camera was processed as having the highest SNR [45]. As seen in the lower part of Fig. 2.4(b) (in the region with higher density of capillaries), the amplitude of pulsations is higher than in the upper part with a single layer of capillaries, and it has almost uniform distribution. Despite the significant amplitude of heart-related pulsations of the reflected green light, no modulation of the capillaries' diameter in the time course was observed in magnified capillary images [45]. However, it should be noted that the RBC speed in each capillary was found to be strongly modulated following the heartbeats. Nevertheless, the modulation of RBC speed cannot directly lead to the light-intensity modulation due to the relatively high speed of RBCs and the strong difference of the average speed (more than ten times) between adjacent capillaries. It should be noted that the observed speed of erythrocytes in capillaries is so high that it leads to their significant displacement in capillaries during the timescale of heartbeats. Typical frame-exposure time accepted for the signal acquisition in imaging PPG systems is about 25 ms. During this period, the number of erythrocytes in a region of interest is independent of their speed, if we assume their density is almost constant.

Under this assumption, the number of photons absorbed by moving erythrocytes is proportional to the exposure time and mean cross-section of RBCs. Therefore, modulation of RBC speed cannot directly lead to the light intensity modulation [45]. In addition, significant diversity of the RBC speed observed in various different capillaries cannot lead to the in-phase modulation of the remitted light intensity in the whole area of observation.

While the conventional BV-model fails to explain the features of RBC dynamics observed under a microscope, the alternative TC-model [33] does this successfully. It is worth noting that Volkov et al. [45] also suggested the presence of one more mechanism capable of modulating the absorption of light by RBCs due to their speed modulation in nonpulsatile capillaries. It is based on the experimental evidence of the light-absorption dependence on the RBCs orientation and aggregation that varies with their speed [62,63]. With an increase in RBCs speed, a decrease in light absorption and scattering is usually observed [63]. However, such a dependence means that the light intensity measured in the

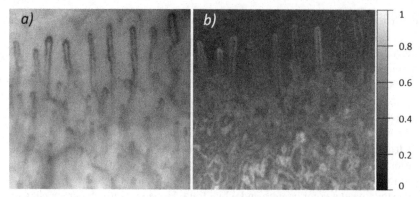

FIGURE 2.4 Mapping the amplitude of PPG waveform. (a) Microscopic image of capillaries. (b) Spatial distribution of the PPG-signal amplitude calculated for the green channel of the image. The color scale on the right side shows the amplitude of the normalized PPG signal (maximum value is 1).

systole (when the RBC speed is higher) will be higher than in the diastole. In other words, this modulation is in counter phase to that predicted by either BV or TC model. Therefore, the mechanism of light modulation due to changes in the RBCs shape/orientation could just diminish experimentally observed APC.

2.8 Irregularity of RBC motion

The predictive ability of the TC-model was demonstrated recently in a new method for capillary visualization in any part of the body [64]. This method is based on the difference of PPG waveforms recorded under a microscope in areas with and without visible skin capillaries containing moving RBCs. It was shown [64] that the modulation of light in the regions between the capillaries has a much more regular shape than in the capillary itself. Fig. 2.5 shows PPG waveforms and their spectra recorded in four regions of interest (ROI) with a size of 4.1×4.1 μm^2. The color of the curves in Fig. 2.5 identifies the ROI position: Red (mid gray in print version) and pink (light gray in print version) were located inside the capillary with moving RBCs, whereas blue (dark gray in print version) and green (gray in print version) were outside the capillaries.

As seen in Fig. 2.5, PPG waveforms outside capillaries (blue and green curves) have regular modulation at the heartbeats, whereas the light intensity modulation is much more random in ROIs selected within capillaries (red and pink curves). We explain this difference as follows. Random, noise-like modulation originates from the inhomogeneity of RBC density in capillaries and their motion. Some of the RBCs are joined together forming aggregates of various lengths [65]. Motion of these aggregates separated by plasma produces noise-like light flickering. Since RBC-plasma sequences and their speeds are different in different capillaries, multiple light interaction with various capillaries smooths over the flickering and leads to more regular modulation of the heart

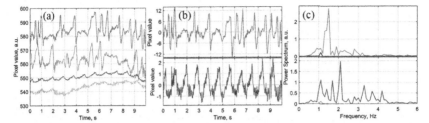

FIGURE 2.5 Particular realizations of PPG waveforms and their spectra. (a) Evolution of the mean pixel value (without any filtration) estimated in four selected ROIs: red (mid gray in print version) and pink (light gray in print version) are inside the capillaries with moving RBCs, whereas blue (dark gray in print version) and green(gray in print version) are outside. (b) PPG waveforms after removal of slowly varying changes in the red and blue ROIs. (c) Power spectra of PPG waveforms in all the selected ROIs.

rate. We suggest that the origin of the heart-related light modulation outside the capillaries is the compression/decompression of the dermis by the pulsatile arteries. Therefore, the superficial capillary layer can be considered as a distributed sensor for monitoring the parameters of blood pulsations occurring in the arteries.

2.9 Occlusion plethysmography

The TC-model served also as a starting point for the development of new applications of the PPG system as a monitor for peripheral blood flow changes during venous occlusion. In this physiological maneuver, return of venous blood is briefly interrupted for four–ten s by inflating a brachial cuff still allowing arterial blood flow into the limb. The rate of blood flow is assessed by estimating the changes in the limb's volume caused by the incoming blood. Usually it is done with either a strain gauge (which measures variations of limb's circumference) [66] or air plethysmography (which measures pressure changes in an air chamber contacting the limb) [67]. There is contact between the skin and sensor in both plethysmography techniques. Evidently, the arterial blood inflow during venous occlusion leads to the increase of the limb's volume and may compress the tissue between the skin and veins. Taking into account the TC-model, we hypothesized that such compression can lead to an increase in the green-light absorption in the capillary layer and therefore to a decrease in the signal measured by an imaging PPG system. This hypothesis was experimentally confirmed [68], resulting in a proposal for a contactless method for the assessment of blood flow reactivity in the limbs [69]. In fact, during the venous occlusion, the blood is accumulating in venules and veins, which is observed by an imaging PPG system as a linear decrease in the DC component.

The typical evolution of the PPG signal during multiple venous occlusions is shown in Fig. 2.6. Note that it was obtained under green illumination of the forearm [69]. Green arrows (dark gray in print version)indicate the beginning of

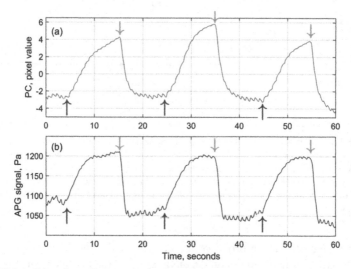

FIGURE 2.6 Comparison of imaging-PPG and air-plethysmography (APG) waveforms in response to triple venous occlusion. (a) PPG waveform, (b) APG waveform. While green arrows (dark gray in print version) show the beginning of each occlusion, the red arrows (gray in print version) indicate the moments of occlusion cuff deflation.

each venous occlusion, whereas the red arrows (gray in print version) show the end of each occlusion. Similarly to the conventional processing of PPG waveforms (see Sect. 2.2), here we subtracted the average value of the signal and changed the sign. One can clearly see in Fig. 2.6(a) that the PPG signal linearly increases immediately after the beginning of the cuff inflation and sharply returns to the initial level after the cuff deflation. This waveform is very similar to the simultaneously recorded, gold standard air-plethysmography (APG) signal shown in Fig. 2.6(b): The beginning and end of each occlusion event are well-synchronized. Moreover, small oscillations at the frequency of about 1 Hz attributed to arterial blood pulsations at the heart rate are in phase in both waveforms. It was found that the heart-related pulsations are resolved even in the occlusion state, but their amplitude (APC) is smaller than before occlusion [69]. The observed diminishment of the heart-related pulsations in the course of occlusion (Fig. 2.6) can be explained by the decrease in the pressure difference applied to a capillary due to the increased pressure in the venules.

Again in this experiment, the venous occlusion was monitored using green light that penetrates only into the superficial layer of the dermis. Therefore, the reason for the observed light-absorption increase is the increase of the blood volume in the superficial layer. This increase can be explained in the framework of the TC-model. Blood accumulating in the veins increases the pressure on the upper layer of the capillaries and squeezes this layer, reducing the distance between the capillaries. Once again, the capillary bed serves as a distributed sensitive network for monitoring the state of deep blood vessels, both arteries and veins. The change in green-light absorption during occlusion can be also explained by an increase in blood volume in small venules that are situated closer to the superficial layer of capillaries than arterioles. Currently, the details of the mechanism for the light modulation in the capillary bed is still under debate. Additional theoretical and experimental studies are required to determine the relationship between the two mechanisms (BV and TC) of green-light modulation by the blood-filling veins during venous occlusion. Nevertheless, recent mathematical simulations of light penetration into the dermis by using the Monte-Carlo method [49,70] demonstrated the low probability that green light interacts with venules or arterioles, thus supporting the predominant role of the TC-model in the formation of a response to venous occlusion. In any case, both mechanisms are able to explain the experimentally observed independence of the PPG-signal-changes rate from the observation point. It was shown that the response of the green-light PPG system to venous occlusion does not depend on the positioning of the observation point along the limb (upper or lower), or surrounding it [69,71].

2.10   Peculiarities of light interaction with cerebral vessels

The previous sections discussed the formation of a PPG waveform during the light interaction with skin capillaries, whereas in this section the PPG signal originating from vessels of the cerebral cortex is considered. Unlike the network of vessels supplying the dermis with blood, the vascular bed of cerebral vessels is arranged differently. As known, the main blood supply to the brain is carried out by meningeal vessels penetrating the brain tissue from the side of the convexital surface [72]. These vessels depart from the Willis circle as paired anterior, middle, and posterior cerebral arteries and further ramify on gyri of the cerebrum, gradually decreasing in diameter from 700 to 40 μm from the conducting to the precortical arteries. Next, from each precortical artery, two–three cortical arteries depart, which penetrate into the cortical space at right angles [72]. Therefore, pulsatile arteries of medium and small caliber are located on the surface of the cerebral cortex and can directly interact with green light. Accordingly, oscillations of the blood volume in the arteries primarily lead to the modulation of green light with the heart rate. This fact was confirmed in the experiments with rats carried out in our group [73]. Fig. 2.7 shows mapping of APC in the open brain cortex.

From the network of blood vessels resolved in the instant, unprocessed frame shown in Fig. 2.7(a), it is difficult to determine which are the arteries and which are the veins. Nevertheless, mapping of the pulsation-amplitude parameter, APC, visualizes exclusively arteries, which is clearly seen in Fig. 2.7(b) where the arteries are highlighted by red (dark gray in print version) as having much higher amplitudes of pulsations. Since in this case green light initially interacts with pulsatile vessels, its heart-related modulation can be explained in frames of the BV-model. However, we cannot completely exclude the presence

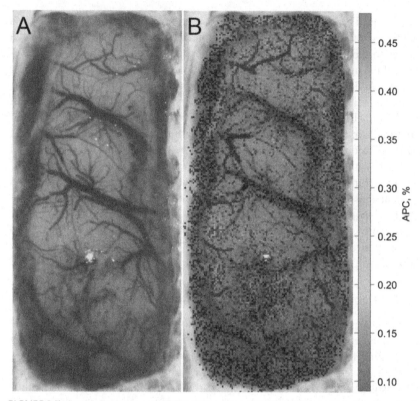

FIGURE 2.7 Spatial distribution of APC in brain cortex. (A) One of the recorded video frames of the open rat's brain, and (B) spatial distribution of APC overlaid with the initial image. The color scale on the right shows AC/DC ratio in percent.

of light modulation due to periodical compression of the brain tissue by large pulsating arteries (TC-model), some of them situated below the visible cortex. The influence of the TC-model becomes more pronounced in an exemplary APC map (see Fig. 2.8) obtained in the human cortex that was opened up during neurosurgical intervention [74].

As can be seen, regions with higher APC (highlighted by red-yellow (dark gray-light gray in print version) in Fig. 2.8B) do not necessarily coincide with the structure of the arteries visible in Fig. 2.8(A). Moreover, the APC is almost uniformly distributed within areas bounded by blue-highlighted, non-pulsating veins in the left part of Fig. 2.8(B), whereas such distribution is more uneven in the upper-right part. Despite the fact that veins, the area of which is approximately double that of the area of arteries, occupy a significant part of the cerebral cortex, these vessels do not contribute in APC. The obvious explanation for this is the significantly lower blood-flow velocity through the veins, the pulsation of which is at least an order of magnitude lower than in the arteries. It is worth

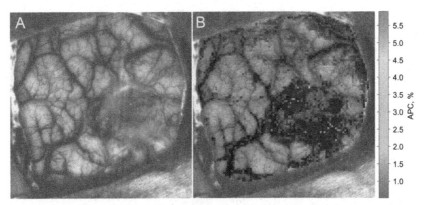

FIGURE 2.8 Mapping the blood pulsations parameters in a brain cortex. (A) One of the recorded video frames of the subject's brain opened up during neurosurgery, and (B) spatial distribution of APC overlaid with initial image. The color scale on the right shows AC/DC ratios in percent.

noting that the mean APC in human cortex (Fig. 2.8 b) is ten times higher than in the rat's cortex (Fig. 2.7B).

2.11   APC as a measure of the arterial tone

Regardless of whether the change in blood volume happens in the pulsating arteries or in the capillary bed when it is compressed by the arteries causes light modulation, the main reason for the modulation is pulse pressure, i.e., the pressure difference in systole and diastole at the point of measurements. However, there are other mechanisms that can affect the amplitude of changes in the lumen of the arteries at the same pulse pressure. For example, the products of brain tissue metabolism, of which CO_2 is of the greatest importance, most actively control the lumen of meningial vessels [75,76]. Despite the fact that large cerebral vessels are involved in autoregulation of cerebral blood flow, the leading role belongs to the meningeal arteries, which determine about 40% of vascular resistance and incorporate specialized sphincters affecting the volumetric flow rate [72]. The tone of the sphincters is determined by metabolic factors and, exerting a regulatory effect, brings into correspondence the trophic need of the brain for blood supply. A decrease in sphincter tone leads to a decrease in local vascular resistance and a corresponding increase in pulse pressure [77]. This increase leads to an increase in the amplitude of arterial lumen oscillations. These oscillations in the lumen determine the variable-in-time component of the PPG waveform. When light interacts directly with arteries, changes in the lumen modulate the absorption of light. In the case of interaction mainly with nonpulsating capillaries, pressure fluctuations change the compression of the tissue, thereby modulating the absorption of light again.

The effect of arterial tone on APC measured in green-light imaging PPG system has been demonstrated in recent experiments to study rats' cerebrovas-

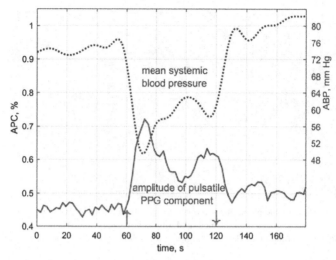

FIGURE 2.9 Typical dynamics of the mean systemic blood pressure (dashed brown curve) and APC (solid blue curve) in cerebral arteries of a rat during painful visceral stimulation. Red (darkgray in print version) and blue (gray in print version) arrows near the X-axes show the beginning and end of the stimulation, respectively.

cular responses to pain stimuli [73]. Simultaneous recording of the systemic blood pressure and imaging PPG in the open rat's brain revealed counter-phase changes in the blood pressure and APC as shown in Fig. 2.9. One can see that painful visceral stimuli (applied at 60 s in the trial) leads to a decrease in systemic arterial pressure accompanied by an increase in APC, whereas the pressure increase at the end of stimuli results in a decrease in APC.

These observations can be explained by a reflex change in the tone of cerebral vessels in response to changes in perfusion pressure [73]. This reaction is in line with the features of the cerebrovascular reflex, the physiological role of which is to maintain the constancy of cerebral blood flow regardless of fluctuations in the systemic blood pressure [78]. In other words, APC follows the local vascular resistance that is varying under the influence of systemic arterial pressure. Therefore, the parameter APC of PPG waveform can be considered as a measure of the cerebral vascular tone.

2.12 Green-light camera-based PPG and cutaneous perfusion

As just discussed, when an imaging PPG system is utilized to measure vascular parameters through the skin, green light interacts predominantly with the superficial layer of capillaries. Given the noncompliance of capillaries with minor change of their size due to blood pressure change (the blood volume in the capillaries does not pulsate) [46], one may ask: Is camera-based PPG suitable for measuring cutaneous blood perfusion? This question received a positive

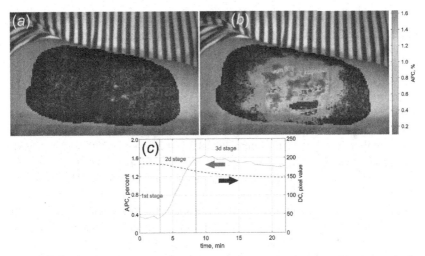

FIGURE 2.10 APC mapping in the area of the capsicum-patch application to the upper arm in the second (a) and twelfth minute (b) after the patch application. The color scale on the right of the maps shows APCs in percent. Evolution of APC averaged within the patch area is shown in the graph (c) by the solid red line, whereas the dashed blue line shows evolution of the mean DC component of the PPG waveform.

answer in experiments on monitoring the effect of capsaicin-patch application on microcirculation [18]. In this experiment, a capsicum patch was attached to a subject's upper arm and illuminated by green light, while video of the forearm was continuously recorded. Video and ECG recording started two minutes prior to the plaster application. The transparency of the capsaicin patch to green light allowed measuring the spatial distribution of APC and its dynamics continuously during the whole period of patch application. APC mapping at the second and twelfth minute after application of the patch are shown in Fig. 2.10(a) and 2.10(b), respectively. In both examples, the APC maps were drawn in pseudo-colors using the same color scale shown on the right side of the images. A fourfold increase in pulsation amplitude after twelve minutes is clearly visible.

The solid red line in Fig. 2.10(c) shows representative examples of the evolution of the APC averaged over the patch area. At the beginning of the patch application, the small APC was observed for all 28 subjects under study [18]. After a certain delay time, APC began to grow rapidly. The delay time varied for each subject, with the mean value of 7.6 ± 3.5 m ranging from 1.1 to 13.9 m. In all cases, the time-course of APC changes after capsaicin application consisted of three clearly distinguishable stages. The first stage was represented by a latent period when the heartbeat related modulation was of rather small amplitude. At the second stage, a sharp increase of APC was observed, and then, in the third stage, APC was saturated.

A typical reaction of the application of capsaicin to human skin is progressive redness of the skin [79], which becomes apparent as an increase of greenlight absorption. Therefore, the skin redness (or hyperemia) could be estimated as the relative change of the DC component of the PPG waveform measured in the location of the patch. The evolution of the DC component is shown in Fig. 2.10(c) by a dashed blue line. As seen, variations of the DC component is significantly smaller than that of APC. In the group of 28 subjects, the relative change in the DC component during the second stage of the patch application is much smaller than in the AC component ($8.3 \pm 4.2\%$ versus $220 \pm 100\%$) [18].

Capsaicin is known to release the calcitonin gene-related peptide (CGRP), which leads to a potent dilatation of cutaneous arteries [80,81]. However, observed capsaicin-induced hyperemia as the change in DC component suggests variations of the RBCs number in the superficial capillary layers to be less than 10% ($8.3 \pm 4.2\%$). Under these conditions, the concomitant four-fold increase in APC (Fig. 2.10(c)) can hardly be explained within the framework of the conventional BV-model. In contrast, the TC-model explains the functional test with capsaicin as follows. As in cerebral vessels, a decrease in sphincter tone (due to release of CGRP) results in an increase in the pulse pressure with a corresponding increase in the amplitude of the arterial lumen oscillation. Therefore, the increased pulsation amplitude of the arteries more strongly compresses the upper capillary layer, thus increasing the AC component of the PPG waveform. At the same time, the decreased arterial tone results in a decrease in local vascular resistance [77] and respective increase of RBC velocity, which causes the increase of cutaneous blood perfusion. Note that the increasing perfusion occurs mainly due to higher RBC speed under conditions of a slightly changing total number of RBC. In other words, capsaicin increases the number of functioning capillaries: Before capsaicin is applied, RBCs fill non-functional capillaries, but their velocity is close to zero, whereas, after the decrease in arterial tone, RBCs begin to move in the capillaries. This is in line with video-capillaroscopy observations showing that the higher mean RBC velocity is associated with higher amplitude of velocity modulation of an almost constant number of RBCs [45]. Therefore, the APC parameter of imaging PPG allows assessment of the cutaneous perfusion and its response to various functional stimuli.

Assessment of the microcirculation response on the local thermal impact using the camera-based PPG system [82] demonstrated also that the temperature-induced increase in the normalized pulsation amplitude is mainly driven by the growth of the pulsatile AC component of the PPG waveform. In this experiment, six–eight independent trials of internal warming of the previously cooled finger were carried out for each of nine subjects. It was found that within ten s of a local increase in the temperature of the finger's skin, the AC component increased by 7.1 ± 2.5 times, whereas the DC component dropped by only $17 \pm 6\%$ [82]. This experiment also underlines the importance of the TC-model of PPG signal formation for implementation of the experimental observations. Moreover, recent experiments on local external heating of the forehead demonstrated the applicability of the camera-based PPG method to measure quantitatively the vasomotor response to local thermal exposure [10]. It was found that APC is a very sensitive marker of changes in skin blood flow that are observed in response to local heating of the forehead skin. Local heating of about 5 °C results in a several-fold APC increase, which is well localized in the heating area. In most cases, the increase in BPA continues after the cessation of heat exposure and the respective diminishment of the skin temperature. Likewise, in the experiment with capsaicin patch application, the increase in cutaneous perfusion is also associated with activation of temperature-sensitive vanilloid receptors of the temporary potential (TRPV) via CGRP. Once again, this activation leads to a decrease in arterial tone, an increase in their pulsation amplitude, and a respective increase in the amplitude of the superficial capillary-layer compression that results in an increase in APC.

2.13 Conclusive remarks

In this chapter, we have discussed the relationship between the two main physiological models (blood-volume, BV, and tissue-compression, TC) for the interaction of light with biological tissue containing *in-vivo* blood vessels. In fact, both models agree that the reason for the modulation of light is the modulation of its absorption due to changes in the volume of blood with which it interacts. The difference is where exactly and for what reasons these changes in blood volume occur. If the conventional BV-model does not specify the source of blood pulsations, proposing that it is most likely arteries and arterioles, then the TC-model pays special attention to the change in blood volume in the superficial capillary layer. We have cited a number of recent experimental studies, the results of which support the TC model and clearly show the need to take into account the mechanical properties of the capillary bed.

One of the conclusions of our analysis is that the superficial capillary layer serves as the distributed network allowing quantitative characterization of the mechanisms of systemic vascular regulation. Moreover, the analysis of light modulation caused by interaction with tissue containing blood vessels makes it possible to reliably identify and objectively assess disorders of local microcirculation and systemic hemodynamics, thereby assessing neurogenic vasomotor reactivity, which is very important for maintaining systemic hemodynamic parameters. Deeper study the basic aspects of light interaction with living tissue is extremely important for both correct interpretation of the experimental data and development of new noninvasive, contactless diagnostic instrument. Further progress in clarifying the physiological mechanism behind PPG will make possible the development of new methods for the early diagnosis of socially significant diseases such as complications of diabetes mellitus; obliterating vascular diseases of the lower limbs; systemic sclerosis; skin, stomach, and intestines cancer; and others.

Acknowledgments

This study was funded by the grant from the Ministry of Science and Higher Education of the Russian Federation (agreement 075-15-2020-800).

References

[1] A.B. Hertzman, The blood supply of various skin areas as estimated by the photoelectric plethysmograph, Am. J. Physiol. 124 (1938) 328–340, https://doi.org/10.1152/ajplegacy.1938.124.2.328.

[2] T. Aoyagi, M. Kishi, K. Yamaguchi, S. Watanabe, Improvement of earpiece oximeter, Osaka (1974) 90–91.

[3] J.W. Severinghaus, Y. Honda, History of blood gas analysis. VII. Pulse oximetry, J. Clin. Monit. Comput. 3 (1987) 135–138, https://doi.org/10.1007/BF00858362.

[4] R-C. Peng, X-L. Zhou, W-H. Lin, Y-T. Zhang, Extraction of heart rate variability from smartphone photoplethysmograms, Comput. Math. Methods Med. 2015 (2015) 516826, https://doi.org/10.1155/2015/516826.

[5] B. Sanudo, M. De Hoyo, A. Munoz-Lopez, et al., Pilot study assessing the influence of skin type on the heart rate measurements obtained by photoplethysmography with the apple watch, J. Med. Syst. 43 (2019) 195, https://doi.org/10.1007/s10916-019-1325-2.

[6] B.P.M. Imholz, W. Wieling, G.A. van Montfrans, K.H. Wesseling, Fifteen years experience with finger arterial pressure monitoring: assessment of the technology, Cardiovasc. Res. 38 (1998) 605–616, https://doi.org/10.1016/S0008-6363(98)00067-4.

[7] M. Nitzan, A. Babchenko, B. Khanokh, D. Landau, The variability of the photoplethysmographic signal – a potential method for the evaluation of the autonomic nervous system, Physiol. Meas. 19 (1998) 93–102, https://doi.org/10.1088/0967-3334/19/1/008.

[8] A.P. Lima, P. Beelen, J. Bakker, Use of a peripheral perfusion index derived from the pulse oximetry signal as a noninvasive indicator of perfusion, Crit. Care Med. 30 (2002) 1210–1213, https://doi.org/10.1097/00003246-200206000-00006.

[9] A.A. Kamshilin, T.V. Krasnikova, M.A. Volynsky, et al., Alterations of blood pulsations parameters in carotid basin due to body position change, Sci. Rep. 8 (2018) 13663, https://doi.org/10.1038/s41598-018-32036-7.

[10] M.A. Volynsky, N.B. Margaryants, O.V. Mamontov, A.A. Kamshilin, Contactless monitoring of microcirculation reaction on local temperature changes, Appl. Sci. 9 (2019) 4947, https://doi.org/10.3390/app9224947.

[11] B. Khanokh, Y. Slovik, D. Landau, M. Nitzan, Sympathetically induced spontaneous fluctuations of the photoplethysmographic signal, Med. Biol. Eng. Comput. 42 (2004) 80–85, https://doi.org/10.1007/BF02351014.

[12] A. Reisner, P.A. Shaltis, D. McCombie, H.H. Asada, Utility of the photoplethysmogram in circulatory monitoring, Anesthesiology 108 (2008) 950–958, https://doi.org/10.1097/ALN.0b013e31816c89e1.

[13] J. Lee, K. Matsumura, T. Yamakoshi, et al., Validation of normalized pulse volume in the outer ear as a simple measure of sympathetic activity using warm and cold pressor tests: towards applications in ambulatory monitoring, Physiol. Meas. 34 (2013) 359–375, https://doi.org/10.1088/0967-3334/34/3/359.

[14] N. Blanik, A.K. Abbas, B. Venema, et al., Hybrid optical imaging technology for long-term remote monitoring of skin perfusion and temperature behavior, J. Biomed. Opt. 19 (2014) 16012, https://doi.org/10.1117/1.JBO.19.1.016012.

[15] T.Y. Abay, P.A. Kyriacou, Reflectance photoplethysmography as noninvasive monitoring of tissue blood perfusion, IEEE Trans. Biomed. Eng. 62 (2015) 2187–2195, https://doi.org/10.1109/TBME.2015.2417863.

[16] A.A. Kamshilin, I.S. Sidorov, L. Babayan, et al., Accurate measurement of the pulse wave delay with imaging photoplethysmography, Biomed. Opt. Express 7 (2016) 5138–5147, https://doi.org/10.1364/BOE.7.005138.

[17] K. Matsumura, K. Shimuzu, P. Rolfe, et al., Inter-method reliability of pulse volume related measures derived using finger-photoplethysmography, J. Psychophysiol. 32 (2018) 182–190, https://doi.org/10.1027/0269-8803/a000197.

[18] A.A. Kamshilin, M.A. Volynsky, O. Khayrutdinova, et al., Novel capsaicin-induced parameters of microcirculation in migraine patients revealed by imaging photoplethysmography, J. Headache Pain 19 (43) (2018), https://doi.org/10.1186/s10194-018-0872-0.

[19] A.B. Hertzman, C.R. Spealman, Observations on the finger volume pulse recorded photoelectrically, Am. J. Physiol. 119 (1937) 334–335, https://doi.org/10.1152/ajplegacy.1937.119.2.257.

[20] J. Nieveen, L.B. van der Slikke, W.J. Reichert, Photoelectric plethysmography using reflected light, Cardiologia 29 (1956) 160–173, https://doi.org/10.1159/000165601.

[21] J. Weinman, A. Hayat, G. Raviv, Reflection photo-plethysmography of arterial-blood-volume pulses, Med. Biol. Eng. Comput. 15 (1977) 22–31, https://doi.org/10.1007/BF02441571.

[22] V.C. Roberts, Photoplethysmography-fundamental aspects of the optical properties of blood in motion, Trans. Inst. Meas. Control 4 (1982) 101–106, https://doi.org/10.1177/014233128200400205.

[23] J.R. Jago, A. Murray, Repeatability of peripheral pulse measurements on ears, fingers and toes using photoelectric plethysmography, Clin. Phys. Physiol. Meas. 9 (1988) 319–329, https://doi.org/10.1088/0143-0815/9/4/003.

[24] S. Loukogeorgakis, R. Dawson, N. Philips, et al., Validation of a device to measure arterial pulse wave velocity by a photoplethysmographic method, Physiol. Meas. 23 (2002) 581–596, https://doi.org/10.1088/0967-3334/23/3/309.

[25] K.H. Shelley, D. Tamai, D. Jablonka, et al., The effect of venous pulsation on the forehead pulse oximeter wave form as a possible source of error in Spo2 calculation, Anesth. Analg. 100 (2005) 743–747, https://doi.org/10.1213/01.ANE.0000145063.01043.4B.

[26] B. Khanoka, Y. Slovik, D. Landau, M. Nitzan, Sympathetically induced spontaneous fluctuations of the photoplethysmographic signal, Med. Biol. Eng. Comput. 42 (2004) 80–85, https://doi.org/10.1007/BF02351014.

[27] J-M. Kim, K. Arakawa, K.T. Benson, D.K. Fox, Pulse oximetry and circulatory kinetics associated with pulse volume amplitude measured by photoelectric plethysmography, Anesth. Analg. 65 (1986) 1333–1339.

[28] W.B. Murray, P.A. Foster, The peripheral pulse wave: information overlooked, J. Clin. Monit. Comput. 12 (1996) 365–377, https://doi.org/10.1007/BF02077634.

[29] N. Selvaraj, A.K. Jaryal, J. Santhosh, et al., Assessment of heart rate variability derived from finger-tip photoplethysmography as compared to electrocardiography, J. Med. Eng. Technol. 32 (2008) 479–484, https://doi.org/10.1080/03091900701781317.

[30] G. Cenini, J. Arguel, K. Aksir, A. van Leest, Heart rate monitoring via remote photoplethysmography with motion artifacts reduction, Opt. Express 18 (2010) 4867–4875, https://doi.org/10.1364/OE.18.004867.

[31] A.A. Kamshilin, S. Miridonov, V. Teplov, et al., Photoplethysmographic imaging of high spatial resolution, Biomed. Opt. Express 2 (2011) 996–1006, https://doi.org/10.1364/BOE.2.000996.

[32] J. Allen, Photoplethysmography and its application in clinical physiological measurement, Physiol. Meas. 28 (2007) R1–R39, https://doi.org/10.1088/0967-3334/28/3/R01.

[33] A.A. Kamshilin, E. Nippolainen, I.S. Sidorov, et al., A new look at the essence of the imaging photoplethysmography, Sci. Rep. 5 (2015) 10494, https://doi.org/10.1038/srep10494.

[34] J.C. de Trefford, K. Lafferty, What does photo-plethysmography measure?, Med. Biol. Eng. Comput. 22 (1984) 479–480, https://doi.org/10.1007/BF02447713.

[35] C-Z. Wang, Y-P. Zheng, Comparison between reflection-mode photoplethysmography and arterial diameter change detected by ultrasound at the region of radial artery, Blood Press Monit. 15 (2010) 213–219, https://doi.org/10.1097/MBP.0b013e328338aada.

[36] P. Zaramella, F. Freato, V. Quaresima, et al., Foot pulse oximeter perfusion index correlates with calf muscle perfusion measured by near-infrared spectroscopy in healthy neonates, J. Perinatol. 25 (2005) 417–422, https://doi.org/10.1038/sj.jp.7211328.

[37] W. Verkruysse, L.O. Svaasand, J.S. Nelson, Remote plethysmographic imaging using ambient light, Opt. Express 16 (2008) 21434–21445, https://doi.org/10.1364/OE.16.021434.

[38] Y. Maeda, M. Sekine, T. Tamura, The advantages of wearable green reflected photoplethysmography, J. Med. Syst. 35 (2011) 829–850, https://doi.org/10.1007/s10916-010-9506-z.

[39] B.A. Fallow, T. Tarumi, H. Tanaka, Influence of skin type and wavelength on light wave reflectance, J. Clin. Monit. Comput. 27 (2013) 313–317, https://doi.org/10.1007/s10877-013-9436-7.

[40] R.R. Anderson, J.A. Parrish, Optical properties of human skin, in: J.D. Reganand, J.A. Parrish (Eds.), The Science of Photomedicine, Springer US, New York, 1982, pp. 147–194.

[41] A.N. Bashkatov, E.A. Genina, V.I. Kochubey, V.V. Tuchin, Optical properties of human skin, subcutaneous and mucous tissues in the wavelength range from 400 to 2000 nm, J. Phys. D, Appl. Phys. 38 (2005) 2543–2555, https://doi.org/10.1088/0022-3727/38/15/004.

[42] H. Gray, Anatomy of the Human Body, Churchill Livingstone Elsevier, Edinburgh, 2008.

[43] F. Mahler, M.H. Muheim, M. Intaglietta, et al., Blood pressure fluctuations in human nailfold capillaries, Am. J. Physiol. 236 (1979) H888–H893, https://doi.org/10.1152/ajpheart.1979.236.6.H888.

[44] S.A. Williams, S. Wasserman, D.W. Rawlinson, et al., Dynamic measurement of human capillary blood pressure, Clin. Sci. 74 (1988) 507–512, https://doi.org/10.1042/cs0740507.

[45] M.V. Volkov, N.B. Margaryants, A.V. Potemkin, et al., Video capillaroscopy clarifies mechanism of the photoplethysmographic waveform appearance, Sci. Rep. 7 (2017) 13298, https://doi.org/10.1038/s41598-017-13552-4.

[46] Y.C. Fung, B.W. Zweifach, M. Intaglietta, Elastic environment of the capillary bed, Circ. Res. 19 (1966) 441–461, https://doi.org/10.1161/01.RES.19.2.441.

[47] L.S. D'Agrosa, A.B. Hertzman, Opacity pulse of individual minute arteries, J. Appl. Physiol. 23 (1967) 613–620, https://doi.org/10.1152/jappl.1967.23.5.613.

[48] G.H. Weiss, R. Nossal, R.F. Bonner, Statistics of penetration depth of photons reemitted from irradiated tissue, J. Mod. Opt. 36 (1989) 349–354, https://doi.org/10.1080/09500348914550381.

[49] S. Chatterjee, K. Budidha, P.A. Kyriacou, Investigating the origin of photoplethysmography using a multiwavelength Monte Carlo model, Physiol. Meas. (2020), https://doi.org/10.1088/1361-6579/aba008.

[50] W. Verkruysse, M. Bartula, E. Bresch, et al., Calibration of contactless pulse oximetry, Anesth. Analg. 124 (2017) 136–145, https://doi.org/10.1016/10.1213/ANE.0000000000001381.

[51] F. Corral, G. Paez, M. Strojnik, A photoplethysmographic imaging system with supplementary capabilities, Opt. Appl. 44 (2014) 191–204, https://doi.org/10.5277/oa140202.

[52] K. Humphreys, T. Ward, C. Markham, Noncontact simultaneous dual wavelength photoplethysmography: a further step toward noncontact pulse oximetry, J. Opt. Soc. Am. Rev. Sci. Instrum. 78 (2007) 44304, https://doi.org/10.1063/1.2724789.

[53] E.K. Chan, B. Sorg, D. Protsenko, et al., Effects of compression on soft tissue optical properties, IEEE J. Sel. Top. Quantum Electron. 2 (1996) 943–950, https://doi.org/10.1109/2944.577320.

[54] E. Nippolainen, N.P. Podolian, R.V. Romashko, et al., Photoplethysmographic waveform as a function of subject's age, Phys. Proc. 73 (2015) 241–245, https://doi.org/10.1016/j.phpro.2015.09.164.

[55] A.A. Kamshilin, O.V. Mamontov, V.T. Koval, et al., Influence of a skin status on the light interaction with dermis, Biomed. Opt. Express 6 (2015) 4326–4334, https://doi.org/10.1364/BOE.6.004326.

[56] I.S. Sidorov, R.V. Romashko, V.T. Koval, et al., Origin of infrared light modulation in reflectance-mode photoplethysmography, PLoS ONE 11 (2016) e0165413, https://doi.org/10.1371/journal.pone.0165413.

[57] A.C.M. Dassel, R. Graaff, A. Meijer, et al., Reflectance pulse oximetry at the forehead of newborns. The influence of varying pressure on the probe, J. Clin. Monit. Comput. 12 (1996) 421–428, https://doi.org/10.1007/BF02199702.

[58] X.-F. Teng, Y.-T. Zhang, The effect of contacting force on photoplethysmographic signals, Physiol. Meas. 25 (2004) 1323–1335, https://doi.org/10.1088/0967-3334/25/5/020.

[59] A.R. Guazzi, M.C. Villarroel, J. Jorge, et al., Non-contact measurement of oxygen saturation with an RGB camera, Biomed. Opt. Express 6 (2015) 3321–3338, https://doi.org/10.1364/BOE.6.003320.

[60] R.F. Schmidt, G. Thews, Human Physiology, Springer-Verlag, Berlin, Heidelberg, 1989.

[61] P.B. Canham, A.C. Burton, Distribution of size and shape in populations of normal human red cells, Circ. Res. 22 (1968) 405–422, https://doi.org/10.1161/01.RES.22.3.405.

[62] L-G. Lindberg, P.Å Öberg, Optical properties of blood in motion, Opt. Eng. 32 (1993) 253–257, https://doi.org/10.1117/12.60688.

[63] A.M.K. Enejder, J. Swartling, P. Aruna, S. Andersson-Engels, Influence of cell shape and aggregate formation on the optical properties of flowing whole blood, Appl. Opt. 42 (2003) 1384–1394, https://doi.org/10.1364/AO.42.001384.

[64] N.B. Margaryants, I.S. Sidorov, M.V. Volkov, et al., Visualization of skin capillaries with moving red blood cells in arbitrary area of the body, Biomed. Opt. Express 10 (2019) 4896–4906, https://doi.org/10.1364/BOE.10.004896.

[65] S. Shin, Y. Yang, J.-S. Suh, Measurement of erythrocyte aggregation in a microchip stirring system by light transmission, Clin. Hemorheol. Microcirc. 41 (2009) 197–207, https://doi.org/10.3233/CH-2009-1172.

[66] I.B. Wilkinson, D.J. Webb, Venous occlusion plethysmography in cardiovascular research: methodology and clinical applications, Br. J. Clin. Pharmacol. 52 (2001) 631–646, https://doi.org/10.1046/j.1365-2125.2001.01495.x.

[67] R.A. Bays, D.A. Healy, R.G. Atnip, et al., Validation of air plethysmography, photoplethysmography, and duplex ultrasonography in the evaluation of severe venous stasis, J. Vasc. Surg. 20 (1994) 721–727, https://doi.org/10.1016/S0741-5214(94)70159-8.

[68] A.A. Kamshilin, V.V. Zaytsev, O.V. Mamontov, Novel contactless approach for assessment of venous occlusion plethysmography by video recordings at the green illumination, Sci. Rep. 7 (2017) 464, https://doi.org/10.1038/s41598-017-00552-7.

[69] V.V. Zaytsev, S.V. Miridonov, O.V. Mamontov, A.A. Kamshilin, Contactless monitoring of the blood-flow changes in upper limbs, Biomed. Opt. Express 9 (2018) 5387–5399, https://doi.org/10.1364/BOE.9.005387.

[70] C.E. Dunn, B. Lertsakdadet, C. Crouzet, et al., Comparison of speckleplethysmographic (SPG) and photoplethysmographic (PPG) imaging by Monte Carlo simulations and in vivo measurements, Biomed. Opt. Express 9 (2018) 4306–4316, https://doi.org/10.1364/BOE.9.004306.

[71] V.V. Zaytsev, O.V. Mamontov, A.A. Kamshilin, Assessment of cutaneous blood flow in lower extremities by imaging photoplethysmography method, Sci. Tech. J. Inf. Technol. Mech. Opt. 19 (2019) 994–1004, https://doi.org/10.17586/2226-1494-2019-19-6-994-1003.

[72] H. Nonaka, M. Akima, T. Nagayama, et al., Microvasculature of the human cerebral meninges, Neuropathology 23 (2003) 129–135, https://doi.org/10.1046/j.1440-1789.2003.00487.x.

[73] O.A. Lyubashina, O.V. Mamontov, M.A. Volynsky, et al., Contactless assessment of cerebral autoregulation by photoplethysmographic imaging at green illumination, Front. Neurosci. 13 (2019) 1235, https://doi.org/10.3389/fnins.2019.01235.

[74] O.V. Mamontov, A.V. Shcherbinin, R.V. Romashko, A.A. Kamshilin, Intraoperative imaging of cortical blood flow by camera-based photoplethysmography at green light, Appl. Sci. 10:6192 (2020), https://doi.org/10.3390/app10186192.

[75] O.B. Paulson, S. Strandgaard, L. Edvinsson, Cerebral autoregulation, Cerebrovasc. Brain Metab. Rev. 2 (1990) 161–192.

[76] C.K. Willie, Y-C. Tzeng, J.A. Fisher, P.N. Ainslie, Integrative regulation of human brain blood flow, J. Physiol. 592 (2014) 841–859, https://doi.org/10.1113/jphysiol.2013.268953.

[77] S. Fantini, A. Sassaroli, K.T. Tgavalekos, J. Kornbluth, Cerebral blood flow and autoregulation: current measurement techniques and prospects for noninvasive optical methods, Neurophotonics 3 (2016) 31411, https://doi.org/10.1117/1.NPh.3.3.031411.

[78] E. Hamel, Perivascular nerves and the regulation of cerebrovascular tone, J. Appl. Physiol. 100 (2006) 1059, https://doi.org/10.1152/japplphysiol.00954.2005.
[79] H. Holst, L. Arendt-Nielsen, H. Mosbech, et al., Capsaicin-induced neurogenic inflammation in the skin in patients with symptoms induced by odorous chemicals, Skin Res. Technol. 17 (2011) 82–90, https://doi.org/10.1111/j.1600-0846.2010.00470.x.
[80] F.A. Russell, R. King, S.J. Smillie, et al., Calcitonin gene-related peptide: physiology and pathophysiology, Physiol. Rev. 94 (2014) 1099–1142, https://doi.org/10.1152/physrev.00034.2013.
[81] B. Fromy, A. Josset-Lamaugarny, G. Aimond, et al., Disruption of TRPV3 impairs heat-evoked vasodilation and thermoregulation: a critical role of CGRP, J. Invest. Dermatol. 138 (2018) 688–696, https://doi.org/10.1016/j.jid.2017.10.006.
[82] A.A. Kamshilin, A.V. Belaventseva, R.V. Romashko, et al., Local thermal impact on microcirculation assessed by imaging photoplethysmography, Biol. Med. 8 (2016) 1000361, https://doi.org/10.4172/0974-8369.1000361.

Chapter 3

Model-based camera-PPG

Pulse-rate monitoring in fitness

Albertus C. den Brinker[a] and Wenjin Wang[b,c]

[a]*AI Data Science & Digital Twin Department, Philips Research, Eindhoven, the Netherlands,*
[b]*Patient Care and Monitoring Department, Philips Research, Eindhoven, the Netherlands,*
[c]*Eindhoven University of Technology, Electrical Engineering Department, Eindhoven,*
the Netherlands

Contents

3.1 Introduction	51	3.5 Discussion 71
3.2 Model-based pulse rate extraction	55	3.6 Conclusions 71
3.3 Fitness application	62	Appendix 3.A PBV determination 72
3.3.1 Experimental setup	63	3.A.1 Optical path descriptors 72
3.3.2 Processing chain	64	3.A.2 Experimental PBV
3.3.3 Performance metric	64	determination 73
3.4 Results	65	Appendix 3.B Pseudocode for
3.4.1 Reference creation	65	model-based PPG 74
3.4.2 System performance	66	References 76

3.1 Introduction

Unobtrusive vital signs monitoring is an actively researched field driven by the notion that it may contribute to novel means for achieving, e.g., a better balance between costs and effectiveness in healthcare, improved care experience, and well-being management. A multitude of technologies can be exploited for this, where camera-based monitoring is attractive in view of the ubiquitousness of low-cost devices. In this chapter, we will focus on camera-based cardiac measurements, in particular, on pulse rate measurements.

Research in the area of unobtrusive pulse rate monitoring is diverse, see Chap. 1, and can be viewed from various angles, such as an application or a technology perspective. Early overview papers include [1–4], the year 2018 brought [5–8], 2019 gave [9–12], and 2020 held [13–16]. In view of this abundance, we will not attempt another overview. Instead, we intend to shed more light on one particular area in this field, namely model-based camera PPG.

Within the field of camera-based monitoring, various approaches can be taken as core processing techniques. There are two broad streams character-

Contactless Vital Signs Monitoring. https://doi.org/10.1016/B978-0-12-822281-2.00011-1
51

ized as model-based and those without a physiological model. The latter include blind source separation (BSS) techniques, as well as supervised machine learning. Model-based approaches have been intensively studied over the last decade, where different variants were created and tested over the years. It is currently being adopted in products, e.g. [17], an app featuring Philips technology.

Model-based in this chapter does not refer to elaborate models of optical properties for skin layers, blood vessels, etc., like those studied for instance in [18,19]. Instead, we will use it as a term for mathematical descriptions of physical phenomena, with the famous example of descriptions of planetary orbits as conic sections. These models intend to capture the joint and dominant characteristics of such systems in the most parsimonious way. It is usually contrasted with data-driven models where the data itself drive a machine-learning stage. There are a number of reasons to highlight the model-based approach. In our view, the most notable reason is that modeling provides insights into the system under study. As such, it typically starts from fairly simple notions or descriptions, while leading to efficient implementations, as well. Efficient implementations allow application on low-cost devices, are less error prone, and easier to keep up-to-date.

The insights and results from the modeling typically fuel new ideas and insights, thus increasing our knowledge and understanding. As a coarse exaggeration, data-driven approaches are not primarily seeking insights [20] because the process is preferably treated as a black box with the promise of ever-increasing accuracy with an ever-increasing amount of data. However, this is debatable because training and testing a system based on data may be sensitive to biases in the data that has been collected for that purpose, leading to a system and conclusions that do not allow for generalization. The explainability and associated tractability of the causes of deteriorated performance are often poor, except for stating that some effects were apparently insufficiently captured in the training set. Also, in machine-learning approaches, a change in some of the optics or camera settings (e.g., sampling rate) may require retraining of the data-driven model, whereas the underlying physiology remains unchanged. Model-based approaches are typically more generalizable, meaning that they can handle these unseen cases by resetting model parameters and some tweaking.

The added value of insights and the simplicity of the algorithms in the model-based PPG approaches have been recognized in the community as witnessed by the high citation rates of some of the model-based papers. On top of that, the performance, as evidenced by the latest competition [21], has shown that model-based approaches [22] rank among the top performers. Therefore, this chapter is devoted to recapturing the model-based work. It also serves to create more clarity and understanding by using a uniform terminology and notation.

For the sake of clarity, we start with discussing two terms: the pulse signal and the PPG signal. The *pulse signal* we define as any signal component in a measured physiological signal caused by the pulsation of the blood circulation. Signals containing a pulse signal encompass blood pressure, blood volume, blood vessel dilation, light absorption by skin tissue, among others. It is clear that the pulse signal observed in the different physiological variables will be different and, even for one physiological parameter, the waveform depends on the location of the measurement. The pulsating rhythm of the cardiac activity is what they have in common. The *PPG signal* is defined as the outcome of a PPG measurement. It is associated with a spot in the body, and it is coupled with the measurement method. Difference in either may lead to different signals (waveforms). A PPG signal does also not necessarily directly reflect a single physiological property; it may integrate over different effects making its origin debatable, see also Chap. 2.

In camera-based PPG, a number of stages can be identified. The first one is the detection of skin or facial landmarks in an image and creating one or more skin patches. From the skin patch(es), the color signals are created typically as the mean value over the patch. The patches may give rise to additional signals such as a movement signal, as will be discussed later. The second step is the construction of a PPG signal, typically from cleverly combining several signals; these signals may come from image patches and involve various color and near-infrared (NIR) channels, movement signals, etc. If there are multiple patches, each patch gives rise to a PPG signal. The third step is therefore a combination of these PPG signals into a single PPG signal. This may be done by averaging, a weighted combination, or a selection mechanism. This final PPG signal can be analyzed for, e.g., the pulse rate and pulse rate variability, or, in combination with other signals, be used for extraction of other information.

The focus of this chapter is on the second step, usually called the core processing, i.e., the downmixing of the various temporal signals derived from the camera image into a single, clean pulse signal. The input to the core processing consists of at least the color channel signals denoted as C. Occasionally, additional information is used in the core process. In particular, information on disturbance signals (like motion) can be exploited, as will be discussed later. These optional additional channels are denoted as D. The output of the core processing is a PPG signal segment associated with a certain patch and denoted as P. The core processing as discussed here is always operating over a finite number of frames, without knowledge of previous data. The number of video frames the processing takes into account is called N, and, as a kind of default, $N = 64$ at a frame rate of 20 frames per second (fps) has been often used as a balance between algorithmic delay and accuracy [23]. With these settings, an interval of about three seconds, meaning on average about three cardiac cycles, is covered. Taking these relatively short segments limits not only the algorithmic delay but also the diversity of the distortions that appear in each segment. Cutting repeatedly a stretch of N frames from a data stream provides each time an N-samples-long PPG-signal segment. The full PPG signal stream can be created by an overlap-add procedure, but this is not considered here.

The concept of model-based PPG extraction is shown in Fig. 3.1. A radiation source (visible light or near-infrared) illuminates a piece of human tissue.

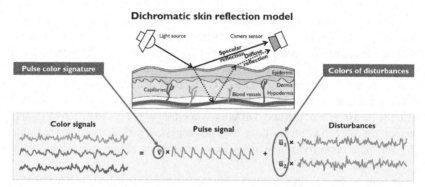

FIGURE 3.1 Schematic illustration of the camera-based PPG extraction and the most relevant signals for the RGB case. A piece of skin tissue is illuminated by a light source, and a camera records the image. The camera collects color signals describing the skin tissue as an average across the skin or a skin patch. From the skin area, other characterization can be collected that are indicative of sources of signal disturbance such as motion. Knowledge and assumptions on relationships between these signals are exploited in model-based PPG extraction methods.

The dichromatic model states that there is a direct reflection from the surface, the specular reflection, and a reflection associated with light penetrating the tissue, which is called the diffuse reflection. The diffuse reflection is influenced by the state of the cardiac blood circulation. Therefore, the diffuse component is treated as a constant component and a zero-mean, pulsating component due to the pumped blood circulation. The latter is called the pulse signal, and the steady component will be called the diffuse skin component or DC-component. The reflected light can be recorded with a sensor, in the figure represented on the bottom left as the red, green, and blue signal from a standard RGB camera, but the concept also holds for infrared (IR) light. Model-based approaches make use of the relationships between pulse components appearing in the different color signals to remove signal components not attributable to the cardiac activity. In particular, it assumes the same waveform for the pulse signal in each color channel, but with different amplitudes.

Moreover, the camera may generate additional signals that can be incorporated in the model-based framework. At the pixel level, the light variations are so small that they are drowned in the sensor/quantization noise. Therefore, spatial averaging is applied, and color information is gathered over an image mask or over skin patches at specific spots determined from facial landmarks. These patches can be further characterized, for instance, by their motion and spatial color homogeneity; this information is also carried in the camera images. Making assumptions on the relationships between these signals and the color signals is another step in model construction. This is illustrated on the bottom-right of Fig. 3.1 as the disturbance signals. In terms of signals, we assume that the pulse is contained in each of three color channels, but with unequal strengths, but also disturbances appear in all different channels with different strengths: In other words, the recorded color channels are the pulse and disturbance signals combined by a mixing matrix. The relative contribution of the pulse signal to the color channels is expressed in Fig. 3.1 by $\bar{v}$, while the effect of the different disturbances (typically motion-induced) is captured in $\bar{u}_i$. Note that the bottom part of the figure illustrates only the assumed relationship between color signals, on the one hand and pulse and disturbance sources, on the other; the effect of sensor noise is not illustrated because it is not part of the optical pathway to the sensor nor is it associated with a particular color direction.

In a series of papers, the development of these models has been described. An overview and introduction to the most relevant model-based approaches is provided in the next section. The latest model-based core algorithm was proposed in [24] and has been tested in NIR. It was suggested that the same methodology would be effective in the case of visible light. This is considered here by a test in one of the toughest RGB scenarios: fitness. The outcome of the test (Sect. 3.4) is used to reflect the current status in model-based approaches, Sects. 3.5–3.6.

3.2   Model-based pulse rate extraction

The model-based approach starts from a simple optical model of the skin. Consider a rigid piece of tissue, a diffuse light source with fixed spectrum illuminating the tissue and a camera focused on the tissue. According to the dichromatic model, two reflections will evolve. The first one is the specular reflection: the mirror-like reflection of the light at the surface of the skin. Second, the light penetrates the skin, being partly absorbed, scattered and reflected, and part of the light reaches the layers containing blood vessels. The amount of light absorption by the blood depends on the amount of blood. This variation is also expressed in the amount of reflected light. Thus, the model-based approach discerns, in the first instance, two components: specular and diffuse, where the latter comprises a steady component and the pulse-related one [25].

In Fig. 3.2, we depict the piece of skin to explain how different spectral components propagate through the system. The different color ranges are associated with a different effective penetration depth, red reaching deeper layers than the green and the blue as indicated by the difference in length of the arrows pointing down from the top layer. The absorption is also wavelength dependent but in a different ranking; as can be seen, the extinction coefficient is larger for blue than for green and red. In the final signal obtained by a normal camera under white-light conditions, where the camera is integrating the entire color range into a three channel representation, the typical amplitude ordering of the pulse strength in the channels of the camera is green, blue, and red (going from high to low amplitude; see also Appendix 3.A).

Displacement of the tissue affects the reflections. Changes in the horizontal and vertical directions can in principle be tracked and compensated for such that always the same piece of tissue in being observed. In practice, this is successful

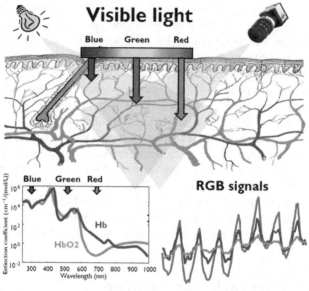

FIGURE 3.2 Schematic illustration of the diffuse light propagation. Top illustration: a piece of skin tissue is illuminated by a light source, and a camera records the image. Different wavelengths have different penetration depths as indicated by the different lengths of the arrows. Bottom left: light absorption by Hb and HbO2 depends on the wavelength as well. Bottom right: the pulse signals as present in the different color channels have different amplitudes.

with limited accuracy. The additional strategy is to measure the velocity and assume that light variations will occur depending on the speed or displacement. Effects of changes of the orientation and non-rigidity of the tissue are much more difficult to capture.

In models developed so far, the following components appear. There are the specular, the diffuse skin, and the pulse components, where the specular reflection is of less importance in NIR [24]. We also have motion signals that we can decompose into motion in the camera plane (movement left/right and up/down) and motion perpendicular to the image plane (depth). Motion in the plane can be determined relatively accurately, while the depth motion is more difficult to capture. This latter motion leads, in a first approximation, to intensity changes and can be suppressed based on this assumption rather than on measurement, as is discussed later. Next to motion, spatial color inhomogeneity is a signal that can be easily obtained (e.g., characterized by the standard deviation) and has been shown to be of value [24,26].

The three components (specular, diffuse, and pulse) are associated with different color directions in the RGB space. The specular color direction essentially reflects the illumination source with the source typically being assumed a broadband, white-light illumination. The diffuse component reflects the skin color. The skin color direction under white light shows remarkably little variation across skin types [27]. Similarly, the pulse color direction shows little variation, and thus the relative color directions are almost fixed.

In the simple model sketched, the skin operates as a filter effectively changing the external illumination to another spectrum before reaching the skin layers containing the blood. This filtering is obviously subject-dependent as different subjects have different skin properties, e.g., pigmentation. However, the light absorption by blood is assumed to be subject-independent, and therefore the effect of the blood volume changes on the spectrum are identical. This reasoning motivates a preprocessing that is intended to compensate for the filtering of skin by normalizing the input color channels.

The normalization is done as follows. Consider a color channel denoted as $C_m(t)$, where m is the channel index ($m = 1, 2, \cdots, M$), and t denotes time. The normalized color channel signals $\widetilde{C}$ are defined as

$$\widetilde{C}_m(t) = C_i(t)/\mu_m - 1, \tag{3.1}$$

where μ_m is the average value of the mth color signal, in practice determined by a short-time averaging operation. The research started was, in the first instance, focused on visible light and low-cost cameras, thus $M = 3$ is dominantly occurring. The normalized color signals are DC-free, and the skin-color direction is mapped onto the vector with direction $[1, 1, 1]$ (in this case, $M = 3$). We also note that light variations associated with motion perpendicular to the image plane are assumed to occur in that direction [24].

From this point onward, various model-based approaches have been developed. Though not the first one published, the PBV [27] is the logical starting point. PBV is the acronym for Pulse derived from Blood Volume. Based on measurements of a large pool of subjects in a white-light condition [27], it was concluded that the orientation of color directions of the skin and that are associated with pulsatility of the blood were largely constant and is therefore referred to as the pulse *signature*. This finding is essentially captured in the earlier description of the skin operating as a spectral filter and the associated preprocessing to normalize the color channels based on the skin. Since the concept of a color signature is fundamental to the modeling, Appendix 3.A provides more details on how this signature can be determined.

Disturbances are assumed to appear in other directions, i.e., their direction in the color space (if any) is unlikely to be identical to that of the pulse. A least-squares optimization can then be defined based on the knowledge of the pulse-color direction as introduced in [27] and more elaborately treated in [24,25]. This defines the PBV method and acts as a first, basic concept from which we will evolve the other methods. The various models discussed in this section are provided in Table 3.1, ordered according to their appearance in the text.

The pulse signature is dependent on the illumination source, implying that, in principle, a color vector needs to be determined for each spectrum and that the spectrum is not allowed to change over time. Fortunately, most situations

TABLE 3.1 Selected model-based publications on PPG core algorithms. Columns (left to right): acronym, year of publication, reference, and aim.

Name	Year	Ref	Aim
PBV	2014	[27]	General method based on pulse color signature
SoftSig	2019	[28]	Relax PBV requirement
CHROM	2013	[29]	Relax PBV requirement
POS	2017	[25]	Relax PBV requirement
PSC$_c$	2020	[30]	Improved α-tuning
SB	2017	[23]	Motion suppression
DIS	2019	[24]	Motion compensation

will have a near diffuse white-light setting allowing to operate with an average white-light PBV. We note that extreme spectral lighting conditions (e.g., only red illumination) cannot be handled because the assumption that the pulse waveform appears in all color channels is then violated. Various ways were conceived to make the PPG extraction less dependent on the specific spectrum of the light source. These are SoftSig, CHROM, and POS which are abbreviations for Soft assumption on the color vector Signature, Chromaticity mapping, and mapping to a color Plane Orthogonal to Skin, respectively.

SoftSig [28] can be viewed as a color-direction probing method which is obviously valuable in case the PBV-knowledge is uncertain (the assumptions are violated, e.g., by improper skin selection or in uncertain lighting conditions). It was proposed within the context of single-element sensor. Such a set-up was motivated by privacy considerations and addresses the challenges in case facial analysis would not be enabled and, therefore, the image frames contain a non-negligible amount of non-skin pixels. Instead of considering a single color direction for the pulse signals, the proposed method probes a number of different directions and creates candidate PPG signals. Next, a selection mechanisms picks the best one out of these candidates. In [28], several selection mechanisms were discussed and tested within the context of a single-element PPG sensor. These methods included metrics based on energy distributions in the frequency domain, skewness, and kurtosis. We note that both PBV and SoftSig can handle arbitrary number of color channels. In contrast, the methods discussed next (CHROM and POS) are restricted to three-channel configurations.

The CHROM, POS, and PSC$_c$ models also have as their starting point a color direction, but not primarily that of the pulse. Their starting point is the direction of a particular disturbance. CHROM projects the signals on the chromaticity plane and, in doing so, eliminates specular reflections. To find the pulse in this plane and to remove diffuse light variations, it uses two directions described by two temporal signals where, for all normal lighting conditions, the pulse is

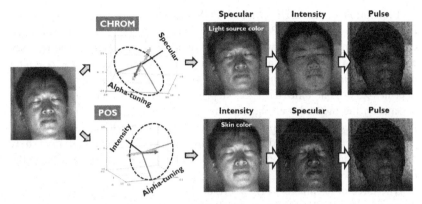

FIGURE 3.3 An example of image decompositions along the different color direction mappings in CHROM and POS. On the right, a single frame from a video sequence is shown. Next to that, normalized color-space decompositions are illustrated. The three right-hand side images (top: CHROM; bottom: POS) show the decomposed images according to the pertinent color directions.

expected to be in anti-phase and the diffuse disturbance in phase, giving rise to a greedy method for removing the disturbance called α-tuning. POS, on the other hand, projects the signals orthogonal to the diffuse (skin) color direction and, thus, in a second step, has to extract the pulse signal from this plane and eliminate the projected specular reflection. This is done in a similar way as in CHROM by using two directions in the plane for which, for normal conditions, in the associated time domain signals, the disturbance is expected to appear in anti-phase. In the pseudocode provided in Appendix 3.B, the projection is implemented as a matrix multiplication, where the down-mix matrix is denoted by $\bar{v}$ and the signals in the α-tuning by $\hat{S}$. It may be clear that the projection orthogonal to the diffuse (skin) color direction (POS) can be applied generically to remove motion-related disturbances, whereas the use of the specular direction as in CHROM to define a first projection plane is coupled to RGB cameras but not to NIR.

In PSC_c [30], the incoming signal is first classified as being clean or disturbed based on expected amplitude of the pulse signal [31]. The signal is divided into a low-frequency (LF) band and a high-frequency (HF) band where the HF band is assumed to contain only negligible amount of pulse-signal content. In case of clean signal, a down-mixing procedure is executed where the ratio of the LF and HF energy content in the PPG signal is maximized. In case of disturbance on the measured signal, the color direction of the maximum energy is used to define a plane on which a projection is performed, a procedure similar to CHROM and POS. Based on the HF content, the best noise disturbance cancellation direction is determined and used for downmixing. As such, it can be viewed as an alternative to the α-tuning procedure. The method is called PSC_c, which stands for Projection vector based on Spectral Characteristics.

Illustration of the CHROM and POS projection principles are given in Figs. 3.3 and 3.4. In Fig. 3.3, the RGB signal is decomposed into the three color directions associated with CHROM and POS, respectively. The primary color direction in CHROM and POS is given by the black arrow, the α-tuning plane by the dashed circle with secondary axes in this plane as the red (mid gray in print version) and blue (dark gray in print version) lines. The α-tuning planes depend on the method, i.e., these planes are different for CHROM and POS. Actual data vectors of the normalized colors are given by the yellow dots (light gray in print version), each dot representing a frame. Note that these dots show their most variation along the intensity direction, and exactly this motivated the POS concept.

The three color images (Fig. 3.3 right) are created by using only one color vector to represent the images. For CHROM, these are the specular color direction and the two directions from the α-tuning. One direction corresponds to the intensity direction as found from the data, and the direction orthogonal to that is the direction used to extract the pulse signal. For simplicity, these are denoted as Specular, Intensity, and Pulse in Fig. 3.3, but more properly should be termed specular direction, direction of the intensity variations in the α-plane, and the remaining orthogonal direction used for pulse extraction. For POS, the three color directions are indicated as the Intensity, Specular, and Pulse in Fig. 3.3. Again, the latter two are a shorthand notation. The specular direction is the specular direction in the α-plane as determined by the tuning, and the pulse image is created using the color vector corresponding to remaining orthogonal direction. In Fig. 3.4, we illustrate the results of the projections of the color-normalized channel signals onto the axes defined in the α-tuning planes, and the outcome of the α-tuning procedure. The vector **w** in the figure indicates the projection direction as effectively used in the two-step optimization and is data-driven.

A next step in the model-based approach was to exploit not only color directions but knowledge from other signals as well. Especially for applications like fitness, the motion is expected to consist of rather steady, quasi-periodic signals, while it is also expected that different movements are coupled to different color directions. In a three-dimensional color space, only two disturbances can be canceled. This raises the question how to remove multiple disturbances, where each disturbance may be assumed a quasi-periodic signal. To address this, signal cleaning was proposed in multiple frequency bands, a concept that indeed improved the performance for fitness applications [23], and can be combined with any of the color-direction based methods. This method SB is a restricted solution since the assumption of steady, quasi-periodic motion is only satisfied in specific use-cases.

Like SB, the latest model called DIS [24] targets motion as an additional source of information in the modeling, but, instead of making assumptions about the motion, it actually measures motion as appearing in the sequence of camera images. The assumption is then that light variations that correlate with this motion are likely to be motion-induced disturbances and unlikely to be attributable

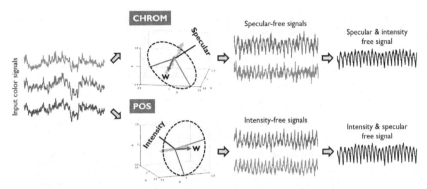

FIGURE 3.4 Illustration of the use of the different color-direction mappings in CHROM and POS. On the right, a single frame from a video sequence is shown. Next to that, normalized color-space decompositions are illustrated. The next column contains the signals in the α-tuning planes (along blue (dark gray in print version) and red (mid gray in print version) directions of the plane) that, due to the projection, are free from specular (CHROM) or intensity (POS) variation. On the right, the signal obtained after α-tuning is shown. Recordings were made in a lab, with white light and still-sitting second author.

to cardiac activity. This idea was not novel in the sense that many proposed camera-PPG systems had some form of motion suppression incorporated either in pre- or post-processing, with an elaborate example of post-processing being [32]. Note that the motion that is taken into account is the motion in the image plane; as already explained, the motion in the depth direction can be handled by projections in a particular color direction. [24] showed that motion removal and color downmixing can be jointly captured in a single-step optimization procedure. A side-effect of good motion removal is that the assumption of the pulse color direction becomes much less critical. In this respect, the method is reminiscent of the proposal in [33] to define an optimal projection as one orthogonal to all motion. This disturbance-removal method is applicable to all signals that are expected to have a color disturbance linearly related to the measurement signal. The method is applicable to multiple disturbances and, unlike SB, does not require an assumption on the character of the disturbing signal.

An overview of the discussed methods and the main requirement is provided in Table 3.2. More in-depth mathematical modeling is provided in [24,25]. Pseudocode describing the various core algorithms in their simplest form are provided in Appendix 3.B.

Model-based approaches are intended to capture the main components of the observed world. As such, they favor generalization at the expense of accuracy, which is in contrast to data-driven methods that are more vulnerable to overfitting and selection bias. The model-based approach still has various factors not yet taken into account, and model refinements may potentially lead to improvements. The main factors not taken into account comprise the non-uniformity of skin patches (not the same tissue everywhere, varying depth of blood vessel,

TABLE 3.2 Selected model-based publications on the core algorithm: acronym, method, and requirement.

Name	Method	Requirement
PBV	LS optimization	Illumination known (pulse color direction)
SoftSig	Color direction selection	A range of color directions
CHROM	2-step optimization	Dominating pulse or disturbance (in α-tuning)
POS	2-step optimization	Dominating pulse or disturbance (in α-tuning)
PSC_C	2-step optimization	LF and HF congruence of disturbance (in α-tuning)
SB	Multi-band optimization	Periodic motion
DIS	LS optimization	Illumination known; access to disturbance signals

etc.), non-rigidity of the tissue, and movements leading to (partial) occlusions (especially rotations).

3.3 Fitness application

The assessment of pulse rates by a noncontact method has been recognized as an attractive option for fitness applications. Research on the topic has however been limited [23,27,34–36]. Measuring pulse rate for fitness is a challenging task. This is due not only to the motion inherent in a workout, but also to the fact that in fitness centers the lighting condition are far from standardized, both in level and spectrum even within one center. To obtain our first insights, we therefore recorded videos in a fitness scenario varying the light condition, light level, and skin type. This study was approved by the Internal Committee Biomedical Experiments of Philips Research, and informed consent has been obtained from each subject.

In the first two conditions (light source and level), the same subject was exercising; for the skin type variation, this was obviously not true. In total, we report on seven light-source conditions, eight illumination levels, and six different subjects for skin-type variation. In total, this makes 21 sessions with the duration of the sessions given in Fig. 3.5, on average about six minutes. The same data was used in [23]. It is clear that the number of variants that is probed in this study does not reflect the full variability that exists in an actual indoor fitness center in terms of subjects, nor in terms of exercising devices.

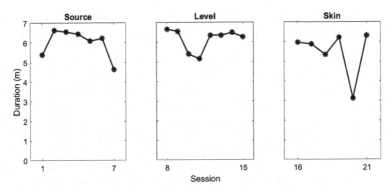

FIGURE 3.5 Duration of the sessions for the 21 sessions divided into three categories.

3.3.1 Experimental setup

During the fitness sessions, the subject varies the running speed on a tread-mill in an indoor environment between low-intensity (three km/h) and high-intensity (12 km/h) within five–eight minutes, depending on the subject's endurance. The background was a skin-contrasting cloth to facilitate the skin-detection/segmentation. Until recently, robust skin detection in fitness was considered a major challenge. However, the performance of algorithms for face detection and face tracking has improved considerably over the last couple of years, especially fueled by AI approaches, e.g. [37].

The following sessions were recorded.

Light-source variation (Sessions 1–7) Three different light sources were used: oblique fluorescent light (from the ceiling), frontal fluorescent light, and frontal halogen light. With these, seven different illumination condition were created: three cases of single illumination sources, three cases involving two light sources, and one case using all three light sources simultaneously. The subject had skin-type III, and the light level was medium, i.e., comparable to Level 4 of the light-level test.

Light-level variation (Sessions 8–15) Light level can influence the outcome: High intensities may cause clipping, while low intensities may lead to low pulsatile amplitudes in RGB channels buried in the camera sensor noise. To vary the intensity, we adjusted the camera aperture to increase/decrease the amount of light entering the camera shutter. A total of eight intensity-levels, from level-1 (low intensity) to level-8 (high intensity), were defined to study this challenge. The subject had skin-type III, and a fluorescent light source was used.

Skin type variation (Sessions 16–21) Six subjects (ages 25–45 yrs, five male and one female) with various skin tones participated in the experiment with three different skin-types according to the Fitzpatrick scale: two Western European subjects (skin-type I–II), two Eastern Asian subjects (skin-type III), and two Southern Asian subjects (skin-type IV–V). The light source was the fluorescent office ceiling light with an illumination direction oblique to the skin-normal, which is the typical illumination condition in a fitness setting. The intensity was set to medium intensity, corresponding to level 4 from the light-level variation test.

In each session, the camera was placed around 2.0 m in front of the subject running on a treadmill located in the lab. We used the global shutter RGB CCD camera USB UI-2230SE-C from IDS, with 640 × 480 pixels, eight-bit depth, and 20 fps. With the used optics, this results in approximately 20,000 skin-pixels. All auto-adjust functions (auto-exposure, auto-gain, auto-white balance, etc.) were turned off and the video was saved in uncompressed format. For ECG data collection, we used a wireless physiological monitoring and feedback device, type NeXus-10 MKII, and synchronized these with the video frames.

3.3.2 Processing chain

The processing to extract the pulse rate from a video includes a number of processing steps. The major algorithmic processing steps include: a global face tracker including a skin mask; a short sliding window for extracting the color and motion signals including bandpass filtering; a core PPG algorithm to construct a short PPG segment; and an overlap-add procedure to generate a long PPG signal from these short PPG sequences. As a final stage, a long sliding window is used to generate the pulse-rate trace from the measured PPG signal, where the pulse rate is calculated from the frequency index of the maximum spectral peak in the frequency domain using segments of about 13 s (256-samples at 20 fps). In line with [38], we used exactly the same frequency extraction mechanism for both reference (ECG) and camera PPG, where the binning of the FFT was such that the pulse-rate quantization step is one BPM.

From the model-based approaches, we selected a limited number of systems to keep the discussion tangible. First of all, we selected DIS as being the latest model not yet evaluated in RGB, also motivated by the fact that this incorporates motion compensation and therefore qualifies as a candidate for the fitness scenario. The second choice is obviously the latest algorithm that showed a good performance in fitness: SB, where we took the POS variant because this was used in the previous study [23]. Both of these algorithms, DIS and SB, have motion correction means on top of a former algorithm. To see the effectiveness of incorporating motion information into the PPG model, we included their predecessors as well, PBV and POS, respectively.

3.3.3 Performance metric

For each session, we determined the deviation d between the camera pulse R_c rate and reference R_r and counted the fraction of frames for which the absolute difference was less or equal to a certain criterion T_γ. This is called the coverage γ. In other words, we have

$$d(k) = |R_c(k) - R_r(k)|, \qquad (3.2)$$

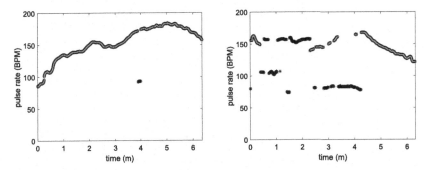

FIGURE 3.6 Examples of reference signal and screening results. Black asterisks indicate the result of peak picking: The red line (gray in print version) indicates the data that is accepted as reference in the performance evaluation. Left plot: Session 12; Right plot: Session 21.

with k the index referring to the sequence of sliding windows used to estimate the pulse rate ($k = 1, 2, \cdots, K$, where K is the total number of these windows in the considered session), and

$$
b_k(T_\gamma) = \begin{cases} 0, & \text{if } d(k) > T_\gamma, \\ 1, & \text{if } d(k) \le T_\gamma. \end{cases} \tag{3.3}
$$

The coverage is then formally defined as

$$
\gamma(T_\gamma) = \frac{\sum_k b_k(T_\gamma)}{K}. \tag{3.4}
$$

With a window hop size of one frame, the number K is slightly lower than the number of frames due to boundary effects (i.e., no zero-padding was used on either side of the PPG signal).

3.4 Results

3.4.1 Reference creation

The ECG provided the reference signal. However, for Session 21, the ECG signal was heavily polluted and, from visual inspection of the graphs, we found for other sessions the procedure of peakpicking was occasionally off. Instances of this are shown in Fig. 3.6. The left plot shows data from Session 12, where the pulse rate has the typical pattern of a one-round workout by running: The pulse rate increases from the start and goes down at the end. Around minute four, there is a short drop in the pulse rate, which we discarded because we doubt that this reflects the real heart rate. The situation in Session 21 was different. From the context, we knew that Session 21 starts at a high pulse rate going down before the normal work-out characteristic. We see that for this session about half of the

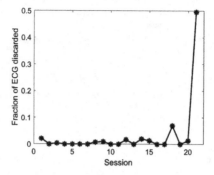

FIGURE 3.7 Fraction of discarded ECG data per session.

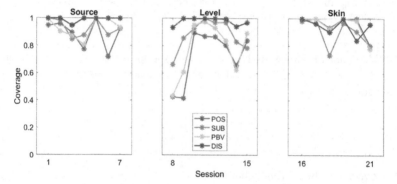

FIGURE 3.8 Fraction of time that estimated rate is within 5 BPM from the reference.

time the pulse rate estimated from the PPG signal has a value actually corresponding to movement frequency. Instead of trying to build another estimator for the pulse rate, it was decided to exclude the doubtful parts in the evaluation. Over all sessions, the amount of data that was discarded was small, except for Session 21, as shown in Fig. 3.7.

3.4.2 System performance

For $T_\gamma = 5$ BPM, the measured fractions over the multiple sessions for the four different algorithms are shown in Fig. 3.8. We observe that the coverage depends on the light source and algorithm: Certain light condition are less favorable. The performance of the DIS algorithm is rather insensitive to the light source. A similar observation holds for the light level condition: DIS is clearly less sensitive to the light level than the other algorithms. Looking at the skin challenge, we observe that the different algorithms have different performance levels for different skin types, but that the amount of variation in performance over the skin types is about the same: roughly ranging from 0.8 to 1.

The performance of the various algorithms over all sessions is shown as a box plot in Fig. 3.9. It is clear that the DIS algorithm provides a much more

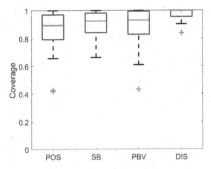

FIGURE 3.9 Boxplot of the coverage per algorithm. Red lines (gray in print version) indicate the median, blue boxes the 50% range, whiskers the range, and a red cross is an outlier.

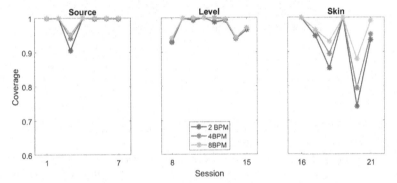

FIGURE 3.10 Fraction of time that estimated rate by the DIS algorithm is within a certain range of the reference value.

constant and higher coverage than the other three. We therefore zoom in some more into the DIS algorithm by considering other coverage thresholds. This is shown in Fig. 3.10, where for $T_\gamma = 2, 4, 8$ BPM the coverage is presented. We see that, for the light source and level condition, the coverage is high (> 0.9), and there is hardly a variation dependent on the criterion. This latter means that, if the difference exist, it is typically a large difference, i.e., a clear outlier. This implies that options for identifying such occasions may exist and most likely can be constructed in a post-processing step. For the skin variation, this is different: The coverage is directly related to the criterion. This indicates that the obtained pulse rate is noisier for particular skin types.

The coverage overviews as presented provides limited insight because it reflects the whole processing change and not just the core algorithm. Therefore, we also look into the spectrograms to highlight the differences between these approaches. We consider three selected cases to illustrate various aspects.

In Fig. 3.11, the spectrograms of the PPG signals derived by the algorithms are shown. All four of them contain the pulse-rate trace typical for an exercise session with an increasing pulse rate. The pulse rate increases from about 60 to

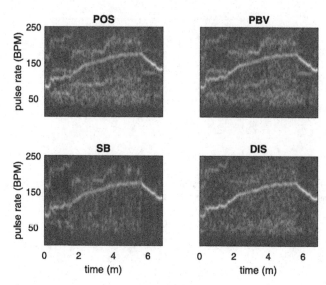

FIGURE 3.11 Spectrograms of the camera-PPG signal obtained by the four different core algorithms in Session 02.

160 BPM, and then the exercise stops and the pulse rate goes down. Especially in the SB and DIS spectrograms, the recovery phase is not only characterized by decreasing pulse rate but also by a cleaner spectrogram. In POS and PBV, we observe (suppressed) motion traces above and below the pulse-rate trace, which are nearly absent in SB and DIS. During exercise, the pulse-rate trace in DIS looks steadier than for SB, while the SB spectrogram may be considered cleaner. The cleanliness of the SB spectrogram is however a direct consequence of the normalization within different subbands, so drawing conclusions from these differences between SB and DIS is not straightforward.

A second example is presented in Fig. 3.12. We observe motion traces at 180 and 130 BPM. These appear better suppressed in DIS than in the other methods (see Fig. 3.12).

As a last example, we present the results of the last session in Fig. 3.13, that is, the session where about half of the reference signal is missing. The pulse-rate trace is the trace going down from 160 to 120, then rising again to about 170 around minute four, followed by a decaying rate. We also observe clear motion traces, especially between minute two and four at around 150 to 160 BPM. This trace is most suppressed in the DIS method. However, at the moment that pulse rate and motion rate coincide (i.e., around minute 4), DIS is not able to recognize the pulse rate.

As an overall conclusion from comparing the spectrograms of SB and DIS, the latter appears to suppress the motion better. In [23], it has been shown that the SB performance profits from an increased window length, implying an increased number of subbands. We therefore consider the effect of the window

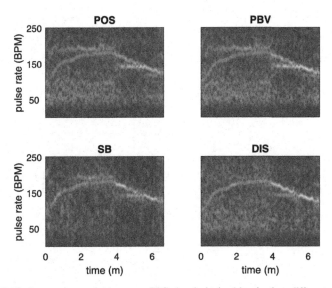

FIGURE 3.12 Spectrograms of the camera-PPG signal obtained by the four different core algorithms in Session 04.

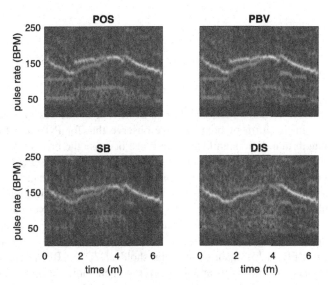

FIGURE 3.13 Spectrograms of the camera-PPG signal obtained by the four different core algorithms in Session 21.

length in the following: We shortened and lengthened the window to 32 and 128 frames, respectively, for the PPG-signal generation. Note that the frequency extraction mechanism remains operating at window length 256.

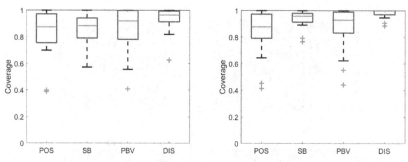

FIGURE 3.14 Boxplots of the coverage at $T_\gamma = 5$ BPM for the four algorithms operating at a length of 32 (left) and 128 frames (right).

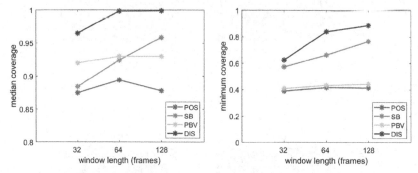

FIGURE 3.15 Median and minimum coverage at $T_\gamma = 5$ BPM for the four algorithms operating at three window lengths in the core algorithm.

The statistics of the coverage at $T_\gamma = 5$ BPM over the 21 sessions is shown in Fig. 3.14 in the form of boxplots. We observe that for POS and PBV the window length makes no significant difference because the boxes interquartile range overlap to a high degree. There is a jump in performance for SB and, to a lesser degree, for DIS as well. This suggest that we should more carefully look into the window-length setting for optimal performance of the motion-oriented PPG optimization algorithms. Fig. 3.14 seems to suggest that we may need to strike a compromise between accuracy and latency for the fitness application.

To gain more insight, the median and minimum coverage over the sessions are plotted in Fig. 3.15, again using a threshold of $T_\gamma = 5$ BPM. The median is taken because it informs us about the average behavior, the minimum as an indicator for the worst-case scenarios. These plots clearly show that, in this application scenario, BPV and POS are not sensitive to window length, whereas SB and DIS are. Since both SB and DIS have extended means for disturbance suppression, it suggests that a prolonged observation helps in getting a better grip on the disturbing signals.

For SB, it holds that the performance in terms of coverage steadily increases with the window length, as shown from both median and minimum coverage characterization. This indeed requires a compromise setting to balance delay and accuracy. On the other hand, DIS seems to saturate: Beyond size 64 (3.2 s), there is no clear performance improvement, while a smaller size (32 frames; 1.6 s) does not give optimal performance. We note that such a short window gets close to a single cardiac cycle on which the optimization is performed, presumably preventing sufficient averaging. Fig. 3.15 shows that the performance of DIS operating at 64 frames is better than SB using doubled algorithmic delay.

3.5 Discussion

The performance analysis and results as reported here were obtained with a limited number of participants and a limited number of fitness devices, and involved a number of lighting conditions, but all in the lab, not in a real application setting. As such, any results can be considered indicative of the current state only.

In line with this, we state that the results presented for the RGB monitoring, together with the results obtained in NIR [24], are considered as indications that the progress in model-based core PPG algorithms has culminated in a promising all-purpose, model-based method. However, new ideas are still popping up in this area, and even a well-defined benchmark including all recent proposals still has to be done. Most likely, model-building with prior knowledge of the specifics of the conditions (illumination, skin types, camera settings) may result in improved core performance for particular applications.

It must also be clear that only the core processing was considered, while many more aspects deserve attention for a properly functioning camera-based PPG monitoring system. Aspects not covered include the preprocessing (skin mask definition and facial landmark tracking) and post-processing (refined frequency extraction), nor the construction of a quality indicator to signal the reliability of the obtained result.

3.6 Conclusions

An overview of the model-based camera-PPG research has been given reflecting research in the last decade. In the course of these years, various flavors have been proposed with different goals in mind, and these have been treated here in an introductory, educational manner to give insights into how they are interrelated, their strengths and weaknesses. The latest model-based variant was proposed and tested in NIR. By reviewing tests of it in the fitness application using RGB, this chapter provides evidence that it qualifies as a high-performance general-purpose PPG method. The PPG core processing has evolved and matured over the years and in its current state enables multiple real-world applications. At the same time, new insights are still being generated and are expected to improve existing systems and broaden the application scope.

Appendix 3.A PBV determination

There are two ways to arrive at the PBV vector. The first one is using spectral characteristics along the optical pathway from the light source to camera sensitivities; the other one is experimental using one or more volunteers in a well-defined setting.

In both cases, similar conditions prevail. For the simulation, a homogeneous piece of skin tissue is assumed uniformly illuminated with diffuse light. This condition is mimicked in the experimental setting by the setup of the light source and the selection of a region of interest in the image. For best results, subject are asked to refrain from any motion.

3.A.1 Optical path descriptors

Consider a homogeneous piece of skin tissue under steady illumination with spectrum $I(\lambda)$, where λ denotes the optical wavelength. The light enters the skin and is scattered and partly reaches the skin areas affected by the blood volume. The reflected light is denoted as $\rho(\lambda, t)$ and modeled as

$$\rho(\lambda, t) = \rho_s(\lambda)\{1 + \rho_b(\lambda)p(t)\}, \tag{3.A.1}$$

i.e., a constant term associated with the average skin, and a time-varying, pulsating component due to the changes of optical character resulting from blood circulation.

When illuminated by a single spectral component λ, the observed signal is $C(t) = I(\lambda)\rho_s(\lambda)\{1 + \rho_b(\lambda)p(t)\}$, and this signal can be characterized by its mean $\mu(\lambda)$ and standard deviation $\sigma(\lambda)$. The resulting characteristic $f(\lambda)$ is then

$$f(\lambda) = \frac{\sigma(\lambda)}{\mu(\lambda)} = \rho_b(\lambda)\sigma_p, \tag{3.A.2}$$

i.e., a function proportional to ρ_b. It is called the relative PPG amplitude [27].

The light reflected from the skin is input to a camera having multiple color channels C_m, $m = 1, \cdots, M$. Each channels has its specific sensitivity $S_m(\lambda)$. The mth channel therefore has as output

$$C_m(t) = \int I(\lambda)\rho_s(\lambda)\{1 + \rho_b(\lambda)p(t)\}S_m(\lambda)d\lambda. \tag{3.A.3}$$

The skin color is the given by the vector $\mathbf{C} = [\mu_1, \cdots, \mu_M]$, with

$$\mu_m = \int I(\lambda)\rho_s(\lambda)S_m(\lambda)d\lambda. \tag{3.A.4}$$

The color variation over time is given by

$$v_m p(t) = p(t)\int I(\lambda)\rho_s(\lambda)\rho_b(\lambda)S_m(\lambda)d\lambda. \tag{3.A.5}$$

The relative (unnormalized) color PPG direction vector is defined as the vector with entries

$$r_m = v_m/\mu_m \propto \sigma_m/\mu_m, \qquad (3.A.6)$$

where σ_m is the standard deviation associated with $C_m(t)$. Finally, the PBV vector is defined as the ℓ^2 normalization of this vector [27]; its entries are

$$q_m = r_m / \sum_{k=1}^{M} r_k^2. \qquad (3.A.7)$$

In view of the normalization, q_m can be determined without knowing the strength of p.

For the simulations, we therefore need to specify all spectral characteristics. Simulations with different skin types are needed because they can differ substantially, e.g. [18,39,40]. Since not all characteristics may be known or only with limited certainty, the experimental PBV determination is the more common approach.

3.A.2 Experimental PBV determination

Also, for the experimental determination, (3.A.8) is used, however now the sensor noise needs to be accounted for:

$$C_m(t) = \int I(\lambda)\rho_s(\lambda)\{1 + \rho_b(\lambda)p(t)\}S_m(\lambda)d\lambda + n(t). \qquad (3.A.8)$$

We assume that $C_m(t)$ has been measured over at least several cardiac cycles, i.e., several seconds. In principle, the experimental procedure could follow the procedure of the simulations by measuring μ_m and σ_m from C_m. However, in view of the fact that such a measurement will absorb the standard deviation of the standard deviation into the estimate of σ_m, a slightly different procedure is followed. Since $p(t)$ is quasi-periodic time series, we assume it may be approximated as a Fourier series when considering $p(t)$ over a not too long time interval. The procedure assumes that we identify the fundamental frequency and determine the amplitude A_m associated with that frequency instead of σ_m. We then take r_m as $r_m = A_m/\mu_m$ and proceed to construct q_m as before.

A slightly different variant is to construct from $C_m(t)$ the normalized color signals $\tilde{C}_m(t)$ as defined by (3.1). In this way, we have upfront normalized by the mean level in mth channel. We then proceed to measure the amplitude $\tilde{a}_m$ of the fundamental frequency associated with $\tilde{C}_m$, and thus have

$$r_m \propto \tilde{a}_m. \qquad (3.A.9)$$

In Fig. 3.A.1, the amplitude spectra of three signals $\tilde{C}_m(t)$ are shown, and $\tilde{a}_m$ can be found at the fundamental frequency of $p(t)$.

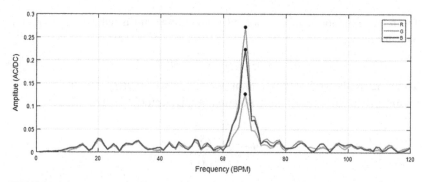

FIGURE 3.A.1 Illustration of the experimental procedure to determine the PBV. The spectra of $\widetilde{C}_m$ are determined, and the amplitudes $\widetilde{a}_m$ associated with the fundamental frequency are taken, here at around 65 BPM. The amplitudes are indicated by the black dots and can directly be used to determine the PBV vector by applying ℓ^2 normalization.

Appendix 3.B   Pseudocode for model-based PPG

In this appendix, the main model-based core algorithm are provided in pseudocode. This concerns the CHROM, POS, SB, PBV and DIS algorithms. The notation of the variables is consistent over the different algorithms and with that in the main text. For clarity, we repeat the definition of the notation and illustrate it in the context of the full processing system as shown in Fig. 3.B.2. Input to the system is a video stream, i.e., a sequence of video frames. In a first processing step, region(s) of interest (RoI) are defined, which we will refer to as (image) patches. Each patch is characterized by a number of signals, most notably the color signals denoted as C and disturbance signals denoted as D. Typically, these are the motion signals associated with the patch in the image plane. The second step is the core processing where the signals over a number of video frames are taken per patch and combined into a PPG segment. The length of these segments is denoted as N, the PPG signal segment as $\mathbf{P}$, and the input segments as $\mathbf{C}$ and $\mathbf{D}$ that are cuts from the signals C and D. In the third processing step, the PPG signal segments and patches are combined to create the full discrete-time PPG signal, typically by pooling across patches and overlap-adding across time. This signal can then be used for further feature analysis like, e.g., pulse rate and pulse rate variability.

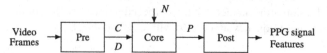

FIGURE 3.B.2 Processing system consisting of pre-, core- and post-processing with in- and output signals of the core processor as indicated. For details, see text.

Algorithm 1 PBV (2014).

Input: $\mathbf{C} := 3 \times N$ RGB signals (e.g., $N = 64$ for 20 fps camera);
1: **Initialize:** $\bar{\mathbf{e}} = [0.3, 0.8, 0.5]$ (PBV signature)
2: $\tilde{\mathbf{C}} = \text{diag}(\text{mean}(\mathbf{C}, 2))^{-1} * \mathbf{C} - 1; \rightarrow$ DC normalization
3: $\hat{\mathbf{C}} = \text{bandpass}(\tilde{\mathbf{S}}, [36, 240]\text{bpm}); \rightarrow$ Band-pass filtering
4: $\mathbf{P} = \bar{\mathbf{e}} * \text{pinv}(\hat{\mathbf{C}}^{\top}); \rightarrow$ Least-squares projection
Output: $\mathbf{P} := 1 \times N$ pulse signal

Algorithm 2 CHROM (2013).

Input: $\mathbf{C} := 3 \times N$ RGB signals (e.g., $N = 64$ for 20 fps camera);
1: **Initialize:** $\bar{\mathbf{v}} = [3, -2, 0; 1.5, 1, -1.5]$ (CHROM-projection plane)
2: $\tilde{\mathbf{C}} = \text{diag}(\text{mean}(\mathbf{C}, 2))^{-1} * \mathbf{C} - 1; \rightarrow$ DC normalization
3: $\hat{\mathbf{C}} = \text{bandpass}(\tilde{\mathbf{C}}, [36, 240]\text{bpm}); \rightarrow$ Band-pass filtering
4: $\hat{\mathbf{S}} = \bar{\mathbf{v}} * \hat{\mathbf{C}}; \rightarrow$ Projection
5: $\mathbf{P} = [1, -\text{std}(\hat{\mathbf{S}}(1, :))/\text{std}(\hat{\mathbf{S}}(2, :))] * \hat{\mathbf{S}}; \rightarrow$ Alpha-tuning
Output: $\mathbf{P} := 1 \times N$ pulse signal

Algorithm 3 POS (2016).

Input: $\mathbf{C} := 3 \times N$ RGB signals (e.g. $N = 64$ for 20 fps camera);
1: **Initialize:** $\bar{\mathbf{v}} = [0, 1, -1; -2, 1, 1]$ (POS-projection plane)
2: $\tilde{\mathbf{C}} = \text{diag}(\text{mean}(\mathbf{C}, 2))^{-1} * \mathbf{C} - 1; \rightarrow$ DC normalization
3: $\hat{\mathbf{C}} = \text{bandpass}(\tilde{\mathbf{C}}, [36, 240]\text{bpm}); \rightarrow$ Band-pass filtering
4: $\hat{\mathbf{S}} = \bar{\mathbf{v}} * \hat{\mathbf{C}}; \rightarrow$ Projection
5: $\mathbf{P} = [1, \text{std}(\hat{\mathbf{S}}(1, :))/\text{std}(\hat{\mathbf{S}}(2, :))] * \hat{\mathbf{S}}; \rightarrow$ Alpha-tuning
Output: $\mathbf{P} := 1 \times N$ pulse signal

Algorithm 4 SB (2017).

Input: $\mathbf{C} := 3 \times N$ RGB signals (e.g., $N = 64$ for 20 fps camera); $\epsilon = 1e - 3(bias)$;
1: **Initialize:** $\bar{\mathbf{v}} = [0, 1, -1; -2, 1, 1]$ (POS-projection as showcase)
2: $\tilde{\mathbf{C}} = \text{diag}(\text{mean}(\mathbf{C}, 2))^{-1} * \mathbf{C} - 1; \rightarrow$ DC normalization
3: $\hat{\mathbf{C}} = \text{bandpass}(\tilde{\mathbf{C}}, [36, 240]\text{bpm}); \rightarrow$ Band-pass filtering
4: $\mathbf{F} = \text{fft}(\hat{\mathbf{C}}, [], 2); \rightarrow$ Time to frequency
5: $\mathbf{S} = \bar{\mathbf{v}} * \mathbf{F}; \rightarrow$ Sub-band projection
6: $\mathbf{Z} = [1, \text{abs}(\mathbf{S}(1, :))./(\epsilon + \text{abs}(\mathbf{S}(2, :)))] * \mathbf{S}; \rightarrow$ Sub-band alpha-tuning
7: $\hat{\mathbf{Z}} = \mathbf{Z}. * \text{abs}(\mathbf{Z})./(\epsilon + \text{abs}(\text{sum}(\mathbf{Z}, 1))); \rightarrow$ Energy normalization
8: $\mathbf{P} = \text{real}(\text{ifft}(\hat{\mathbf{Z}}, [], 2)); \rightarrow$ Frequency to time
Output: $\mathbf{P} := 1 \times N$ pulse signal

Algorithm 5 DIS (2019).

Input: C := $3 \times N$ RGB signals (e.g., $N = 64$ for 20 fps camera); **D** := $2 \times N$ disturbance signals;

1: **Initialize:** $\bar{\mathbf{e}} = [0.3, 0.8, 0.5, 0, 0]$ (DIS signature)
2: $\mathbf{S} = [\mathbf{C}; \mathbf{D}]$;
3: $\tilde{\mathbf{S}} = \text{diag}(\text{mean}(\mathbf{S}, 2))^{-1} * \mathbf{S} - 1$; $\rightarrow$ DC normalization
4: $\hat{\mathbf{S}} = \text{bandpass}(\tilde{\mathbf{S}}, [36, 240]\text{bpm})$; $\rightarrow$ Band-pass filtering
5: $\mathbf{P} = \bar{\mathbf{e}} * \text{pinv}(\hat{\mathbf{S}}^{\top})$; $\rightarrow$ Least-squares projection
Output: P := $1 \times N$ pulse signal

References

[1] D.J. McDuff, J.R. Estepp, A.M. Piasecki, E.B. Blackford, A survey of remote optical photoplethysmographic imaging methods, in: 2015 37th Annual International Conference of the IEEE Engineering in Medicine and Biology Society (EMBC), 2015, pp. 6398–6404.

[2] A. Sikdar, S.K. Behera, D.P. Dogra, Computer-vision-guided human pulse rate estimation: a review, IEEE Reviews in Biomedical Engineering 9 (2016) 91–105.

[3] Y. Sun, N. Thakor, Photoplethysmography revisited: from contact to noncontact, from point to imaging, IEEE Transactions on Biomedical Engineering 63 (3) (2016) 463–477.

[4] A. Al-Naji, K. Gibson, S. Lee, J. Chahl, Monitoring of cardiorespiratory signal: principles of remote measurements and review of methods, IEEE Access 5 (2017) 15776–15790.

[5] F. Bousefsaf, C. Maaoui, A. Pruski, Remote sensing of vital signs and biomedical parameters: a review, Modelling, Measurement and Control C 79 (4) (2018) 173–178.

[6] P.V. Rouast, M.T.P. Adam, R.R. Chiong, D. Cornforth, E. Lux, Remote heart rate measurement using low-cost RGB face video: a technical literature review, Frontiers of Computer Science 12 (2018) 858–872.

[7] R. Spetlík, J. Cech, J. Matas, Non-contact reflectance photoplethysmography: progress, limitations, and myths, in: 2018 13th IEEE International Conference on Automatic Face Gesture Recognition (FG 2018), 2018, pp. 702–709.

[8] S. Zaunseder, A. Trumpp, D. Wedekind, H. Malberg, Cardiovascular assessment by imaging photoplethysmography – a review, Biomedizinische Technik/Biomedical Engineering 63 (5) (2018) 617–634.

[9] C.H. Antink, Simon S. Lyra, M. Paul, X. Yu, S. Leonhardt, A broader look: camera-based vital sign estimation across the spectrum, Yearbook of Medical Informatics 28 (1) (2019) 102–114.

[10] X. Chen, J. Cheng, R. Song, Y. Liu, R. Ward, Z.J. Wang, Video-based heart rate measurement: recent advances and future prospects, IEEE Transactions on Instrumentation and Measurement 68 (10) (2019) 3600–3615.

[11] M. Harford, J. Catherall, S. Gerry, J.D. Young, P. Watkinson, Availability and performance of image-based, non-contact methods of monitoring heart rate, blood pressure, respiratory rate, and oxygen saturation: a systematic review, Physiological Measurement 40 (6) (2019) 06TR01.

[12] F. Khanam, A. Al-Naji, J. Chahl, Remote monitoring of vital signs in diverse non-clinical and clinical scenarios using computer vision systems: a review, Applied Sciences 9 (20) (2019) 4474.

[13] Y. Deng, A. Kumar, S.S. Agaian, S.P. DelMarco, V.K. Asari, Standoff heart rate estimation from video – a review, in: SPIE Defense + Commercial Sensing, Anaheim (CA), 27 April–8 May 2020.

[14] M.A. Hassan, A.S. Malik, D. Fofi, B. Karasfi, F. Meriaudeau, Towards health monitoring using remote heart rate measurement using digital camera: a feasibility study, Measurement 149 (2019) 106804.

[15] M. Leo, P. Carcagnì, P.L. Mazzeo, P. Spagnolo, D. Cazzato, C. Distante, Analysis of facial information for healthcare applications: a survey on computer vision-based approaches, Information 11 (3) (2020) 128.

[16] R. Sinhal, K. Singh, M.M. Raghuwanshi, An overview of remote photoplethysmography methods for vital sign monitoring, in: M. Gupta, D. Konar, S. Bhattacharyya, S. Biswas (Eds.), Computer Vision and Machine Intelligence in Medical Image Analysis: International Symposium, ISCMM 2019, vol. 992, Springer, Singapore, 2020, pp. 21–31.

[17] WinSocial app., https://play.google.com/store/apps/details?id=com.winsocial.underwriting. (Accessed 24 December 2020).

[18] R.R. Anderson, J.A. Parrish, The optics of human skin, Journal of Investigative Dermatology 77 (1981) 13–19.

[19] A.V. Moco, et al., New insights into the origin of remote PPG signals in visible light and infrared, Scientific Reports 8 (1) (May 2018).

[20] Q. Zhan, W. Wang, G. de Haan, Analysis of cnn-based remote-ppg to understand limitations and sensitivities, Biomedical Optics Express 11 (3) (Mar. 2020) 1268–1283.

[21] X. Li, H. Han, H. Lu, X. Niu, Z. Yu, A. Dantcheva, G. Zhao, S. Shan, The 1st challenge on remote physiological signal sensing (RePSS), in: 2020 IEEE/CVF Conference on Computer Vision and Pattern Recognition Workshops (CVPRW), Seattle, WA, USA, June 2020, pp. 1274–1281.

[22] A. Woyczyk, V. Fleischhauer, S. Zaunseder, Skin segmentation using active contours and Gaussian mixture models for heart rate detection in videos, in: 2020 IEEE/CVF Conference on Computer Vision and Pattern Recognition Workshops (CVPRW), Seattle, WA, USA, June 2020, pp. 1265–1273.

[23] W. Wang, A.C. den Brinker, S. Stuijk, G. de Haan, Robust heart rate from fitness videos, Physiological Measurement 38 (6) (May 2017) 1023–1044.

[24] W. Wang, A.C. den Brinker, G. de Haan, Discriminative signatures for remote-PPG, IEEE Transactions on Biomedical Engineering 67 (5) (2020) 1462–1473.

[25] W. Wang, A.C. den Brinker, S. Stuijk, G. de Haan, Algorithmic principles of remote PPG, IEEE Transactions on Biomedical Engineering 64 (7) (July 2017) 1479–1491.

[26] W. Wang, A.C. den Brinker, G. de Haan, Full video pulse extraction, Biomedical Optics Express 9 (8) (Aug 2018) 3898–3914.

[27] G. de Haan, A. van Leest, Improved motion robustness of remote-PPG by using the blood volume pulse signature, Physiological Measurement 35 (9) (Oct. 2014) 1913–1922.

[28] W. Wang, A.C. den Brinker, G. de Haan, Single-element remote-PPG, IEEE Transactions on Biomedical Engineering 66 (7) (July 2019) 2032–2043.

[29] G. de Haan, V. Jeanne, Robust pulse rate from chrominance-based rPPG, IEEE Transactions on Biomedical Engineering 60 (10) (Oct. 2013) 2878–2886.

[30] K. Zhou, S. Krause, T. Blöcher, W. Stork, Enhancing remote-PPG pulse extraction in disturbance scenarios utilizing spectral characteristics, in: 2020 IEEE/CVF Conference on Computer Vision and Pattern Recognition Workshops (CVPRW), Seattle, WA, USA, June 2020, pp. 1130–1138.

[31] W. Wang, A.C. den Brinker, S. Stuijk, G. de Haan, Amplitude-selective filtering for remote-PPG, Biomedical Optics Express 8 (3) (Mar 2017) 1965–1980.

[32] B. Wu, P. Huang, C. Lin, M. Chung, T. Tsou, Y. Wu, Motion resistant image-photoplethysmography based on spectral peak tracking algorithm, IEEE Access 6 (2018) 21621–21634.

[33] C.S. Pilz, I.B. Makhlouf, V. Blazek, S. Leonhardt, On the vector space in photoplethysmography imaging, in: 2019 IEEE/CVF Conference on Computer Vision and Pattern Recognition Workshops (CVPRW), Seoul, Korea (South), June 2019, pp. 1580–1588.

[34] W. Wang, B. Balmaekers, G. de Haan, Quality metric for camera-based pulse rate monitoring in fitness exercise, in: Proc. IEEE Int. Conf. Image Process. (ICIP), Phoenix, AZ, USA, Sept. 2016, pp. 2430–2434.

[35] B. Wu, C. Lin, P. Huang, T. Lin, M. Chung, A contactless sport training monitor based on facial expression and remote-PPG, in: 2017 IEEE International Conference on Systems, Man, and Cybernetics (SMC), Oct 2017, pp. 846–851.

[36] C. Zhao, C.-L. Lin, W. Chen, M.-K. Chen, J. Wang, Visual heart rate estimation and negative feedback control for fitness exercise, Biomedical Signal Processing and Control 56 (2020) 101680.

[37] J. Li, Y. Wang, C. Wang, Y. Tai, J. Qian, J. Yang, C. Wang, J. Li, F. Huang, DSFD: dual shot face detector, in: Proceedings of the IEEE/CVF Conference on Computer Vision and Pattern Recognition (CVPR), June 2019, pp. 5060–5069.

[38] Y. Mironenko, K. Kalinin, M. Kopeliovich, M. Petrushan, Remote photoplethysmography: rarely considered factors, in: The IEEE/CVF Conference on Computer Vision and Pattern Recognition (CVPR) Workshops, June 2020.

[39] Y. Kanzawa, T. Naito, Y. Kimura, Human skin detection by visible and near-infrared imaging, in: Proc. IAPR Conf, in: Mach. Vision Appl., vol. 12, June 2011, pp. 14–22.

[40] M.J. Mendenhall, a.S. Nunez, R.K. Martin, Human skin detection in the visible and near infrared, Applied Optics 54 (35), 10559–10570.

Chapter 4

Camera-based respiration monitoring

Motion and PPG-based measurement

Wenjin Wang[a,c] and Albertus C. den Brinker[b]
[a]*Patient Care and Monitoring Department, Philips Research, Eindhoven, the Netherlands,*
[b]*AI Data Science & Digital Twin Department, Philips Research, Eindhoven, the Netherlands,*
[c]*Eindhoven University of Technology, Electrical Engineering Department, Eindhoven, the Netherlands*

Contents

4.1 Introduction

Monitoring of the respiration rate is ubiquitous for many healthcare applications because it is one of the most important vital signs to indicate a person's health. Changes in spontaneous respiration rate can be used as an early indicator of physiological deterioration or delirium in patients, while the average respiration rate can provide insights into a person's well-being [2], e.g., level of stress and quality of sleep. To measure the respiration rate, contact-based sensors have been developed and matured, in both clinical and home-based applications [6]. These include systems based on capnography to measure nasal or oral airflow and electrical impedance tomography or motion sensors (e.g., accelerometers) to measure respiratory motion from the chest or abdomen. However, contact-based measurements are usually cumbersome, uncomfortable, and inconvenient to use, and they may require effort from trained personnel (e.g., to attach electrodes to the skin or to strap an inductive belt around the chest). As an alternative, contactless respiration monitoring has been explored, and the camera is one of the investigated sensing modalities alongside Doppler radar, acoustic sensors, and WiFi [14,16,30].

Camera systems have many benefits for health monitoring. It does not require direct/mechanical contact between sensors and skin, thus reducing the interference and potential infection/contamination caused by contact measurement, especially improving user experience in the scenario like sleep monitoring and baby monitoring (with fragile skin). It also enables automatic detection and tracking of a Region of Interest (RoI) in a video (e.g., face, chest, or belly) that reduces the need for user compliance and/or personnel intervention, which is suitable for long-term and continuous monitoring over day and night (24/7). It has been shown that cameras can measure a broad range of physiological signals and body signs/activities (e.g., pulse rate (variability), respiration rate, blood-oxygen saturation, pulse transit time, skin temperature, atrial fibrillation, body movement, emotions, pain/discomfort, etc.) [31]. Among those measurements, respiration is an essential component of a camera-based health-monitoring system. It has been extensively researched in the last decade and matured into products such as baby monitoring [11,12], sleep monitoring [24], Philips Vital Signs App for portable monitoring [8], and Philips VitalEye for respiratory triggering/gating in Magnetic Resonance (MR) [27].

For camera-based respiration measurement, various proposals have been made to extract a respiratory signal and rate from a video sequence. Fundamentally, all those proposals exploit one of the following three principles for measurement: motion sensing, thermography and photoplethysmography:

- Motion sensing. Respiration is characterized by movement of chest and abdomen and thus causes displacement of these areas in an image [2,10,23]. It uses the same principle as contact-based motion sensors, and its performance has been reported in various imaging sensors, such as RGB [21], Near Infrared (NIR) [4], and depth (time-of-flight) cameras [1,20] to measure either 2D or 3D respiratory motion.
- Thermography. The exchange of air causes temperature changes that can be measured at the nose and mouth [3]. The feasibility has been demonstrated in both high-resolution thermal cameras and low-resolution thermopiles [9,17].
- Photoplethysmography (PPG). Respiratory effort modulates the blood circulation [22], i.e., it compresses the blood vessels and varies the amount of arterial blood at a peripheral skin site.

Among these three principles, motion-based measurement attracts the most attention in research due to its simplicity and reproducibility, i.e., it has been extensively researched and recently applied in products for baby monitoring and respiratory triggering/gating for MR. In comparison, thermography requires dedicated/costly thermal sensors that are less ubiquitous; photoplethysmography has more constraints in practical use-cases, i.e., PPG-based respiration measurement requires a high-quality PPG signal, which leads to higher demands on the camera sensor and more restrictions on the environment than motion-based approaches. However, since motion-based and PPG-based measurements

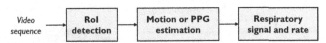

FIGURE 4.1 General pipeline for motion-based or PPG-based respiration measurement from a video.

are feasible with a single camera sensor (RGB or NIR), by acquiring motion signals from the chest area and PPG signals from the (facial) skin area simultaneously, we investigate both approaches in this chapter with the use case of a NIR camera.

A typical motion-/PPG-based respiration monitoring system has the following three ingredients (see Fig. 4.1): Region-of-Interest (RoI) detection, motion or PPG estimation, and respiratory signal construction and rate calculation. RoI detection automatically finds a sub-region in an image that contains the respiratory signal, such as the chest or abdomen (motion-based) or facial area (PPG-based). The motion or PPG estimation determines the motion or PPG signal inside the RoI. In the last step, the motion or PPG signal is translated into a respiratory signal, and, from this signal, the length of individual respiratory cycles is determined or a repetition rate is calculated. The RoI-detection algorithm can be constructed by machine learning (e.g., supervised detection of face, chest, or belly), while the last step calls for rather generic feature-extraction methods. We consider the motion or PPG estimation to be the unique and core step of the pipeline because it determines the fundamental behavior of a monitoring system, i.e., whether the system is able to capture small displacements (at the level of sub-pixel distances) or whether the measurement is sensitive to respiratory modulation of blood-volume changes.

In PPG, the pixel intensity changes due to skin absorption variations are determined, which is an indirect measurement of respiratory motion. Motion-based measurements determine the displacement of the chest or abdomen and thus directly reflect respiratory effort. Two motion-estimation approaches [2,10] are commonly used as the basis (engine) for most proposed solutions: optical flow and profile correlation. Optical flow measures the displacement at a pixel level between two video frames [18,19]. Profile correlation determines the translational movement of a group of pixels (e.g., the image area like a patch or template) by considering the cross correlation between subsequent video frames [7,29]. In this chapter, we review and discuss two measurement modalities (PPG-based and motion-based) and two common algorithms for the motion-based modality. Specifically, we hope to answer following two questions:

- Which measurement modality is more suitable for respiration monitoring: motion-based or PPG-based?
- Which core algorithm is more suitable in motion-based measurement: optical flow or correlation filter?

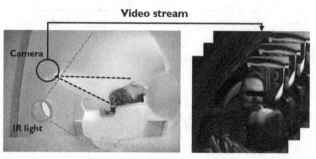

FIGURE 4.2 Philips VitalEye camera setup installed in the high-end MR system (Philips 3T Ingenia Ambition and Ingenia Elition) for respiratory triggering/gating [25]. It consists of an infrared camera and an in-bore infrared light source. The infrared image sequence is obtained with the VitalEye camera.

To address these two questions, we need to specify the context (use case) of respiration monitoring because different applications may render very different requirements and challenges, i.e., the best possible option (if any) depends on the use case. At the moment, there is very little insight into this matter. To carry out a direct comparison, we opted for respiration monitoring in a healthcare setting, in particular, in an MR environment. This application area constitutes interesting options for increased usage of camera-based contactless monitoring, and no thorough benchmarking of respiration algorithms for this use case is available. It restricts the conclusions drawn from the benchmarking to a large degree to this particular use case, but we envision this work as the first stepping stone in a much larger benchmarking effort.

In MR imaging, respiration monitoring is used for gating and triggering of the scans. This is necessary because the movement of chest and abdomen due to the respiration would otherwise lead to artifacts (motion blur) in the MR images, especially the edge sharpness of moving structures is reduced [15]. To solve this problem, respiratory triggering and gating techniques have been developed and applied in clinical settings using contact-based motion sensors (e.g., a respiratory belt) [5]. More recently, camera-based contactless sensing was introduced by Philips in the VitalEye product [26,27]. Fig. 4.2 shows the VitalEye camera setup installed in an MR system. It is an excellent showcase for how to use camera-based vital signs monitoring to optimize the clinical operation and workflow. It has been shown that VitalEye camera-based respiratory triggering/gating not only improves patient comfort and clinical efficiency during the MR scanning, but also improves the MR imaging quality (see the example in Fig. 4.3), i.e., its triggering is fast (20-ms temporal resolution) and detailed (sub-mm motion detection) depending on the camera's sampling rate and image resolution.

In this chapter, we stress that the goal is not to investigate the triggering/gating function for MRI nor the medical image quality, but rather the performance of core algorithms for respiratory signal measurement in MR-specific settings.

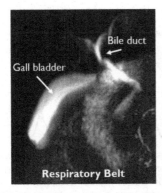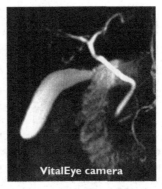

FIGURE 4.3 The MR images obtained by the respiratory triggering of using contact-based respiratory belt (left) and VitalEye camera (right). The result is from a clinical study at University of Bonn, Germany. It is clear that the MR image acquired with the assistance of VitalEye camera has less motion blur.

As already mentioned, each application case has its own challenges. For MR, the monitored subjects are stabilized as they are lying on the scanning table and asked to remain motionless. Therefore, this scenario would in the first instance appear to be a relatively simple case for accurate sensing. However, the monitoring issues arise from the environmental setting where positioning of cameras and light sources have restrictions and only part of the chest or face is visible.

The remainder of this chapter is structured as follows. In Sect. 4.2, we introduce the camera setup in the clinical setting used for collecting the benchmark dataset. In Sect. 4.3, we revisit PPG-based and motion-based respiration measurement and two core algorithms commonly used for motion-based measurement, with a focus on detailing their algorithmic steps and principles. In Sect. 4.4, we evaluate the core algorithms in the MR dataset and analyze their merits and limitations. Finally, in Sect. 4.5, we draw our conclusions.

4.2 Setup and measurements

In this section, we introduce the camera setup, the recording setting, and the synchronized data collection from cameras and contact-based reference sensors.

4.2.1 Camera setup in the MR system

A clinical MR system that already provides camera-based respiratory-triggered MRI (VitalEye, 3T Ingenia Elition X, Philips Healthcare, The Netherlands) [13] is used for the experiments. It is equipped with a camera that provides a view into the MR bore as shown in Fig. 4.4. A mirror is used to display content for patient entertainment. The target monitoring area is the chest because this contains the relevant information, but this area is largely occluded (by the mirror arm), implying that only a limited RoI is available. Without considerable adaptation

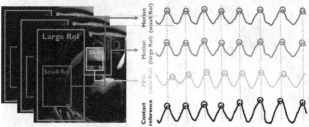

FIGURE 4.4 The camera view in the clinical MR system. Both the camera video data and contact reference data are acquired simultaneously for benchmarking. For each video, three fixed RoIs are predefined for respiratory signal extraction and are indicated by: yellow (light gray in print version) bounding-box for PPG-based measurement, and red (mid gray in print version) and blue (dark gray in print version) bounding-boxes for motion-based measurement. The reason for defining two RoIs (small and large) for motion-based measurement is to study the robustness to RoI selection.

to the MR system itself, the occlusion is an unavoidable challenge within an MR setting. The camera (IDS UI-1220SE, IDS Imaging Development Systems, Obersulm, Germany) has a monochrome CMOS sensor (global shutter, 752 × 480 pixels, 8 bits), operated at 20 frames per second (fps). The exposure time was manually set per session to maximize the dynamic range. All automatic adjustment functions of the camera (e.g., auto-exposure, auto white-balance, auto-focus, and auto-gain) were switched off. The camera has a daylight cut-filter that eliminates the visible light disturbance (e.g., fluorescent light from the ceiling). An active near-infrared light source is applied to provide constant and stable infrared illumination.

4.2.2 Data collection and preparation

The collected data consists of simultaneous acquisition of the signals from camera and contact sensors (i.e., a respiratory belt) for eleven healthy adult subjects (mean age of 46.8 yrs., ranging from 32 to 64 yrs.). The scanning protocol was that for routine MR scanning. The respiratory motion sensor of the clinical MR system sampled at 496 Hz. The camera video data and contact motion sensor data were synchronized by time stamps. This study was approved by the Internal Committee Biomedical Experiments of Philips Research, and informed consent was been obtained from each participant.

As the purpose of this study is to investigate the core algorithms for respiration measurement, we manually set the respiratory RoI. In practice, automatic RoI detection can be implemented based on machine learning technology, but this is not the focus of this chapter. For video processing, we define three RoIs (see Fig. 4.4): one face RoI (for PPG measurement), one small chest RoI positioned at the respiratory area, and one large chest RoI including respiratory and non-respiratory areas (i.e., including a large part of the MR machine). The reason for using two chest RoIs is to validate the sensitivity of motion estimation on RoI definition, assuming that automatic RoI detection may not always find an accurate respiratory area in practice. In this sense, we may gain insight into the influence of an inaccurate RoI on the sensitivity of the estimation of tiny motions.

4.3 Methods

As mentioned in Sect. 4.1, this chapter focuses on the investigation of PPG-based and motion-based respiratory signal extraction and two basic motion-based algorithms: optical flow and profile correlation. This chapter focuses on the investigation of PPG- and motion-based respiratory signal extraction. We discuss only briefly the PPG method as more details are provided in Chap. 3. We provide in more depth the two motion-based algorithms: optical flow and profile correlation. Lastly we introduce the performance measures used to compare the various algorithms. We use the following mathematical conventions in this chapter: Italic characters denote scalars; boldface characters denote vectors and matrices.

4.3.1 PPG-based

The PPG-based respiratory signal extraction is rather intuitive. It measures the skin-pixel intensity changes due to blood-absorption variations at a peripheral skin site. Considering a 2D image captured by a monochrome video camera at time t, we manually select an RoI (e.g., face) and assume the pixel intensity at the position (x, y) as $I(x, y, t)$. The signal component obtained at this frame is derived by averaging the skin pixels:

$$C(t) = \mu(I(x, y, t)), \qquad (4.1)$$

where $\mu(\cdot)$ denotes the spatial-averaging operator. By using a band-pass filter (e.g., [0.1, 0.6] Hz for a healthy adult), we can reduce non-respiratory components from $\mathbf{C}$, such as high-frequency pulsatile components, low-frequency Mayer waves, and sensor noise. The filtered signal, denoted as $\mathbf{S}$, is considered to be the respiratory signal measured from the intensity-variation signal. We note that since a monochrome camera is used, there is no multiwavelength combination in this process to reduce disturbance (e.g., non-respiratory body motion or lighting changes) as is typical in camera-based PPG-extraction approaches. But we also mention that, in the MR setup with an active IR light source and a patient in a stable position, we do not expect significant disturbances from the patient or environment.

4.3.2 Motion-based: optical flow

For the motion-based approach, we elaborate on two core algorithms: optical flow and profile correlation (see their mechanims in Fig. 4.5). We first revisit the

(a) Optical flow pipeline

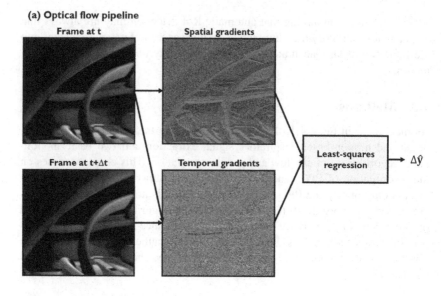

(b) Profile correlation pipeline

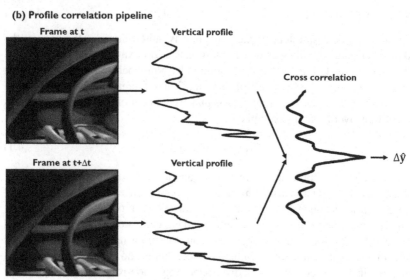

FIGURE 4.5 The essence of two core algorithms used for motion estimation: (a) optical flow uses the least-squares regression between an image's spatial gradients and temporal gradients to estimate the vertical motion; and (b) profile correlation uses the cross-correlation between spatial representations (profiles) to estimate the vertical motion.

optical flow-based motion estimation. Given a manually defined RoI, we denote the pixel intensity at the position (x, y) as $I(x, y, t)$. When motion occurs, the entire intensity profile is assumed to be located at another position (at Δx and

Δy distance from the current location) after Δt:

$$I(x, y, t) = I(x + \Delta x, y + \Delta y, t + \Delta t). \tag{4.2}$$

For our case of monitoring respiratory motion where the motion is measured from a distance and at a high sampling rate (relative to the respiration rate), the pixel movement is considered to be small. $I(x + \Delta x, y + \Delta y, t + \Delta t)$ can be approximated by the Taylor-series expansion:

$$I(x + \Delta x, y + \Delta y, t + \Delta t) \approx I(x, y, t) + \frac{\partial I}{\partial x}\Delta x + \frac{\partial I}{\partial y}\Delta y + \frac{\partial I}{\partial t}\Delta t. \tag{4.3}$$

Combining (4.2) and (4.3), we have:

$$\frac{\partial I}{\partial x}\Delta x + \frac{\partial I}{\partial y}\Delta y + \frac{\partial I}{\partial t}\Delta t = 0. \tag{4.4}$$

Dividing (4.4) by Δt ($\Delta t \neq 0$) gives:

$$\frac{\partial I}{\partial x}\frac{\Delta x}{\Delta t} + \frac{\partial I}{\partial y}\frac{\Delta y}{\Delta t} + \frac{\partial I}{\partial t} = 0, \tag{4.5}$$

where $(\frac{\Delta x}{\Delta t}, \frac{\Delta y}{\Delta t})$ denotes the velocity of the pixel movement, which is called "optical flow", written as **v**. $(\frac{\partial I}{\partial x}, \frac{\partial I}{\partial y})$ denotes the partial derivatives of the pixel in the x and y directions in an image (i.e., the spatial gradient), written as $\nabla \mathbf{I}$. $\frac{\partial I}{\partial t}$ denotes the temporal gradients between two images, written as I_t, and, in practice, is approximated by the difference between the intensities in two frames.

Thus (4.5) can be expressed as vector multiplication:

$$\nabla \mathbf{I} \cdot \mathbf{v}^\top + I_t = 0. \tag{4.6}$$

The optical flow or motion vector at position (x, y) can be solved by least-squares regression based on multiple pixel correspondences, which is the essence of classical method of Lucas–Kanade [19]. Here we assume that the vertical direction contains the strongest respiratory energy based on the targeting application, which is also the assumption made by prior art [2,28]. Thus we only consider the motion estimated in the y-direction and have:

$$v_y = \left(\frac{\partial I}{\partial y}\right)^{-1} \frac{\partial I}{\partial t}. \tag{4.7}$$

For comparison with the displacement that is estimated by the cross correlation, it is convenient to translate the velocity v_y into displacement using the same delay Δt as deployed in the cross-correlation algorithm $\Delta \hat{y}$. We mention that, in order to understand the fundamental behavior of optical flow in pixel motion analysis, we use the most basic version without dedicated steps (e.g., spatial or temporal smoothing, multiscale analysis, heuristics) that may change the characteristics of pixel movement. We are aware of alternatives, such as recent Convolutional Neural Network based methods, which we consider beyond the scope of this study.

4.3.3 Motion-based: profile correlation

The second motion-based approach relies on correlation techniques to estimate the displacement between spatial representations of two frames. The displacement is used to generate the motion signal. The spatial representation can be a 2D image patch (i.e., template) or 1D profile vector. Here we elaborate on the case of 1D profile because it is used by a well-recognized core algorithm—Profile Correlation (ProCor) [2].

Similar to the optical flow case, we consider the images patches (RoI) denoted as I at two time instances: t and $t + \Delta t$. Since respiratory motion is typically generated in the vertical direction, a vertical profile is constructed described by a vector $\mathbf{P}$ containing signal values associated with the vertical positions of the RoI. The signal values in $\mathbf{P}$ are taken as a projection of I on the vertical axis, i.e., by averaging the pixels per image row. We thus have created the two vectors $\mathbf{P}_t$ and $\mathbf{P}_{t+\Delta t}$.

To estimate the shift between these two profiles, we calculate their cross-correlation using Fourier transformations. First, the spatial representation is converted into the frequency domain by Fourier transformation ($\mathcal{F}(\cdot)$):

$$\mathbf{F}_t = \mathcal{F}(\mathbf{P}_t), \tag{4.8}$$

and then the cross-correlation spectrum is calculated between $\mathbf{F}_t$ and $\mathbf{F}_{t+\Delta t}$:

$$\mathbf{R} = \mathbf{F}_t \odot \mathbf{F}^*{}_{t+\Delta t}, \tag{4.9}$$

where $\odot$ denotes the entry-wise product and $*$ denotes conjugation. Next, we apply the inverse Fourier transform ($\mathcal{F}^{-1}(\cdot)$) of $\mathbf{R}$ to convert it back to the spatial domain:

$$\mathbf{r} = \mathcal{F}^{-1}(\mathbf{R}). \tag{4.10}$$

Finally, we determine the location of correlation peak in $\mathbf{r}$ by:

$$\Delta \hat{y} = \arg \max_y \{\mathbf{r}\}, \tag{4.11}$$

where $\Delta \hat{y}$ denotes the displacement of the profile on the vertical direction.

4.3.4 Respiratory signal and rate

For motion-based approaches, we concatenate the vertical shifts measured between two video frames in the time domain (by either optical flow or profile

Two-frame interval (50 ms)   **Two-frame interval (200 ms)**

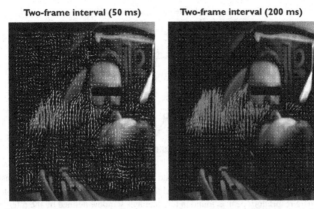

FIGURE 4.6 An example of motion vectors estimated by optical flow from the dense image blocks, where the two-frame interval used to compute motion vectors is set to 50 ms and 200 ms, respectively. The motion vectors with an amplitude larger than 0.2 sub-pixels are denoted in red (gray in print version).

correlation) to generate a long-term signal that represents motion velocity:

$$\mathbf{Y} = \{\Delta \hat{y}_1, \Delta \hat{y}_2, ..., \Delta \hat{y}_N\}. \tag{4.12}$$

Based on the definition, we use cumulative summation (i.e., discrete-time integration) to transform the velocity signal into a respiratory motion signal:

$$S_i = \sum_1^i Y_i. \tag{4.13}$$

We note that the PPG-based approach does not require this step. In the end, a simple peak detector[1] is applied to find peaks in the respiratory signal. The detected peaks can be used to estimate Inter-Beat Interval (IBI), and the inverse of IBI represents the instantaneous respiration rate.

To evaluate the algorithms, we measure the beat-to-beat accuracy between the contact respiratory signal and camera respiratory signal based on their detected respiratory peaks (see Fig. 4.4). The first screening of the data revealed that direct comparison of distances between peaks of the camera-based respiration signals and the reference in terms of squared error is extremely sensitive to outliers. Therefore, we introduce the following robust metrics. For each respiratory peak from the contact device, we set a tolerance window centered around the peak, where the window radius is 50% of IBI with respect to its preceding and proceeding peaks (i.e., no overlap between the windows in time). If a single respiratory peak is detected from the camera signal within the tolerance window, it is counted as a valid measurement (see Fig. 4.7), denoted as a True Positive

---

[1] The Matlab® function **findpeaks**($\cdot$) with default settings is used to detect peaks in a time signal.

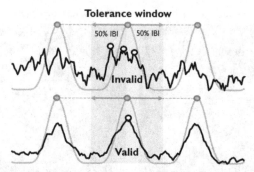

FIGURE 4.7 Illustration of defining valid/invalid respiratory peaks for evaluation metrics. The black curve is the camera respiratory signal, and the gray curve is the reference phantom signal. If a single-camera respiratory peak is found in the tolerance window defined by the reference signal, it is a valid measurement for the camera.

(TP) peak. If none or multiple peaks are detected in the tolerance window, it is counted as a False Negative (FN). The camera peaks that are not identified as TP peaks are denoted as False Positive (FP) peaks. Next, we use the metrics, precision, and recall to quantify the evaluation. The precision is calculated by:

$$precision = \frac{TP}{TP + FP}, \tag{4.14}$$

and the recall is calculated by:

$$recall = \frac{TP}{TP + FN}, \tag{4.15}$$

where precision and recall indicate the accuracy and sensitivity of the camera system, respectively. Finally, we use F_1-score to combine precision and recall into a single metric value (the harmonic mean):

$$F_1 = 2 \cdot \frac{precision \cdot recall}{precision + recall}. \tag{4.16}$$

For completeness, we provide the following recording details. The PPG-based algorithm uses the face RoI and two motion-based algorithms use two different predefined chest RoIs (i.e., small RoI 150×200 pixels and large RoI 250×450 pixels). For motion-based algorithms, a frame interval of 200 ms (corresponding to four frames of a 20-,fps camera) is used to generate the motion vectors. We mention that, if the time interval between two video frames is too small (e.g., adjacent frames with 50-ms delay), the motion vector becomes difficult to estimate because the intensity changes due to the small displacements start to drown in the sensor noise (see Fig. 4.6). All algorithms use the same peak detector to locate respiratory peaks in their measured respiratory signals.

TABLE 4.1 Averaged metric values of PPG- and motion-based (OptFlow and ProCor) algorithms obtained from eleven test subjects. Two respiratory RoIs are used for benchmarking motion-based algorithms.

Metric	Face RoI	Small motion RoI		Large motion RoI	
	PPG	OptFlow	ProCor	OptFlow	ProCor
Precision (%)	68.5	80.7	83.3	82.4	83.0
Recall (%)	83.3	89.2	89.6	88.6	90.2
F_1-score (%)	75.1	84.4	85.9	85.0	86.2

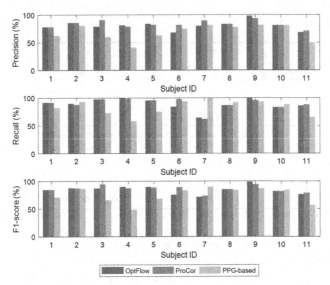

FIGURE 4.8 The precision, recall and F_1-score obtained by motion-based (OptFlow and Pro-Cor) and PPG-based algorithms from eleven test subjects, where motion-based algorithms use the small RoI.

4.4 Results and discussion

In this section, we discuss the results obtained by PPG-based and motion-based approaches in the MR dataset with various RoI settings.

Table 4.1 shows averaged metric values obtained by PPG-based and motion-based (OptFlow and ProCor) approaches. The first impression is that PPG-based approach is worse than motion-based, i.e., the F_1-score is around 10% less. This is largely due to the lower precision of the PPG-based approach (69.5%), where more false positives are generated. Its recall is also lower but not significantly lower than motion-based approaches. Table 4.1 also shows that the two motion-based algorithms have very similar performance independent of RoI setting. The F_1-score is around 85%.

Figs. 4.8–4.9 show the results on individual basis. In line with the statistics from Table 4.1, we see that motion-based approaches generally outperform the

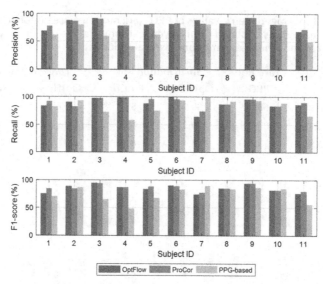

FIGURE 4.9 The precision, recall, and F_1-score obtained by motion-based (OptFlow and Pro-Cor) and PPG-based algorithms from eleven test subjects, where motion-based algorithms use the large RoI.

PPG-based approach on a subject level, and there is no major difference between the two motion algorithms. However, some variations can be observed between individuals, i.e., the motion-based approach is clearly better than the PPG-based approach for Subject 4, whereas PPG-based seems better for Subject 7. To obtain more insights into the causes for these differences, we plot the respiratory spectrograms obtained by three algorithms in Fig. 4.10. For Subject 4, we observe that, in the middle of the recording, the PPG signal is lost. Inspection of the video material suggest that this is due to head movements during that period. Such head movements do not disrupt the motion measurements at the chest. For Subject 7, disturbances appear in the respiration traces at specific parts in the recording. Inspection of the original video content suggest that these are attributable to involuntary body motions not interfering with the stable head position. This comparison shows that the two modalities can be hampered by independently occurring sources of disturbances and, thus, that hybrid methods can be designed to improve the (robustness of the) performance.

There are multiple reasons why performance of the methods differs over the modalities. The motion-based approach measures pixel movement at the chest, which is a direct measurement of respiratory source. In contrast, PPG-based approach measures pixel-intensity changes due to respiration-modulated blood volume changes, which is an indirect measurement of respiratory motion. It relies on the assumption that respiratory motion has sufficient modulation intensity in arterial blood pulsation that allows it to be measured at a distant peripheral skin site, whereas such modulation effect is presumably subject-dependent,

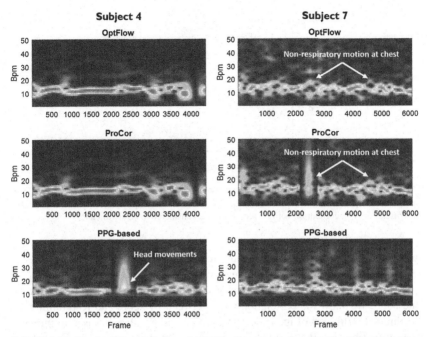

FIGURE 4.10 Respiratory spectrograms obtained by three approaches from two subjects: motion-based measurement is better for Subject 4, while PPG-based measurement is better for Subject 7.

posture-dependent, and respiratory-effort dependent (e.g., deep breathing or shallow breathing). In addition, PPG- and motion-based approaches have different quantizations in measuring the signal, meaning that the same motion source is quantized differently, i.e., PPG-based is quantized by pixel depth (e.g., eight bit, 256 integer values), while motion-based is quantized by pixel resolution. The error propagation depends on the algorithmic steps and is obviously different in both schemes.

We also notice that PPG- and motion-based approaches have different measurement delays. The motion-based approach has an accurate alignment/synchronization with the respiratory motion (inhaling and exhaling phases) because it directly measures the origin of the source. The PPG-based approach measures the blood-volume changes at a skin site (e.g., face or forehead), which has a distance with respect to the motion source at chest. Thus modulation of the blood volume introduced at the thorax or abdomen will be visible some time later in other parts of the body. This causes delays between the peaks and valleys appearing in these signals. This delay may be less relevant for respiration rate monitoring, but it is very critical for MR applications where accurate synchronization between MR data acquisition and motion at the chest is a necessity. Any time delay between the measurement and true body motion will degrade the

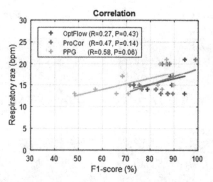

FIGURE 4.11 Correlation between F_1-score and respiration rate for three approaches. Results for motion-based approaches are obtained from the small RoI. R-value and P-value are used to indicate the correlation, i.e., correlation with P-value < 0.05 is considered to be significant.

quality of motion synchronization and removal. In view of this, motion-based respiration measurement is recommended for the use case of MR triggering.

To investigate if the performance of measurements is related to respiration rate itself, we show the correlation plots between F_1-score and respiration rate for three algorithms in Fig. 4.11. The R-value and P-value are used to indicate the correlation, i.e. if the P-value is smaller than the significance level (0.05 is used by default), it suggests that the underlying correlation (R-value) is significant. As can be seen, all three methods have positive correlations between the F_1-score and respiration rate, suggesting accurate measurements are more difficult at lower respiration rates. However, the P-values of motion-based approaches (OptFlow and ProCor) are clearly larger than 0.05, meaning that the correlation is not significant, and thus motion-based measurement is practically insensitive to respiration rate. In comparison, the PPG-based approach has an R-value of 0.58 and P-value of 0.06, though slightly larger than 0.05 (not significant); it is less clear whether or not the performance is related to the respiration rate in view of the limited data points and their uneven spread across the performance axis.

Though the present results favor the application of motion-based respiratory analysis over a PPG-based one, we mention that a physical limitation always exists for camera-based measurement: If the respiratory motion is so small/tiny (due to extremely shallow breathing) that the induced intensity changes are below the sensor noise level, the motion-based algorithms will fail. At the same time, if the motion is so limited, it is unlikely to cause pronounced modulation in the blood-volume signal. Fortunately, if the chest motion is limited, it is unlikely to cause pronounced artifacts in the MR imaging, but a good quantitative analysis of such an effect (e.g., extremely shallow breathing) is currently missing. For more general applications in video health monitoring (e.g., sleep monitoring or baby monitoring), we may still want to improve the measurement sensitivity, especially to cope with shallow breathing. In principle, one could consider options in either the hard- or software domain, e.g., by using a sensor with higher resolution, more bit depth, and lower sensor noise, or denoising techniques to improve the image quality.

4.5 Conclusions

Camera-based respiration monitoring is an emerging research topic for video health monitoring. It has been extensively explored in the last decade, and its maturity is evident from its presence in products. The motion-based approaches have been widely applied to measure the respiratory signal and rate from a video, where the essential issue is "tiny motion estimation". Measuring respiration from modulation induced in the blood-volume changes has also been proposed, though it is less ubiquitous than motion-based approach. Since benchmarking of these methods is limited, we started with addressing this issue in a specific healthcare application where the challenges and objectives are well-defined—respiratory triggering/gating for magnetic resonance (MR). The measurement and algorithmic principles are discussed in detail and their performances are compared in a clinical MR setup equipped with a camera monitoring system. The results show that motion-based modality is better than PPG-based modality, with a higher measurement sensitivity and less measurement delay with respect to the ground-truth (i.e., chest respiratory motion). The two benchmarked motion-based algorithms have rather similar performances within this application (e.g., with an average F_1-score around 85%) and a performance largely insensitive to the RoI setting. This latter presumably holds due to the stationary of the background in this application. There are variations between individuals that are mainly due to motion disturbance at different measurement sites per subject (facial expression, involuntary body motion). The results suggest that the performance is independent of the respiration rate for the motion-based approach; this is currently less clear for the PPG-based approach. We stress that these conclusions are restricted to the MR case. Other use cases (e.g., baby monitoring, sleep monitoring) may have different requirements and challenges. Further improvements in the measurement sensitivity should be considered in future studies and applications.

References

[1] Paul S. Addison, et al., Continuous respiratory rate monitoring during an acute hypoxic challenge using a depth sensing camera, Journal of Clinical Monitoring and Computing (2019) 1–9.

[2] M. Bartula, T. Tigges, J. Muehlsteff, Camera-based system for contactless monitoring of respiration, in: 2013 35th Annual International Conference of the IEEE Engineering in Medicine and Biology Society (EMBC), 2013, pp. 2672–2675.

[3] S.L. Bennett, R. Goubran, F. Knoefel, Comparison of motion-based analysis to thermal-based analysis of thermal video in the extraction of respiration patterns, in: 2017 39th Annual International Conference of the IEEE Engineering in Medicine and Biology Society (EMBC), 2017, pp. 3835–3839.

[4] F. Deng, et al., Design and implementation of a noncontact sleep monitoring system using infrared cameras and motion sensor, IEEE Transactions on Instrumentation and Measurement 67 (7) (2018) 1555–1563.

[5] Richard L. Ehman, et al., Magnetic resonance imaging with respiratory gating: techniques and advantages, American Journal of Roentgenology 143 (6) (1984) 1175–1182.

[6] Mia Folke, et al., Critical review of non-invasive respiratory monitoring in medical care, Medical and Biological Engineering and Computing 41 (4) (2003) 377–383.

[7] Manuel Guizar-Sicairos, Samuel T. Thurman, James R. Fienup, Efficient subpixel image registration algorithms, Optics Letters 33 (2) (2008) 156–158.

[8] Philips app measures vitals using iPad camera, https://www.mobihealthnews.com/14848/philips-app-measures-vitals-using-ipad-camera, 2011.

[9] Jagadev Preeti, Lalat Indu Giri, Non-contact monitoring of human respiration using infrared thermography and machine learning, Infrared Physics & Technology 104 (2020) 103117.

[10] Rik Janssen, et al., Video-based respiration monitoring with automatic region of interest detection, Physiological Measurement 37 (1) (2015) 100.

[11] Joao Jorge, et al., Data fusion for improved camera-based detection of respiration in neonates, in: Gerard L. Coté (Ed.), Optical Diagnostics and Sensing XVIII: Toward Point-of-Care Diagnostics, vol. 10501, International Society for Optics, SPIE, 2018, pp. 215–224.

[12] J. Jorge, et al., Non-contact monitoring of respiration in the neonatal intensive care unit, in: 2017 12th IEEE International Conference on Automatic Face Gesture Recognition (FG 2017), 2017, pp. 286–293.

[13] S. Krueger, et al., Optical unobtrusive physiology sensor for respiratory-triggered MRI acquisitions, in: Proceedings of the Annual Meeting of the International Society of Magnetic Resonance in Medicine (ISMRM), France, 2018, p. 2529.

[14] Yee Siong Lee, et al., Monitoring and analysis of respiratory patterns using microwave Doppler radar, IEEE Journal of Translational Engineering in Health and Medicine 2 (2014) 1–12.

[15] C.E. Lewis, et al., Comparison of respiratory triggering and gating techniques for the removal of respiratory artifacts in MR imaging, Radiology 160 (3) (1986) 803–810.

[16] Xuefeng Liu, et al., Contactless respiration monitoring via off-the-shelf WiFi devices, IEEE Transactions on Mobile Computing 15 (10) (2015) 2466–2479.

[17] Lorato Ilde, et al., Unobtrusive respiratory flow monitoring using a thermopile array: a feasibility study, Applied Sciences 9 (12) (2019) 2449.

[18] Bruce D. Lucas, Generalized Image Matching by the Method of Differences, PhD thesis, Pittsburgh, PA, July 1984.

[19] Bruce D. Lucas, Takeo Kanade, An iterative image registration technique with an application to stereo vision, in: Proceedings of the 7th International Joint Conference on Artificial Intelligence – vol. 2, IJCAI'81, Vancouver, BC, Canada, Morgan Kaufmann Publishers Inc., San Francisco, CA, USA, 1981, pp. 674–679.

[20] Vishnu Vardhan Makkapati, Sai Saketh Rambhatla, Camera based estimation of respiration rate by analyzing shape and size variation of structured light, in: 2016 IEEE International Conference on Acoustics, Speech and Signal Processing (ICASSP), IEEE, 2016, pp. 2219–2223.

[21] C. Massaroni, et al., Measurement system based on RBG camera signal for contactless breathing pattern and respiratory rate monitoring, in: 2018 IEEE International Symposium on Medical Measurements and Applications (MeMeA), 2018, pp. 1–6.

[22] Leila Mirmohamadsadeghi, et al., Real-time respiratory rate estimation using imaging photoplethysmography inter-beat intervals, in: 2016 Computing in Cardiology Conference (CinC), IEEE, 2016, pp. 861–864.

[23] A.V. Mogo, et al., Camera-based assessment of arterial stiffness and wave reflection parameters from neck micro-motion, Physiological Measurement 38 (8) (July 2017) 1576–1598.

[24] T. Nochino, Y. Ohno, S. Okada, Development of noncontact respiration monitoring method with web-camera during sleep, in: 2017 IEEE 6th Global Conference on Consumer Electronics (GCCE), 2017, pp. 1–2.

[25] N.V. Koninklijke Philips, Philips VitalEye, https://www.philips.com/a-w/about/news/media-library/20191201-Philips-VitalEye.html, 2019.

[26] N.V. Koninklijke Philips, Philips VitalEye for MR systems, https://www.philips.co.uk/c-dam/b2bhc/gb/resource-catalog/landing/brightontender/mr-vital-eye-product-specification-lr.pdf, 2018.

[27] Philips Philips, VitalEye helps to simplify MR workflow, https://www.philips.com/a-w/about/news/media-library/20191201-Philips-VitalEye.html, 2019.

[28] M. Rocque, Fully automated contactless respiration monitoring using a camera, in: 2016 IEEE International Conference on Consumer Electronics (ICCE), 2016, pp. 478–479.

[29] Jignesh N. Sarvaiya, Suprava Patnaik, Salman Bombaywala, Image registration by template matching using normalized cross-correlation, in: 2009 International Conference on Advances in Computing, Control, and Telecommunication Technologies, IEEE, 2009, pp. 819–822.

[30] Tianben Wang, et al., C-FMCW based contactless respiration detection using acoustic signal, Proceedings of the ACM on Interactive, Mobile, Wearable and Ubiquitous Technologies 1 (4) (2018) 1–20.

[31] Wenjin Wang, Robust and automatic remote photoplethysmography, PhD thesis, Eindhoven University of Technology, The Netherlands, 2017.

Chapter 5

Camera-based blood oxygen measurement

Izumi Nishidate

Graduate School of Bio-applications and Systems Engineering, Tokyo University of Agriculture and Technology, Tokyo, Japan

Contents

5.1 Introduction

Quantitative evaluation of biological chromophores is useful for detecting various skin diseases including cancers, monitoring health status and tissue metabolism, and assessing clinical and physiological vascular functions. The major chromophores in biological tissues are oxygenated hemoglobin and deoxygenated hemoglobin, which show distinctive optical absorption properties in the visible to the near-infrared (NIR) wavelength region [1]. They show more distinctive contrast at 650 nm and 940 nm in red and near-infrared region, respectively. When the concentration of each chromophore varies, the corresponding change may be observed in diffusely reflected light from the tissues in this wavelength range. Peripheral tissues and cells demand a continuous supply of oxygen, which are delivered via blood circulation. The absorption spectra of oxygenated hemoglobin and deoxygenated hemoglobin differ since the binding of oxygen to hemoglobin changes the light absorption spectrum of hemoglobin [2]. Delivery of oxygen to peripheral tissues can be evaluated from the diffuse reflectance spectrum based on the absorption spectra of oxygenated hemoglobin and deoxygenated hemoglobin. Hemoglobin oxygen saturation is the fraction of oxygenated hemoglobin relative to total hemoglobin in whole blood. It is expressed as a percentage and abbreviated SaO_2 for arterial oxygen saturation and $SvO2$ for venous oxygen saturation. The arterial oxygen saturation measured by the pulse oximeter is called percutaneous arterial oxygen saturation (SpO_2). On the other hand, the mixture of SaO_2 and SvO_2 in a given section of tissue is called tissue oxygen saturation (StO_2) [3]. It is a useful indicator for monitoring peripheral tissue oxygen consumption, hypoperfusion, cyanosis, and tissue viability. Diffuse reflectance and transmittance spectra of peripheral tissue are affected by the periodic temporal variation in blood volume due to the cardiac pulse traveling through the body. Photo-plethysmography (PPG) has been widely used to assess systemic physiological parameters such as heart rate, blood pressure, cardiac output, vascular compliance, and percutaneous oxygen saturation (SpO_2) [4,5].

Diffuse reflectance spectra with continuous-wave light can be easily produced by using a light source with a broad-band spectrum (fluorescent tube, incandescent or light emitting diode (LED)), affordable optical components, and a spectrometer. Analysis of diffuse reflectance spectra provide useful information about tissue activities and functions that are related to biological chromophores. Thus, diffuse reflectance spectroscopy (DRS) has been widely used for the evaluating skin chromophores [6–16]. Multi-spectral imaging based on diffuse reflectance has been employed for visualizing the spatial distribution of chromophore contents in living tissue using a series of discrete optical interference filters, a liquid-crystal wavelength tunable filter [17–19], and a hyperspectral imaging system [20–23]. A simple method for quantitative imaging of melanin and hemoglobin concentrations in *in vivo* skin tissue, based on diffuse reflectance images and using multiple regression analysis aided by Monte Carlo simulations, has been previously proposed [24]. NIR diffuse-reflectance imaging can be achieved with a combination of a general-purpose digital monochromatic image sensor, a few narrowband optical filters, and a light source with a broadband spectrum or using a hyperspectral camera. It can be used for visualizing the spatial distribution of blood oxygen saturation, lipids, and water in tissues.

For practical uses, more simple, cost-effective, and portable equipment is needed. Imaging with a digital red–green–blue (RGB) camera is one of the promising tools to satisfy those demands. Several techniques with an RGB camera-based imaging have been used for the non-invasive characterization of biological tissues where contrast is obtained from the absorption of light by biological chromophores [25–27]. This can be achieved using a general-purpose digital color camera and a white-light source such as a white LED, which enable cost-effective, easy-to-use, battery-powered, portable, remotely administrable, and/or point-of-care solutions. Therefore, an RGB camera-based diffuse-reflectance imaging technique meets medical and healthcare needs. In this chapter, a simple and affordable imaging technique to evaluate blood oxygen saturation and hemoglobin concentration in biological tissues by using a digital RGB camera is described. Several examples for potential applications to biomedical imaging are presented.

5.2 Principle

Let us briefly review the basic principle of the method for estimating spatial maps of concentrations of oxyhemoglobin (C_{HbO}), deoxyhemoglobin (C_{HbR}), total hemoglobin (C_{HbT}), and tissue oxygen saturation (StO_2) in skin tissue [28, 29]. The responses of RGB channels (R, G, and B) in each pixel of the skin-tissue color image acquired by a digital RGB camera can be expressed as

$$\begin{bmatrix} R \\ G \\ B \end{bmatrix} = \mathbf{L_1} \begin{bmatrix} X \\ Y \\ Z \end{bmatrix}, \qquad (5.1)$$

where $\mathbf{L_1}$ is a transposition matrix to convert the tristimulus values in the Commission Internationale de l' Éclairage XYZ (CIEXYZ) color system X, Y, and Z to corresponding responses of R, G, and B. The tristimulus values X, Y, and Z in this equation are defined as

$$X = \kappa \sum E(\lambda)\bar{x}(\lambda)\Theta(\lambda), \qquad (5.2)$$

$$Y = \kappa \sum E(\lambda)\bar{y}(\lambda)\Theta(\lambda), \qquad (5.3)$$

$$Z = \kappa \sum E(\lambda)\bar{z}(\lambda)\Theta(\lambda), \qquad (5.4)$$

where λ, $E(\lambda)$, and $\Theta(\lambda)$ are the wavelengths, the spectral distribution of the illuminator, and the diffuse-reflectance spectrum of living tissue, respectively, whereas $\bar{x}(\lambda)$, $\bar{y}(\lambda)$, and $\bar{z}(\lambda)$ are color matching functions in the CIEXYZ color system. The values of constant κ that result in Y being equal to 100 for the perfect diffuser is given by

$$\kappa = \frac{100}{\sum E(\lambda)\bar{y}(\lambda)}. \qquad (5.5)$$

In Eqs. (5.2)–(5.5), the summation can be carried out using data at ten nm intervals, from 400 to 700 nm. Assuming that the skin tissue mainly consists of epidermis containing the melanin and dermis containing oxygenated hemoglobin, deoxygenated hemoglobin, and bilirubin, the diffuse reflectance of skin tissue Θ can be expressed as

$$\Theta = \frac{I}{I_0} = \left[\int_0^\infty P_e(\mu'_{s,e}, l_e) \exp\left(-\mu_{a,m} l_e\right) dl_e \right]$$
$$\times \left[\int_0^\infty P_d(\mu'_{s,d}, l_d) \exp[-(\mu_{a,HbO} + \mu_{a,HbR} + \mu_{a,bil}) l_d] dl_d \right], \qquad (5.6)$$

where I and I_0 are the detected and incident-light intensities respectively, $P(\mu_s', l)$ is the path length probability function that depends on the scattering properties, as well as on the geometry of the measurements, μ_s', μ_a, and l are the reduced scattering coefficients, absorption coefficient, and the photon path length, respectively. In addition, the subscripts m, HbO, HbR, e, and d indicate melanin, oxygenated hemoglobin, deoxygenated hemoglobin, epidermis, and dermis, respectively. The absorption coefficient μ_a of each chromophore can be expressed as the product of its concentration C and the extinction coefficient ϵ, and can be defined as

$$\mu_a = C \times \varepsilon. \tag{5.7}$$

Therefore, the responses of I_R, I_G, and I_B can be expressed as a function of concentrations of melanin (C_m), oxygenated hemoglobin (C_{HbO}), and deoxyhemoglobin (C_{HbR}).

First, the responses of the RGB channels I_R, I_G, and I_B in each pixel of the image are transformed into XYZ-values by a matrix $\mathbf{N_1}$ as

$$\begin{bmatrix} X \\ Y \\ Z \end{bmatrix} = \mathbf{N_1} \begin{bmatrix} R \\ G \\ B \end{bmatrix}. \tag{5.8}$$

The matrix $\mathbf{N_1}$ can be determined based on measurements of a standard color chart that has 24 color chips and is supplied with data giving the CIEXYZ values for each chip under specific illuminations and corresponding reflectance spectra. In principle, the inverse matrix of $\mathbf{L_1}$ in Eq. (5.1) can be used as a matrix $\mathbf{N_1}$ in Eq. (5.8). The values of X, Y, and Z are then transformed into C_m, C_{HbO}, and C_{HbR} by matrix $\mathbf{N_2}$. It is difficult to determine the matrix $\mathbf{N_2}$ based on $\mathbf{L_1}$ and Eqs. (5.2)–(5.6) because $P(\mu_s', l)$ and l for each layer are usually unknown. Three hundred diffuse-reflectance spectra $\Theta(\lambda)$ in a wavelength range from 400 to 700 nm at intervals of 10 nm are derived by Monte Carlo simulation (MCS) for light transport [30] in skin tissue. The simulation model consisted of two layers representing the epidermis and dermis. In a single simulation of diffuse reflectance at each wavelength, five-M photons are randomly launched. The absorption coefficients of melanin $\mu_{a,m}(\lambda)$, oxygenated hemoglobin $\mu_{a,HbO}(\lambda)$, and deoxygenated hemoglobin $\mu_{a,HbR}(\lambda)$ were obtained from the values of $\epsilon_m(\lambda)$, $\epsilon_{HbO}(\lambda)$, and $\epsilon_{HbR}(\lambda)$ in the literature [2,6]. The absorption coefficient of the epidermis depends on the volume concentration of melanin in the epidermis C_m. The absorption coefficient of a melanosome given in the literature is used as the absorption coefficient of melanin $\mu_{a,m}(\lambda)$. This corresponds to the absorption coefficient of the epidermis for the case in which $C_m = 100\%$. The absorption coefficients of the epidermis for ten lower concentrations of $C_m = 1$–10% at intervals of 1% are subsequently derived by simply proportioning them to that of $C_m = 100\%$, and the absorption coefficients were input for the epidermis. The hemoglobin concentration of blood, having a 44% hematocrit with 150 g/L of hemoglobin, was converted to 100% volume concentration (100 vol.%) of hemoglobin. The sum of the absorption coefficients of oxygenated hemoglobin and deoxygenated hemoglobin, $\mu_{a, \text{HbO}}(\lambda)$ and deoxyhemoglobin $\mu_{a, \text{HbR}}(\lambda)$, represents the absorption coefficient of total hemoglobin $\mu_{a, \text{HbT}}(\lambda)$; the values for $C_{\text{HbT}} = 0.2, 0.4, 0.6, 0.8$, and 1.0 vol.% are used as inputs for the dermis layer in the MCS. Tissue oxygen saturation StO_2 is determined by $\mu_{a, \text{HbO}}(\lambda)/\mu_{a, \text{HbT}}(\lambda)$, and values of 0%, 20%, 40%, 60%, 80%, and 100% were used for the simulation. The refractive index of the epidermis and dermis layers is assumed to be the same and fixed at 1.4 for all simulations. The thicknesses of the epidermis and dermis layer are set to 0.06 and 4.94 mm, respectively. The reduced scattering coefficient $\mu_s'(\lambda)$ calculated from the typical values for the scattering coefficient $\mu_s(\lambda)$ [13] and anisotropy factor $g(\lambda)$ [13] are used for both the epidermis and dermis. The XYZ-values are then calculated based on the simulated $\Theta(\lambda)$. These calculations are performed for the various combinations of C_m, C_{HbO}, and C_{HbR} to obtain the data sets of chromophore concentrations and XYZ-values. Multiple regression analysis with 300 data sets establishes the regression equations for C_{HbO} and C_{HbR}:

$$C_{\text{HbO}} = \alpha_0 + \alpha_1 X + \alpha_2 Y + \alpha_3 Z, \tag{5.9}$$

$$C_{\text{HbR}} = \beta_0 + \beta_1 X + \beta_2 Y + \beta_3 Z. \tag{5.10}$$

The regression coefficients α_i and β_i ($i = 0, 1, 2, 3$) reflect the contributions of the XYZ-values to C_{HbO} and C_{HbR}, respectively, and are used as the elements of a 4×2 matrix $\mathbf{N_2}$ as

$$\mathbf{N_2} = \begin{bmatrix} \alpha_0 & \alpha_1 & \alpha_2 & \alpha_3 \\ \beta_0 & \beta_1 & \beta_2 & \beta_3 \end{bmatrix}. \tag{5.11}$$

The coefficients α_i and β_i are established based on the color values simulated by the Monte Carlo simulation for light transport in skin model. Transformation with $\mathbf{N_2}$ from the tristimulus values to chromophore concentrations is thus expressed as

$$\begin{bmatrix} C_{\text{HbO}} \\ C_{\text{HbR}} \end{bmatrix} = \mathbf{N_2} \begin{bmatrix} 1 \\ X \\ Y \\ Z \end{bmatrix}. \tag{5.12}$$

Once we have determined the matrices $\mathbf{N_1}$ and $\mathbf{N_2}$, images of C_{HbO} and C_{HbR} are reconstructed without the MCS. The total hemoglobin concentration image is simply calculated as $C_{\text{HbT}} = C_{\text{HbO}} + C_{\text{HbR}}$ and tissue oxygen saturation of hemoglobin as $StO_2\% = (C_{\text{HbO}}/C_{\text{HbT}}) \times 100$.

It should be noted that C_{HbO} and C_{HbR} in this case are calculate based only on the direct current (DC) components of skin colors unlike the alternating current (AC) components used in the pulse oximeter. As already described, StO_2

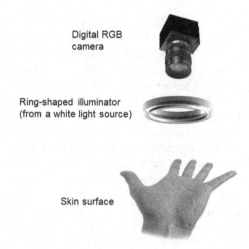

Digital RGB
camera

Ring-shaped illuminator
(from a white light source)

Skin surface

FIGURE 5.1 Schematic diagram of the imaging system.

is calculated from the values of C_{HbO} and C_{HbR} using Eq. (5.11). The values of C_{HbO} and C_{HbR} are estimated from the color values through Eqs. (5.9) and (5.10). Therefore, the color values are not calibrated to the measured or actual values of StO_2 in this method. Instead of the direct calibration of color values to actual value of StO_2, experiments with optical skin phantoms having different StO_2 levels are conducted to validate the estimated value of StO_2.

5.3 Application: monitoring blood oxygen saturation in human skin

Fig. 5.1 schematically shows the imaging system. A metal-halide white-light source with a wavelength range from 380 to 740 nm illuminates the surface of a sample via a light guide with a ring illuminator. Diffusely reflected light was captured by a 24-bit RGB CCD camera with a camera lens to acquire an RGB color image. An IR-cut filter in the camera rejects unnecessary longer-wavelength light (> 700 nm). Therefore, a portion of the light spectra will not be used for the imaging. A standard white diffuser with 99% reflectance was used to regulate the white balance of the camera. The RGB image was stored in a personal computer and analyzed according to the visualization process described earlier.

Fig. 5.2 shows an example of the *in vivo* resultant images of RGB color, StO_2, C_{HbO}, C_{HbR}, and C_{HbT} obtained from a human subject during cuff occlusion at 250 mmHg. Inflation of the cuff to 250-mm Hg prevents blood flow from leaving the measurement site and also hinders arterial inflow. Fig. 5.3 shows the time courses of StO_2, C_{HbO}, C_{HbR}, and C_{HbT} averaged over the ROI for 12 human subjects during cuff occlusion at 250-mm Hg. Since StO_2 measured by this method represents oxygen saturation for the mixture of arterio-venous blood,

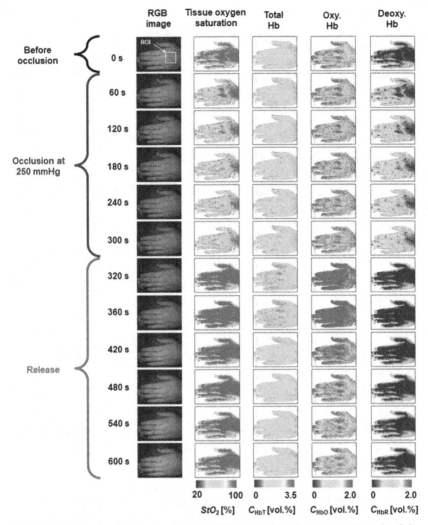

FIGURE 5.2 Example of the *in vivo* results of RGB color images, tissue oxygen saturation StO_2, total hemoglobin C_{HbT}, oxygenated hemoglobin C_{HbO}, and deoxygenated hemoglobin C_{HbR}, obtained from a human subject during cuff occlusion at 250-mm Hg.

the average value of $56 \pm 21\%$ for StO_2 is lower than the typical normal arterial oxygen saturation, which ranges from 98 to 100%. During cuff occlusion, C_{HbO} and C_{HbR} decreased and increased, respectively. The value of StO_2 exhibited the well-known deoxygenation curve, in which the oxygen saturation falls exponentially. The slight increase in C_{HbT} probably has a physiological cause. This is because, during occlusion, the venous outflow is reduced more than the arterial inflow. After the cuff was deflated, both StO_2 and C_{HbT} increase sharply and then gradually return to their normal levels.

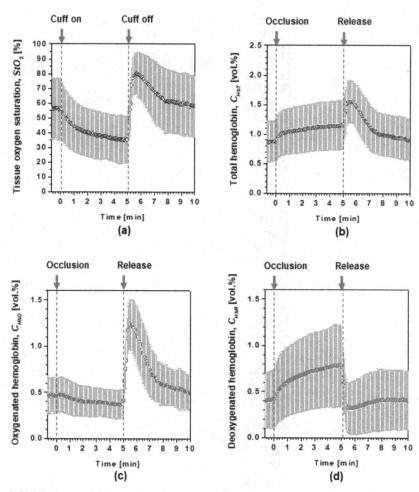

FIGURE 5.3 Time courses of (a) tissue oxygen saturation StO_2, (b) total hemoglobin C_{HbT}, (c) oxygenated hemoglobin C_{HbO}, (d) deoxygenated hemoglobin C_{HbR} averaged over the ROI for 12 human subjects during cuff occlusion at 250-mm Hg. Data are expressed as mean ± SD ($n = 12$).

5.4   Application: monitoring blood oxygen saturation in skin during changes in fraction of inspired oxygen

A patient with a bluish discoloration in the skin known as cyanosis is alarming to a medical doctor. Such cases may be severely hypoxemic or hypoxic, constituting a medical emergency [31]. Cyanosis first appears in the extremities when oxygen demand outstrips the supply, resulting in a high amount of deoxygenated blood accumulating in blood-perfused dermal tissues [32,33]. It may result from a number of medical conditions including reduced cardiac

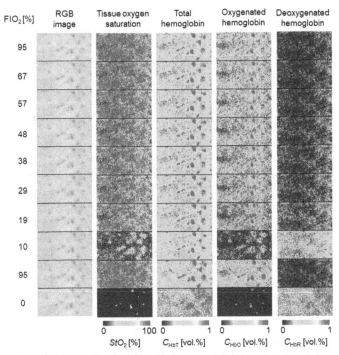

FIGURE 5.4 Typical sequential images obtained from *in vivo* rat dorsal skin during changes in FiO$_2$ including RGB color images, tissue oxygen saturation StO_2, total hemoglobin C_{HbT}, oxygenated hemoglobin C_{HbO}, and deoxygenated hemoglobin C_{HbR}.

output (heart failure), peripheral vasoconstriction (hypothermia), and regional ischemia (arterial thrombosis) [31,34,35]. Therefore, evaluation of blood oxygen saturation is important for clinical applications. Furthermore, information on the spatial distribution of methemoglobin would aid in the diagnosis and treatment of diseases. It should be noted that the bluish appearance of skin color in cyanosis is related to the DC components of skin color and not the AC variation used in the pulse oximeter. As previously described, a diffuse reflectance white standard is used to regulate the white balance of the camera, while a standard color chart is introduced to establish the color transformation matrix N_1 using a specific light source. The influence of environmental light can be reduced if the procedures of white balance and color calibration are performed using a specific light source under environmental lighting.

Fig. 5.4 shows typical sequential images obtained from *in vivo* rat dorsal skin including RGB color images, StO_2, C_{HbT}, C_{HbO}, and C_{HbR} during changes in FiO$_2$ obtained from rat dorsal skin including RGB color images, StO_2, C_{HbO}, C_{HbR}, and C_{HbT} during changes in FiO$_2$. Fig. 5.5 shows the time courses of (a) StO_2, (b) C_{HbO}, (c) C_{HbR} and (d) C_{HbT} averaged over the entire region of each image for 11 rats during changes in FiO$_2$. Values of C_{HbO} and C_{HbR} de-

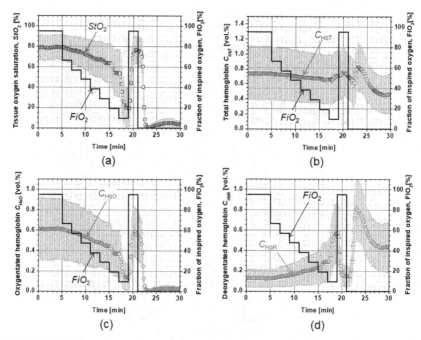

FIGURE 5.5 Time courses of average chromophore concentration and tissue oxygen saturation averaged over the ROI on *in vivo* rat dorsal skin during changes in FiO_2. (a) Tissue oxygen saturation StO_2, (b) total hemoglobin C_{HbT}, (c) oxygenated hemoglobin C_{HbO}, (d) deoxygenated hemoglobin C_{HbR}. Data are expressed as mean ± SD ($n = 11$).

creased and increased, gradually, as FiO_2 decreased, which caused decreases in StO_2. On the other hand, the value of StO_2 dropped dramatically when FiO_2 was below 19%, indicating the onset of hypoxemia due to hypoxia. The value of C_{HbT} gradually increased after the onset of severe hypoxia (FiO_2=9%) and anoxia (FiO_2=0%), implying an increase in blood flow compensating for hypoxia. Time courses of C_{HbO}, C_{HbR}, C_{HbT}, and StO_2, while changing FiO_2, were consistent with well-known physiological responses to changes in FiO_2.

5.5   Application: monitoring blood oxygen saturation in brain

Cerebrovascular diseases (CVDs) affect the blood vessels and blood circulation in brain tissues. Common CVDs include ischemic stroke, transient ischemic attack, and subarachnoid hemorrhage. CVDs can damage and deform cerebral arteries delivering oxygen and nutrients to the brain tissues. Neuronal cells demand a continuous supply of oxygen and glucose, which are delivered via blood circulation. Impaired cerebral blood circulation can cause irreversible cell injury or cell death as a consequence of neuronal energy failure [36]. Various treatment strategies for CVDs, including medication, lifestyle changes, and/or surgery, are available depending on the specific underlying cause. In surgical treatments for

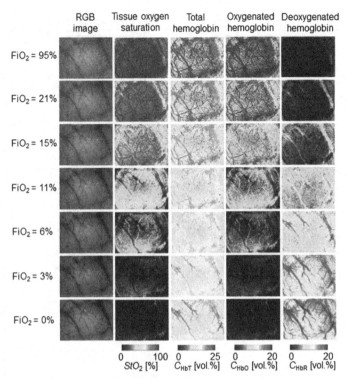

FIGURE 5.6 Typical resultant images obtained from *in vivo* exposed rat brains during changes in FiO$_2$, including RGB color images, tissue oxygen saturation StO$_2$, total hemoglobin C$_{HbT}$, oxygenated hemoglobin C$_{HbO}$, and deoxygenated hemoglobin C$_{HbR}$.

CVDs, cortical hemodynamics are often controlled by bypass graft surgery, temporary occlusion of arteries, or surgical removal of veins. Since the brain is vulnerable to hypoxemia and ischemia, interruption of cerebral blood flow reduces the oxygen supply to tissues and can induce irreversible damage to cells and tissues. Monitoring of cerebral hemodynamics and alterations to cellular structure during neurosurgery is thus extremely important. Cerebral hemodynamics, including cerebral blood flow, cerebral blood volume, and oxygen saturation of hemoglobin in red blood cells (RBCs), affect the light-absorbing properties of brain tissues. In this section, the method described in Sect. 5.2 is modified and extended to imaging for cerebral hemodynamics of *in vivo* brain tissues [37].

Fig. 5.6 shows typical resultant images for C$_{HbO}$, C$_{HbR}$, C$_{HbT}$, and StO$_2$ obtained from *in vivo* exposed rat brains during changes in FiO$_2$. Fig. 5.7 shows the typical time courses of (a) C$_{HbO}$, (b) C$_{HbR}$, (c) C$_{HbT}$, and (d) StO$_2$ averaged over the area for the ROI on the parenchyma during changes in FiO$_2$. Time courses of SpO$_2$ and HR are also compared with StO$_2$ and C$_{HbT}$, respectively, in Fig. 5.7. Values of C$_{HbO}$ and C$_{HbR}$ are decreased and increased, gradually, as FiO$_2$ decreased, which caused decreases in StO$_2$. On the other hand, the value

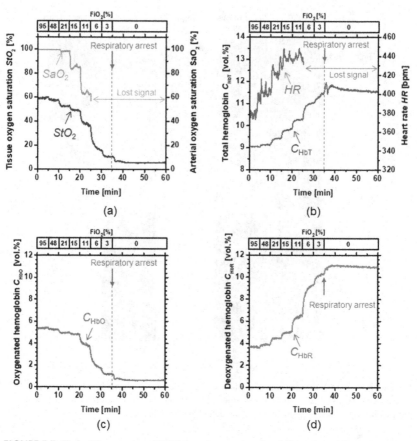

FIGURE 5.7 Typical time courses of (a) tissue oxygen saturation StO_2 and arterial oxygen saturation SaO_2, (b) total hemoglobin C_{HbT} and heart rate HR, (c) oxygenated hemoglobin C_{HbO}, and (d) deoxygenated hemoglobin C_{HbR} averaged over the area for the region of interest (ROI) on the parenchymal region.

of SpO_2 dropped dramatically when FiO_2 is below 21%, indicating the onset of hypoxemia due to hypoxia. The value of C_{HbT} gradually increased according to reductions in FiO_2. The value of HR is gradually increased after the onset of hypoxia and reached a maximum amplitude approximately one m after respiratory arrest (RA), implying an increase in blood flow compensating for hypoxia. Immediately following RA, values of both C_{HbO} and C_{HbT} dropped sharply. Time courses of C_{HbO}, C_{HbR}, C_{HbT}, and StO_2 while changing FiO_2, are consistent with well-known physiological responses to changes in FiO_2. The method can capture hemodynamic responses in brain tissue and thus appears promising for monitoring brain function and tissue vitality in neurosurgery, as well as in the diagnosis of several neurological disorders, such as traumatic brain injury, seizure, ischemic stroke, and cerebral hyperperfusion syndrome. Integrating this method into a surgical microscope system will enable the display of a blood oxygen-saturation map of brain surface in real time on a monitor during neurosurgery.

5.6   Application: monitoring blood oxygen saturation in hepatic ischemia-reperfusion

Ischemia and subsequent reperfusion of liver tissue may occur in a number of clinical settings, such as those associated with low-flow states, or temporary occlusion of the supplying blood vessels during liver transplantation, and to a large extent in liver resection surgery. Hepatic ischemia-reperfusion is characterized by circulatory derangement, including decreases in sinusoidal diameters and blood flow, increases in the heterogeneity of hepatic microvascular perfusion, and even total cessation of blood flow within individual sinusoids [38]. Therefore, ischemia-reperfusion damage to donor liver tissue is thought to be a vital role that may play a role in the decline of posttransplant graft function and ultimate rejection. In liver transplantation, normothermic ischemia-reperfusion causes up to 10% of early graft rejections [39]. Hepatocellular damage caused by ischemia-reperfusion correlates with the extent of microcirculatory reperfusion failure after portal-triad cross-clamping [40]. For liver transplantation, the hepatic pedicle is temporarily clamped [41] to restrict blood inflow and to reduce intraoperative blood loss, which results in deprivation of tissue oxygen and depletion of energy production, thereby shifting cellular metabolism to anaerobic pathways. The resulting hepatic ischemia, along with subsequent reperfusion, causes the ischemia-reperfusion injury.

Measurement of hepatic tissue oxygenation has been shown to correlate significantly with the microcirculatory derangement and hepatic dysfunction induced by ischemia-reperfusion. Reestablishment of blood flow to the liver tissue represents a vital requirement for recovery of cellular and organ function, which typically aggravates ischemia-induced tissue damage. Moreover, during reperfusion, an additional insult is added to the damage sustained during ischemia [42]. Several studies have reported that post-ischemic reperfusion is associated with inadequate energy supply, a subsequent reduction of hepatic tissue oxygen saturation, and impairment of mitochondrial adenosine triphosphate (ATP) regeneration, along with incomplete recovery of hepatocellular excretory function [43]. Hypoxic or ischemic damage to the hepatic sinusoidal endothelium is generally considered to be the first pathological change in the cascade of events resulting in ischemia-reperfusion injury during the manifestation of graft dysfunction [44]. Ischemia-reperfusion injury is closely related to an increase in the incidence of rapid decreases in tissue oxygen saturation StO_2, reduction of heme aa_3 in cytochrome c oxidase indicating mitochondrial dysfunction, and hepatocellular injury such as apoptosis, necrosis, and cell death [45]. On reentry of blood into the organ during reperfusion, there is an overproduction of toxic oxygen radicals followed by endothelial cell damage, causing disturbances in the microcirculation of liver tissue that lead to alterations in hepatic tissue oxygenation [40]. Maintenance of adequate hepatic blood flow and sufficient hepatic oxygenation is one of the most important factors in hepatic surgical success. As liver function depends greatly on hemodynamics and viability of liver tissue, measurements of hemoglobin concentration and hemoglobin saturation in hepatic tissue have been shown to correlate significantly with microcirculatory liver failure induced by ischemia-reperfusion injury [46].

Fig. 5.8 shows typical *in vivo* results obtained from liver tissue, including RGB images, C_{HbO}, C_{HbR}, C_{HbT}, and StO_2 images before ischemia (oxygen breathing, air breathing), and during ischemia, reperfusion, and post-mortem, respectively [47]. In this experiments, hepatic ischemia was induced by clamping the portal triad (the hepatic artery, the portal vein, and the bile duct) supplying the median lobes and left lateral lobe with no occlusion of caudate lobe. A ten-minute ischemic period was followed by reperfusion for a period of 120 minutes. After 120 minutes of reperfusion, an anoxic condition was produced by nitrogen breathing. StO_2, C_{HbT}, C_{HbO}, and C_{HbR} of the normal liver tissue (oxygen breathing and air breathing state) revealed high levels of oxyhemoglobin and tissue oxygen saturation. The large increase in C_{HbR} and decreases in C_{HbO} and StO_2 typically observed in ischemia are consistent with the changes in time courses in StO_2 reported in the literature [48,49]. Successive images recorded five-min post-reperfusion show the spatial changes in C_{HbO} and StO_2 that are indicated by a white arrow on the less perfused portion of the imaged area, which is accompanied with scattered hepatic lobular perfusion. Immediately after reperfusion, a slight increase in C_{HbO} and StO_2 are observed in the imaged area, possibly indicating the gradual compensation of hepatic blood flow. In addition, changes in StO_2, C_{HbT}, C_{HbO}, and C_{HbR} are observed at two-min and five-min after onset of nitrogen breathing and a rapid increase in C_{HbR} and C_{HbT} in post-mortem liver tissue appeared.

Time courses of the changes in StO_2, C_{HbT}, C_{HbO}, and C_{HbR} averaged over the samples and over the entire imaged area during the period of pre-ischemia (oxygen breathing and air breathing), ischemia, reperfusion, and post-mortem [47] are plotted in Fig. 5.9. The symbols i, ii, iii, iv, v, and vi in each figure indicate O_2-breathing, air-breathing, ischemia under air-breathing, reperfusion under air-breathing, N_2-breathing, and post-mortem, respectively. C_{HbO} and C_{HbR} are decreased and increased, respectively, following the onset of ischemia, compared to the preischemic levels. Consequently, StO_2 decreased, thereby reflecting reduced blood flow and oxygen supply to the tissue. Ligation of the hepatic pedicle of the median and left lateral lobes resulted in approximately 70% occlusion or ischemia of the liver [46], as indicated by a rapid decrease of hepatic StO_2 on the initiation of ischemia. Note that the decrease in C_{HbT} found at an ischemia is possibly due to hepatic inflow occlusion and maintenance of outflow. Hepatic tissue oxygen saturation during oxygen breathing and air breathing (baseline) are estimated to be $74.09 \pm 0.70\%$ and $68.12 \pm 1.68\%$,

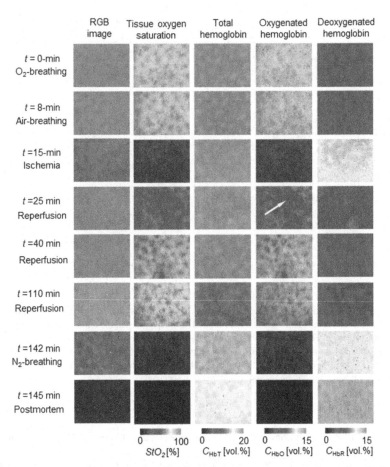

FIGURE 5.8 Typical images of *in vivo* liver tissue at various measurement time points. Left to right: the first column shows images at $t = 0$ (the beginning of measurement and 0 min after start of oxygen breathing); the second column shows images at $t =$ eight min (three min after start of air breathing); the third column shows images at $t = 15$ min (five min after the onset of ischemia); the fourth column shows images at $t = 25$ min (five min after the onset of reperfusion); the fifth column shows images at $t = 40$ min (20-min after the onset of reperfusion); the sixth column shows images at $t = 110$ min (90-min after the onset of reperfusion); the seventh column shows images at $t = 142$ min (two min after the onset of nitrogen breathing); and the eighth column shows images at $t = 145$ min (five min after the onset of nitrogen breathing). Successive images recorded five-min post-reperfusion show the spatial changes in C_{HbO} and StO_2 that are indicated by a white arrow on the less perfused portion of the imaged area, which is accompanied with scattered hepatic lobular perfusion.

respectively. Immediately after reflow of the portal triad, hepatic StO_2 showed a remarkably heterogeneous distribution with an average of $55.48 \pm 8.72\%$ after 120 min of reperfusion. However, this did not return to baseline levels, possibly due to ischemia-reperfusion associated hepatic dysfunction.

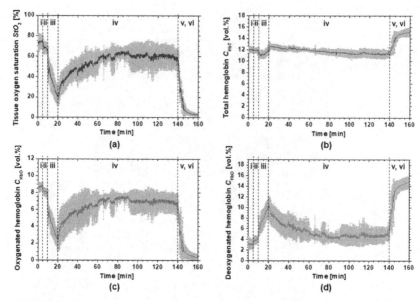

FIGURE 5.9 Typical time courses of tissue oxygen saturation and chromophore concentrations averaged over five rats and averaged over the ROI corresponding to the entire image for (a) tissue oxygen saturation StO_2, (b) total hemoglobin C_{HbT}, (c) oxygenated hemoglobin C_{HbO}, and (d) deoxygenated hemoglobin C_{HbR}. Error bars represent the standard deviation ($n=5$). The symbols i, ii, iii, iv, and v in each figure indicate O_2-breathing, air-breathing, ischemia under air-breathing, reperfusion under air-breathing, and post-mortem, respectively.

References

[1] V. Tuchin, Tissue Optics: Light Scattering Methods and Instruments for Medical Diagnosis, 2nd ed., SPIE, 2007.

[2] S.A. Prahl, Tabulated molar extinction coefficient for hemoglobin in water, http://omlc.org/spectra/hemoglobin/summery.html, 1991. (Accessed 30 September 2019).

[3] D.A. Boas, M.A. Franceschini, Haemoglobin oxygen saturation as a biomarker: the problem and a solution, Philos. Trans. R. Soc. A 369 (2011) 4407–4424.

[4] J. Allen, Photoplethysmography and its application in clinical physiological measurement, Physiol. Meas. 28 (3) (2007) R1–R39.

[5] A.A. Kamal, J.B. Harness, G. Irving, A.J. Mearns, Skin photoplethysmography–a review, Comput. Methods Programs Biomed. 28 (4) (1989) 257–269.

[6] G. Zonios, J. Bykowski, N. Kollias, Skin melanin, hemoglobin, and light scattering properties can be quantitatively assessed in vivo using diffuse reflectance spectroscopy, J. Invest. Dermatol. 117 (2001) 1452–1457.

[7] J.W. Feather, M. Hajizadeh-Saffar, G. Leslie, J.B. Dawson, A portable scanning reflectance pectrophotometer using visible wavelengths for the rapid measurement of skin pigments, Phys. Med. Biol. 34 (1989) 807–820.

[8] D.K. Harrison, S.D. Evans, N.C. Abbot, J.S. Beck, P.T. McCollum, Spectrophotometric measurements of haemoglobin saturation and concentration in skin during the tuberculin reaction in normal human subjects, Clin. Phys. Physiol. Meas. 13 (1992) 349–363.

[9] D.J. Newton, D.K. Harrison, C.J. Delaney, J.S. Beck, P.T. McCollum, Comparison of macro- and maicro-lightguide spectrophotometric measurements of microvascular haemoglobin oxygenation in the tuberculin reaction in normal human skin, Physiol. Meas. 15 (1994) 115–128.

[10] A.A. Stratonnikov, V.B. Loschenov, Evaluation of blood oxygen saturation in vivo from diffuse reflectance spectra, J. Biomed. Opt. 6 (2001) 457–467.

[11] G. Zonios, J. Bykowski, N. Kollias, Skin melanin, hemoglobin, and light scattering properties can be quantitatively assessed in vivo using diffuse reflectance spectroscopy, J. Invest. Dermatol. 117 (2001) 1452–1457.

[12] G.N. Stamatas, N. Kollias, Blood stasis contributions to the perception of skin pigmentation, J. Biomed. Opt. 9 (2004) 315–322.

[13] I. Nishidate, Y. Aizu, H. Mishina, Estimation of melanin and hemoglobin in skin tissue using multiple regression analysis aided by Monte Carlo simulation, J. Biomed. Opt. 9 (2004) 700–710.

[14] P.R. Bargo, S.A. Prahl, T.T. Goodell, R.A. Sleven, G. Koval, G. Blair, S.L. Jacques, In vivo determination of optical properties of normal and tumor tissue with white light reflectance and empirical light transport model during endoscopy, J. Biomed. Opt. 10 (2005) 034018.

[15] S.-H. Tseng, P. Bargo, A. Durkin, N. Kollias, Chromophore concentrations, absorption and scattering properties of human skin in vivo, Opt. Express 17 (2009) 14599–14617.

[16] K. Yoshida, I. Nishidate, Rapid calculation of diffuse reflectance from a multilayered model by combination of the white Monte Carlo and adding-doubling methods, Biomed. Opt. Express 5 (2014) 3901–3920.

[17] M.G. Sowa, J.R. Payette, M.D. Hewko, H.H. Mantsch, Visible-near infrared multispectral imaging of the rat dorsal skin flap, J. Biomed. Opt. 4 (1999) 474–481.

[18] A.K. Dunn, A. Devor, H. Bolay, M.L. Andermann, M.A. Moskowitz, A.M. Dale, D.A. Boas, Simultaneous imaging of total cerebral hemoglobin concentration, oxygenation, and blood flow during functional activation, Opt. Lett. 28 (2003) 28–30.

[19] I. Kuzmina, I. Diebele, D. Jakovels, J. Spigulis, L. Valeine, J. Kapostinsh, A. Berzina, Towards noncontact skin cancer melanoma selection by multispectral imaging analysis, J. Biomed. Opt. 16 (2011) 060502.

[20] B.S. Sorg, B.J. Moeller, O. Donovan, Y. Cao, M.W. Dewhirst, Hyperspectral imaging of hemoglobin saturation in tumor microvasculature and tumor hypoxia development, J. Biomed. Opt. 10 (2005) 044004.

[21] S.F. Bish, M. Sharma, Y. Wang, N.J. Triesault, J.S. Reichenberg, J.X.J. Zhang, J.W. Tunnell, Handheld diffuse reflectance spectral imaging (DRSi) for in-vivo characterization of skin, Biomed. Opt. Express 5 (2014) 573–586.

[22] A. Nkengne, J. Robic, P. Seroul, S. Gueheunneux, M. Jomier, K. Vie, SpectraCam®: a new polarized hyperspectral imaging system for repeatable and reproducible in vivo skin quantification of melanin, total hemoglobin, and oxygen saturation, Skin Res. Technol. 24 (2018) 99–107.

[23] F. Vasefi, N. MacKinnon, R.B. Saager, A.J. Durkin, R. Chave, E.H. Lindsley, D.L. Farkas, Polarization-sensitive hyperspectral imaging in vivo: a multimode dermoscope for skin analysis, Sci. Rep. 4 (2014) 4924.

[24] I. Nishidate, A. Wiswadarma, Y. Hase, N. Tanaka, T. Maeda, K. Niizeki, Y. Aizu, Non-invasive spectral imaging of skin chromophores based on multiple regression analysis aided by Monte Carlo simulation, Opt. Lett. 36 (2011) 3239–3241.

[25] J. O'Doherty, P. McNamara, N.T. Clancy, J.G. Enfield, M.J. Leahy, Comparison of instruments for investigation ofmicrocirculatory blood flow and red blood cell concentration, J. Biomed. Opt. 14 (2009) 034025.

[26] N. Tsumura, H. Haneishi, Y. Miyake, Independent-component analysis of skin color image, J. Opt. Soc. Am. A 16 (1999) 2169–2176.

[27] N. Tsumura, Y. Miyake, F.H. Imai, Medical vision: measurement of skin absolute spectral-reflectance image and the application to component analysis, in: Proceedings of the Third International Conference on Multispectral Color Science, Joensuu, Finland, 2001 June, pp. 25–28.

[28] I. Nishidate, K. Sasaoka, T. Yuasa, K. Niizeki, T. Maeda, Y. Aizu, Visualizing of skin chromophore concentrations by use of RGB images, Opt. Lett. 33 (19) (2008) 2263–2265.

[29] I. Nishidate, N. Tanaka, T. Kawase, T. Maeda, T. Yuasa, Y. Aizu, T. Yuasa, K. Niizeki, Noninvasive imaging of human skin hemodynamics using a digital red-green-blue camera, J. Biomed. Opt. 16 (2011) 086012.

[30] L. Wang, S.L. Jacques, L. Zheng, MCML: Monte Carlo modeling of light transport in multi-layered tissues, Comput. Methods Programs Biomed. 47 (1995) 131–146.

[31] A. Baernstein, K. Smith, J. Elmore, Singing the blues: is it really cyanosis?, Respir. Care 53 (2008) 1081–1084.

[32] C.D. Marple, Cyanosis, Am. J. Nusr. 58 (2) (1958) 222–225.

[33] G. Casey, Oxygen transport and the use of pulse oxymetry, Nurs. Stand. 15 (47) (2001) 46–53.

[34] G. Baranoski, S. Van Leeuwen, T. Chen, Elucidating the biophysical processes responsible for the chromatic attributes of peripheral cyanosis, in: 39th Annual International Conference of the IEEE Engineering in Medicine and Biology Society (EMBC), Jeju Island, South Korea, 2017, pp. 90–95.

[35] S.M. McMullen, W. Patrick, Cyanosis, Am. J. Med. 126 (3) (2013) 210–212.

[36] N.R. Sims, H.Muyderman, Mitochondria, oxidative metabolism and cell death in stroke, Biochim. Biophys. Acta 1802 (1) (2010) 80–91.

[37] A. Mustari, T. Kanie, S. Kawauchi, S. Sato, M. Sato, Y. Kokubo, I. Nishidate, In vivo evaluation of cerebral hemodynamics and tissue morphology in rats during changing fraction of inspired oxygen based on spectrocolorimetric imaging technique, Int. J. Mol. Sci. 19 (2) (2018) 491.

[38] K. Chun, J. Zhang, J. Biewer, D. Ferguson, M.G. Clemens, Microcirculatory failure determines lethal hepatocyte injury in ischemia/reperfused rat livers, Shock 1 (1) (1994) 3–9.

[39] R. Cursio, P. Colosetti, M.C. Saint-Paul, S. Pagnotta, P. Gounon, A. Iannelli, P. Auberger, J. Gugenheim, Induction of different types of cell death after normothermic liver ischemia-reperfusion, Transplant. Proc. 42 (10) (2010) 3977–3980.

[40] M. Goto, S. Kawano, H. Yoshihara, Y. Takei, T. Hijioka, H. Fukui, T. Matsunaga, M. Oshita, T. Kashiwagi, H. Fusamoto, et al., Hepatic tissue oxygenation as a predictive indicator of ischemia-reperfusion liver injury, Hepatology 15 (3) (1992) 432–437.

[41] J.H. Pringle, Notes on the arrest of hepatic hemorrhage due to trauma, Ann. Surg. 48 (4) (1908) 541–549.

[42] M. Kretzschmar, A. Krüger, W. Schirrmeister, Hepatic ischemia-reperfusion syndrome after partial liver resection (LR): hepatic venous oxygen saturation, enzyme pattern, reduced and oxidized glutathione, procalcitonin and interleukin-6, Exp. Toxicol. Pathol 54 (5–6) (2003) 423–431.

[43] W. Kamiike, M. Nakahara, K. Nakao, M. Koseki, T. Nishida, Y. Kawashima, F. Watanabe, K. Tagawa, Correlation between cellular ATP level and bile excretion in the rat liver, Transplantation 39 (1) (1985) 50–55.

[44] J.J. Lemasters, H. Bunzendahl, R.G. Thurman, Reperfusion injury to donor livers stored for transplantation, Liver Transplant. Surg. 1 (2) (1995) 124–138.

[45] J.W. Kupiec-Weglinski, R.W. Busuttil, Ischemia and reperfusion injury in liver transplantation, Transplant. Proc. 37 (4) (2005) 1653–1656.

[46] A.E. El-Desoky, D.T. Delpy, B.R. Davidson, A.M. Seifalian, Assessment of hepatic ischemia reperfusion injury by measuring intracellular tissue oxygenation using near infrared spectroscopy, Liver 21 (1) (2001) 37–44.

[47] S. Akter, S. Kawauchi, S. Sato, S. Aosasa, J. Yamamoto, I. Nishidate, In vivo imaging of hepatic hemodynamics and light scattering property during ischemia-reperfusion in rats based on spectrocolorimetry, Biomed. Opt. Express 8 (2) (2017) 974–992.

[48] S. Akter, S. Maejima, S. Kawauchi, S. Sato, A. Hinoki, S. Aosasa, J. Yamamoto, I. Nishidate, Evaluation of light scattering and absorption properties of in vivo rat liver using a single-reflectance fiber probe during preischemia, ischemia-reperfusion, and postmortem, J. Biomed. Opt. 20 (7) (2015) 076010.

[49] S. Akter, T. Tanabe, S. Maejima, S. Kawauchi, S. Sato, A. Hinoki, S. Aosasa, J. Yamamoto, I. Nishidate, In vivo estimation of optical properties of rat liver using single-reflectance fiber probe during ischemia and reperfusion, Opt. Rev. 22 (6) (2015) 1–6.

Chapter 6

Camera-based blood pressure monitoring

Keerthana Natarajan[a], Mohammad Yavarimanesh[a], Wenjin Wang[b,c], and Ramakrishna Mukkamala[d]

[a]Department of Electrical and Computer Engineering, Michigan State University, East Lansing, MI, United States, [b]AI Data Science & Digital Twin Department, Philips Research, Eindhoven, the Netherlands, [c]Eindhoven University of Technology, Electrical Engineering Department, Eindhoven, the Netherlands, [d]Department of Bioengineering and Department of Anesthesiology and Perioperative Medicine, University of Pittsburgh, Pittsburgh, PA, United States

Contents

High blood pressure (BP) afflicts about one in three adults worldwide [1]. While the incidence increases with age, many people develop hypertension early in adulthood (e.g., more than one in five US adults under 40-years old are hypertensive [2]). The condition is usually asymptomatic, but the risk for stroke and heart disease increases monotonically with BP for a given age [3]. Lifestyle changes and many inexpensive, once-daily medications can lower BP and cardiovascular risk [4]. Yet, only about three in seven people with hypertension are aware of their condition, and just one of these seven has their BP under control [5]. Epidemiological data on hypertension in low-resource settings are more alarming [5]. As a result, hypertension has emerged as the leading cause of disability-adjusted-life years lost [6].

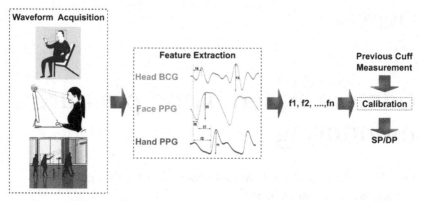

FIGURE 6.1 Camera-based blood pressure (BP) monitoring. This potential contactless approach involves acquisition of photo-plethysmography (PPG) or ballistocardiography (BCG) waveforms via a video camera, extracting features from the waveforms that correlate with BP and calibrating the features to systolic BP (SP) and diastolic BP (DP) using previous cuff measurements.

Auscultatory and oscillometric BP measurement devices have been instrumental in the fight against hypertension. At the same time, these devices may bear responsibility for the abysmal hypertension awareness and control rates due to their reliance on an inflatable cuff. Cuff-based devices are not readily available, especially in low-resource settings and are cumbersome to use. So, most people do not regularly check their BP [7]. Regular measurements during daily life are needed to circumvent white-coat and masked effects in the clinic in which patients present with higher or lower BP than usual and to average out the large variations in BP that occur over time due to stress, physical activity, and other factors [8–10]. If BP could be measured seamlessly, then many people would become aware of their condition or motivated enough to take their medications [11].

Cuff-less BP measurement devices are therefore being widely pursued. Most of these devices are based on contact sensing. Contactless monitoring of BP with video cameras, as shown in Fig. 6.1, is another possible approach that is being increasingly investigated. The goal of this chapter is to facilitate research to draw conclusions on the feasibility of such camera-based BP monitoring. We begin by arguing that this approach, if viable, carries significant advantages in battling hypertension over the other possible cuff-less approaches. We then explain theoretical principles that may allow extraction of BP from video-camera recordings of a person. We next summarize the key experimental studies to date on the camera-based approach, as well as investigations into cuff-less BP measurement via contact sensors that offer insight into the feasibility of the contactless approach. We conclude with our recommendations for future research and outlook on video cameras for contactless BP monitoring.

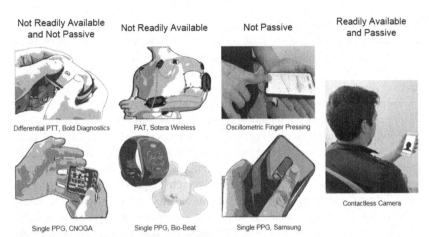

FIGURE 6.2 Competing cuff-less BP measurement devices under investigation. Camera-based BP monitoring is uniquely suitable for hypertension management because it could provide passive and frequent BP measurements during daily life with devices that are already available to many (adapted from [12,14–18]).

6.1 Advantages over other potential cuff-less BP measurement devices

Many other devices are being investigated for BP measurement without a cuff. As shown in Fig. 6.2, these cuff-less devices use contact sensors and come in the form of a wearable (e.g., wristband watch or chest patch) [12,13] or "nearable" (i.e., typically a smartphone [14,15] but also other compact devices that can be carried [16,17]).

Two wearable devices (ViSi Mobile System, Sotera Wireless and BB-613 WP, Biobeat Technologies) are now FDA-approved for cuff-less BP monitoring in between periodic cuff BP measurements [19,20], which are used to calibrate the devices to yield values in units of mmHg. Note that an FDA-approved wristwatch offered by a leading medical equipment manufacturer (HeartGuide, Omron Healthcare) actually employs an inflatable cuff [21]. A major advantage of wearable devices is that they can afford passive or even continuous BP monitoring. However, wearable devices are typically expensive and therefore not as readily accessible as cuff-based devices. Nearable devices designed specifically for BP measurement (see, e.g., left of Fig. 6.2) may likewise be unavailable to most. Hence, since virtually every adult has some nontrivial risk of suffering from hypertension, these devices may only have marginal impact on managing the condition.

Smartphones are general-purpose devices that are widely available. About 45% of people worldwide own a smartphone [22], and smartphones should become common in low-resource settings soon due to lower costs from increased competition in the market [23]. However, the smartphone-based BP measurement devices under investigation require the user to perform an activity. The user must either interact with the device in a certain way (e.g., place a fingertip on a sensor) for cuff-less BP monitoring in between periodic cuff calibrations [15] or perform a finger-pressing method under visual guidance for cuff-less and calibration-free BP monitoring [14,24]. Hence, people may not be compliant in making regular measurements with the devices.

By contrast, the camera-based approach can potentially allow for passive BP monitoring in the mass population due to the ubiquity of video cameras and the contactless paradigm (see Fig. 6.2). For example, as explained in [25], each time a person uses a smartphone (which is a few hours per day on average [26]), their face may be in the field of view of the front camera. It is possible for an application to control this camera to identify suitable periods (e.g., a still user in good lighting conditions) to make BP measurements via facial video processing. Similarly, as a person continually works on a computer, an application controlling the webcam could also measure BP. Such seamless, frequent BP monitoring during daily life with pervasive devices could uniquely help in enhancing global hypertension awareness and control rates.

6.2 Theoretical principles

While camera-based BP monitoring may seem idealistic, there are theoretical underpinnings to support its feasibility. The general theory is to: (6.2.1) acquire arterial waveforms from video camera recordings of a person; (6.2.2) extract features from one or more waveforms that correlate with BP based on known physiology or otherwise; and (6.2.3) map the features to BP via a predefined model formed with cuff BP measurements (i.e., cuff calibration). This theory is basically the same as most other cuff-less BP measurement devices except that the waveforms are obtained in a contactless manner.

6.2.1 Contactless acquisition of arterial waveforms

Arterial waveforms can be acquired from camera recordings of a person via video processing methods. As illustrated in Fig. 6.3, the measurement principles are photo-plethysmography and ballistocardiography. These principles are described in detail in Chap. 4. We present salient points now.

Photo-plethysmography (PPG):

Contact PPG provides excellent balance between effectiveness and convenience in measuring arterial waveforms and is the foundation of widely used pulse oximeters (see [27,28] for reviews). In this approach, a volume of tissue is illuminated with a light emitting diode on one side of the tissue, and the transmitted light is received by a photodetector on the opposite side (i.e., transmission-mode) or the reflected light is received by an adjacent photodetector on the same side (i.e., reflectance-mode). The received light is of lower intensity mainly due to the absorption of light by the various tissue elements, including hemoglobin

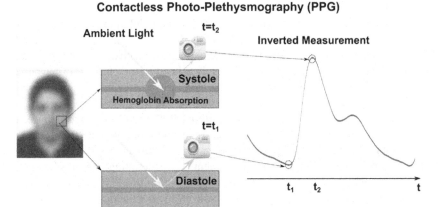

FIGURE 6.3 Contactless PPG and BCG measurement principles. Contactless PPG is a reflectance-mode approach with ambient light illuminating the skin, whereas contactless BCG detects head-to-toe direction displacement of any body part due to the heartbeat.

in blood. Since the amount of hemoglobin is proportional to the volume of blood, the measurement includes a time-varying component that changes inversely with the changes in pulsatile arterial-blood volume about its mean value (via, e.g., a logarithmic relationship according to the Beer–Lambert–Bouguer law). Contact PPG is conventionally applied to a fingertip, earlobe, or toe and uses visible to infrared wavelengths. The longer infrared wavelengths penetrate deeper under the skin and therefore permit more robust measurement under the conditions of dark skin or cold-induced cutaneous vasoconstriction. On the other hand, hemoglobin absorption sensitivity is much lower at infrared than green wavelengths.

Contactless PPG with a video camera is a reflectance modality with ambient light serving as the illuminator (see Fig. 6.3) [29]. For conventional red–

green–blue (RGB) cameras, light in the visible range (450–650 nm wavelengths) is mainly absorbed by the melanin in skin and hemoglobin in cutaneous blood vessels, and the residual light is reflected back to the camera. Each recorded video pixel of the skin therefore includes subtle color changes superimposed on the average skin color that are inversely related to the changes in pulsatile arterial blood volume about its mean value. PPG waveforms at various skin locations on the face and even the hand may then be extracted via video processing. As explained in Chaps. 1 and 4, the resulting PPG waveforms (via the green wavelength, in particular) could represent pulsatile blood volume from deeper than cutaneous arteries. The waveform quality can degrade with dark skin, aging, facial makeup, and a cold environment. By penetrating beneath the skin, infrared cameras may afford more robust acquisition of PPG waveforms via a similar contactless principle [30,31]. However, these cameras including active infrared light sources are not readily available. Note that sensor contact pressure on the skin, which markedly alters the amplitude and shape of PPG waveforms [32], is not a factor in the contactless approach.

The general requirements for the camera are low sensor noise (i.e., large pixel size), stable frame rate > 20–30 fps (PPG waveforms mainly oscillate at frequencies up to 10–15 Hz), and an uncompressed video recording format (to maintain all color information). All auto-adjustment functions provided by the camera (e.g., auto-white balancing, auto-exposure, auto-gain, and auto-focus) should be turned off. While consumer-grade webcams can suffice, industrial RGB cameras offer better sensor quality and, more importantly, provide full control of the camera features. A dedicated light source also improves the results. However, it is important that the light spectrum is stable over time. For infrared applications, a monochrome camera with daylight cutoff filter and an active infrared light source are needed. This equipment is not readily available but essential for specific applications like night-time monitoring. For camera selection, it is important to use global shutter sensors rather than rolling shutter sensors, which will introduce a phase delay between waveforms at different sites [33].

A camera recording of the skin of a person must be taken. Video processing may then be employed to acquire PPG waveforms according to the following steps.

(i) Skin Region of Interest (RoI) Detection: Face detection or living skin detection is applied. Face detection can be achieved by machine learning-based approaches such as the conventional Viola–Jones face detector [34] or recent convolutional neural network-based face detectors [35]. For multi-site measurement on the face, a facial landmark detector can be used to further detect facial corners and segment the face into sub-regions [36]. Alternatively, living-skin detection can be applied [37], which targets the image areas containing blood-volume pulsation. Typical facial areas for contactless PPG are the forehead and cheeks due to stronger skin pulsatility [38].

(ii) Color-Signal Generation: The skin pixels in the RoI are averaged per video frame. The average values are then concatenated over time to generate color traces (e.g., RGB traces). Such spatial averaging reduces quantization noise [39].

(iii) PPG Waveform Extraction: The PPG waveform is extracted from the (multi-channel) color traces, which is considered to be the core algorithmic step. Extraction means removing non-pulsatile components in the color signals, such as motion, sensor noise, and light spectrum changes, or other disturbances. This step is typically achieved by multiwavelength channel combination, which can eliminate non-pulsatile distortions similar to source de-mixing. Various methods have been proposed, such as blind source separation-based [40,41], optic-physiologic model-based [39,42–44], and machine learning-based [45–47] approaches. Taking the POS algorithm [39] as an example for illustration, it first performs DC normalization on color signals to eliminate the dependency on lighting intensity and skin-tone color. Then, it estimates a projection direction in the normalized RGB space such that the projected signal can be separated from the distortions. The assumption is that the pulse source and motion source have different spectral properties and therefore different color variation directions in the space. Additional signal processing steps (e.g., detrending, band-pass filtering) should be applied to further reduce distortions. Note that such processing typically mitigates out-of-band noise (i.e., outside the heart rate (HR) frequency and its harmonics), whereas channel combination can reduce in-band distortion. The PPG waveform may not include as fine details as its contact-sensing counterpart.

Ballistocardiography (BCG):

Contact BCG is a longstanding measurement principle, which has witnessed a resurgence recently due to advances in sensing technology (see [48] for a review). BCG concerns the measurement of the subtle body movements that occur in response to each heartbeat. When blood is ejected in the ascending aorta, the body moves downward to conserve momentum. Conversely, when blood moves through the descending aorta, the body moves upward. Such longitudinal movement, which is significantly larger than the transverse body movements that also occur with each heartbeat, can be measured as a displacement, velocity, or acceleration. Popular measurement devices are a bed-accelerometer system, a sensitive weighing scale or force plate, and accelerometers in wristbands or smartphones. These devices yield BCG waveforms that exhibit "I", "J", and "K" waves. As shown in Fig. 6.4, these waves may arise as the difference in BP gradients in the ascending and descending aorta as follows:

$$F_{BCG}(t) = A_D [P_1(t) - P_2(t)] - A_A [P_0(t) - P_1(t)], \qquad (6.1)$$

where $F_{BCG}(t)$ is the BCG waveform as a force; $P_0(t)$, $P_1(t)$, and $P_2(t)$ are BP waveforms at the inlet of the ascending aorta, aortic arch, and outlet of the descending aorta; and A_A and A_D are the average cross-sectional areas of the ascending and descending aorta [49]. Hence, the BCG waveform is mainly embedded with arterial BP waveform information.

Contactless BCG is an approach for direct measurement of the body displacement in response to each heartbeat (see Fig. 6.3) [50]. Longitudinal displacement of a body part may be readily extracted via conventional video processing. The resulting displacement may be twice differentiated with respect to time to yield the more common acceleration BCG waveform. While contactless PPG measures diffuse skin reflection, contactless BCG measures specular skin reflection, which is the mirror-like reflection variations directly from the skin surface. As a result, shorter wavelengths (e.g., blue wavelength) are preferred for BCG. Furthermore, contactless BCG does not require skin pixels, so it may be more effective than contactless PPG when, for example, skin is dark or sparsely visible. On the other hand, BCG is very sensitive to motion artifact including respiration by virtue of tracking body movement and may also be dependent on lighting conditions [51].

Contactless BCG is based on motion analysis and therefore not wavelength dependent. In fact, the motion tracking may be applied to a gray image. Nevertheless, the camera requirements are similar to contactless PPG, but BCG waveforms do fluctuate at frequencies up to 25 Hz. Video processing of a camera recording of a person may be employed to acquire BCG waveforms according to the following steps [50].

(i) RoI Detection: An RoI where BCG motion may appear is detected. Note that the RoI may be cloth. Empirically, the preferred RoI is a boundary (e.g., shoulder, mouth) with clear/shape edges wherein motion is easier to be registered. This step can likewise be achieved by machine learning-based approaches.

(ii) Motion Signal Generation: Motion estimation methods are applied to the RoI to measure motion signals typically in the vertical direction from pixel displacements in consecutive frames. Optical flow [52] and correlation filters [53] find pixel matches between subsequent frames, and the matching vector represents the displacement. Specifically, optical flow-based methods have been commonly used, such as dense optical flow (Farneback [54]), sparse optical flow (Lucas Kanade [55]), and recently developed convolutional neural network-based methods. The major consideration of selecting an appropriate motion estimation algorithm for BCG is the "motion sensitivity". A secondary consideration can be robustness to disturbances like lighting changes. After extracting the motion vectors between consecutive frames, they are concatenated in time to generate long-term motion traces.

(iii) BCG Waveform Extraction: The local motion signals are combined into a global motion signal that represents the BCG waveform. Additional signal processing steps (e.g., band-pass filtering) should be employed to further reduce non-BCG components (e.g., slower respiration, other body motions, and sensor noise).

Challenges:

While video processing can permit ultra-convenient acquisition of PPG and BCG waveforms, motion, low signal conditions, as well as variable lighting, are serious challenges. It may be most prudent to first develop the contactless approach while the person is still, the camera-person orientation is fixed, and the lighting and temperature conditions are controlled [56]. A camera-based system could perform surveillance to identify when such ideal conditions occur and then acquire the waveforms only in those instances. If this initial approach proves successful, then proceeding towards more frequent measurement in variable, real-world conditions could follow.

6.2.2   Extraction of waveform features that correlate with BP

Our knowledge of the various contactless arterial waveform features that can correlate with BP stems from contact-sensor investigations. As shown in Fig. 6.4, some features are based on physiology, whereas other features may be more data-driven. At least 10 sec segments of the waveform(s) may be needed for robust feature extraction.

Pulse transit time (*PTT*):

PTT is a popular feature with strong roots in physiology and the most supporting evidence (see [57] for a review). When the heart ejects blood, a pressure wave enters and propagates through the arterial system. This wave may be pictured as acute dilation of the arterial wall (see Fig. 6.4) and travels much faster than blood. PTT is the time delay required for the pressure wave to move between proximal and distal sites. It decreases as the artery becomes less compliant (i.e., stiffens) due to fluid dynamic principles. The well-known Bramwell–Hill equation captures this relationship for PTT (τ) in the form of pulse wave velocity ($v = l/\tau$, where l is the wave travel length) as follows:

$$v = \frac{l}{\tau} = \sqrt{\frac{A}{\rho}\frac{dP}{dA}}. \tag{6.2}$$

Here, A is the cross-sectional area of the artery, P is the distending BP, ρ is blood density, and dA/dP is the arterial compliance. The compliance of an artery decreases as BP rises due to the material properties of the arterial wall. That is, while elastin fibers therein may be considered as a linear material, collagen fibers are slack and do not exert tension until the arterial wall is stretched. The relationship between area/compliance and BP in the aorta is captured by

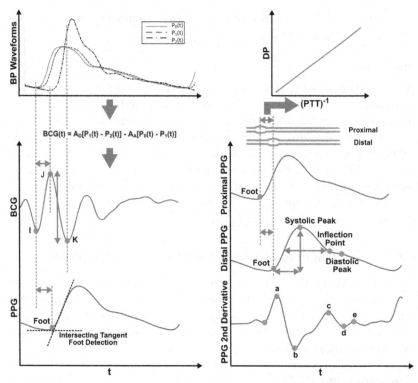

FIGURE 6.4 PPG and BCG waveform features that may correlate with BP. Some features are based on physiology, such as the pulse transit time (PTT) between two waveforms and perhaps the BCG IJ and JK amplitudes as indicated via the BCG model, whereas others including PPG waveform features may be more data-driven.

Wesseling's model as follows:

$$A = A_{max}\left[\frac{1}{2} + \frac{1}{\pi}\tan^{-1}\left(\frac{P - P_M}{P_R}\right)\right] \tag{6.3}$$

$$\frac{dA}{dP} = \frac{A_{max}}{\pi P_R\left[1 + \left(\frac{P-P_M}{P_R}\right)^2\right]}, \tag{6.4}$$

where A_{max} is the maximum cross-sectional area at high BP; P_M is the BP at maximum compliance; and P_R is the BP range for which the compliance is relatively constant [58,59]. Substituting these equations into Eq. (6.2) and assuming higher BP ($P \gg P_M + P_R$) gives the following simple, yet effective, inverse relationship between BP and PTT:

$$P = K_1\frac{l}{\tau} + K_2, \tag{6.5}$$

where $K_1 = \sqrt{\frac{2\rho P_R}{\pi+2}}$ and $K_2 = P_M$ are unknown calibration parameters.

This PTT–BP relationship ignores confounding physiology. Firstly, the arterial wall also includes smooth muscle fibers. Smooth muscle contraction causes arterial compliance and thus PTT to decline (e.g., K_1 decreases with smooth muscle contraction). Such a change in PTT can occur independently of BP and acutely [57,60]. However, smooth muscle is less abundant in the aorta. Secondly, elastin fibers are replaced by collagen fibers with aging so that PTT declines throughout life (e.g., K_1 decreases with aging). However, such a change in PTT occurs slowly. Hence, Eq. (6.5) is most tenable for PTT through the aorta and over a period of time in which aging is not a factor.

PTT can be measured as the time delay between proximal and distal arterial waveforms (see Fig. 6.4). Conventionally, for arterial stiffness quantification, PTT is detected as the trough-to-trough or foot-to-foot time delay between two BP waveforms. The reason is that the foot occurs early in systole before the pressure wave, which is reflected mainly at the microcirculation, returns to the heart. In fact, because of wave reflection, the peak-to-peak time delay between proximal and distal BP waveforms can be negative [60]. The PPG foot (which follows the BP foot as implied by Eq. (6.6)) has also been shown to be superior to the PPG peak for tracking BP changes [32,61]. A robust way to detect the foot is the intersecting tangent method (see Fig. 6.4). In contact sensor studies, ECG is often used to obtain a surrogate of the proximal waveform. The time delay between ECG and arterial waveforms (i.e., pulse arrival time (PAT)) is contaminated by the pre-ejection period (PEP), which can change independently of PTT and BP [62]. However, PEP is not a problem in camera-based BP monitoring, as cardiac electrical activity is not observed. BCG can provide a true and convenient but less robust measurement of the proximal waveform. The BCG model of Eq. (6.1) predicts that the onset/peak of the I-wave coincides with the foot of the ascending aortic/aortic arch BP waveform (see Fig. 6.4). However, the BCG J-wave via a wristband accelerometer may also be effective in denoting proximal timing [63]. BCG wave detection is difficult without ECG assistance, and the J- and I-waves are often defined as the peak of the BCG waveform within 300 ms following the R-wave and the local minimum immediately prior to the J-wave [48,64]. Note that PTT corresponds to diastolic BP (DP), in particular when it is detected via the waveform foot, which is at the level of diastole.

Relatively unique PTTs can be detected with video cameras. Facial video recordings allow detection of the time delay between BCG and face PPG waveforms [65], two PPG waveforms from different locations on the face (e.g., cheek and forehead) [56,66–68], or possibly two PPG waveforms of different wavelengths [69]. Video recording of the face and hand (see Fig. 6.1) permit detection of the time delay between BCG and hand PPG waveforms [65] or face-and-hand PPG waveforms [67,70–72]. Note that the latter time delay is actually the difference between PTT to the hand and PTT to the face. These camera-based PTTs indicate time delays through smaller arteries, wherein smooth muscle is most abundant, and are short (e.g., < 50 ms for PTT via face and hand PPG waveforms) [68,71]. Since various types of noise are a major problem in the contactless approach, the time delays may be detected more robustly from the entire waveforms (via, e.g., cross-correlation) [72]. However, such PTTs may be contaminated by wave reflection. Note that, for relatively low frame rates (e.g., 30–60 fps), the waveforms should be interpolated prior to PTT detection.

Single PPG waveform features:

PPG waveform features are under study as BP markers (see [73,74] for reviews). However, there is currently no supporting theory.

The Kelvin–Voigt model of viscoelasticity provides a simple relationship between same site BP and PPG waveforms in the frequency-domain ($P(\omega)$ and $V(\omega)$) as follows:

$$V(\omega) = \frac{1}{j\omega\eta + E} P(\omega), \qquad (6.6)$$

where E and η are the elastic modulus and the coefficient of viscosity of the arterial wall [75]. The transfer function here is a lowpass filter with gain of $1/E$ and cutoff frequency of E/η. Assuming constant E and η, the model indicates that pulse pressure (PP = systolic BP (SP)–DP), and the peak-to-peak PPG amplitude are proportional for heart rates in the filter passband. The BP and PPG pulse widths would be similarly related. Assuming an exponential DP decay with RC time constant (as predicted by the popular Windkessel model) and that BP increases with this time constant via vasoconstriction, then the PPG pulse width increases with BP. PPG amplitude and pulse widths are commonly used features (see Fig. 6.4). Note that widths representing the shorter systolic period (e.g., time interval from PPG foot to peak) may be inversely related to cardiac contractility [76] and thus BP. However, these assumptions are not valid. Firstly, E is inversely related to arterial compliance and therefore increases with increasing BP and smooth muscle contraction, whereas E/η may decrease (which would increase the PPG pulse width) with increasing BP [32]. In fact, PPG amplitude was shown to increase with increasing SP/DP induced by mental arithmetic (which increases PP via stroke volume) but decrease with similar BP increases caused by a cold pressor test (which causes smooth muscle contraction) [61]. PPG amplitude may also vary with the intensity of the light source and temperature, whereas PPG pulse widths decrease with increasing HR. However, these PPG amplitude and width variations are correctable (e.g., by normalizing the PPG amplitude the mean PPG value). Secondly, BP also increases via cardiac output, and vasoconstriction may not necessarily increase BP (e.g., it can compensate for falling cardiac output so as to maintain BP [77]).

Another model of the PPG waveform that has been proposed is a superposition of forward and reflected waves [78]. In this model, the systolic peak is due to the forward wave, while a diastolic peak is caused by the reflected wave. The time interval between the systolic and diastolic peaks (see Fig. 6.4) may then indicate twice the PTT. One of the FDA-approved devices claims to extract a PTT from a single PPG waveform [20]. However, the model may be at odds with conventional arterial viscoelastic principles.

While there is no theory supporting PPG features as specific markers of BP, it is possible that certain physiologic patterns are common in daily life and that PPG features are useful in tracking BP changes in these limited circumstances. Exhaustive data-driven features obtained via all combinations of amplitudes and time intervals between all local extrema of the PPG waveform and its derivatives (see Fig. 6.4) and via the frequency-domain [79–83] may ultimately help shed light on a currently hidden theory.

Single BCG waveform features:

Studies of BCG waveform features as markers of BP are not as popular, but there may be a viable theory based on the BCG model of Eq. (6.1). According to this model (see Fig. 6.4), the IJ interval corresponds to PTT through the aorta, whereas the J and JK amplitudes may indicate aortic and peripheral PP to within an A_D scale factor. A key point is that the arterial area may change only little over the typical BP range and far less than arterial compliance [58,59]. While these relationships are actually approximations (see Fig. 6.4), the model at least indicates that both aortic PTT and PP information are somehow embedded in a single BCG waveform. Data-driven BCG features [84] may therefore better correlate with BP.

Other arterial waveform features:

HR and HR variability are other arterial waveform features. While increases in HR increase BP via cardiac output, increases in BP reduce HR via cardiovascular reflexes. A large subject study showed little correlation between BP and HR [85]. HR variability derived from an ECG waveform quantifies autonomic nervous-system function [86]. For example, high frequency (0.15–0.40 Hz) HR spectral power indicates parasympathetic nervous-system function (which, e.g., increases during sleep), whereas the ratio of low-frequency (0.04–0.15 Hz) to high-frequency HR spectral power indicates the balance between sympathetic and parasympathetic nervous-system function (which, e.g., increases upon standing). However, due to the sigmoidal reflex relationship from BP to HR [87], HR variability may be small under either low or high BP. A large subject study showed little correlation between BP and HR spectral powers [85]. That said, HR and HR variability could possibly provide value in tracking BP when combined with other waveform features, such as PTT [88]. Note that arterial waveforms actually provide pulse-rate variability, which differs from HR variability due to variations in PAT.

Other video camera features:

Video cameras offer richer information than contact PPG and BCG sensors. Additional information arises from the spatial component. On the other hand,

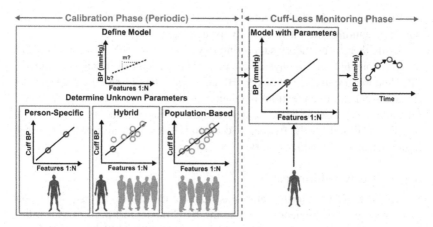

FIGURE 6.5 Cuff-based methods for calibrating arterial waveform features to BP. A calibration model is periodically formed using waveform feature-cuff BP measurement pairs from the person (more accurate) or a population of similar people (more convenient), and cuff-less BP may be obtained thereafter via the model.

spatial averaging over a pixel array is often used to improve waveform signal quality. Contextual information in video recordings such as facial expression, body motion and posture, and environment, could also help. However, such potential value of video recordings remains untapped.

6.2.3   Calibration of features to BP using cuff BP measurements

None of the video camera features is in units of mmHg, so the features must be calibrated to yield absolute BP. Calibration, in fact, may represent the greatest challenge in realizing any cuff-less BP monitoring device. As a representative example, Eq. (6.5) relates PTT to BP with K_1 and K_2 as unknown parameters. Since these parameters vary from person-to-person and over time within a person, the challenge is to determine the two (or more) parameters for a person so that PTT (and/or other features) can be continually converted to a value in BP units.

As shown in Fig. 6.5, calibration is generally based on cuff BP measurements (see [89] for a book chapter on calibration or "initialization" of PTT-based BP monitors). First, a parametric model is established to relate cuff-less features to BP. Then, the model parameters are determined using simultaneous measurement of the features and cuff BP. Thereafter, BP is measured without a cuff by applying the features to the fully defined calibration model. The model parameters would have to be updated periodically to account for vascular aging effects.

Parametric models:

Various models have been proposed to relate arterial waveform features to BP. Models with fewer features can mitigate the burden of obtaining cuff BP measurements for determining the model parameters. However, simpler models may limit accuracy. For example, a single feature would not allow independent determination of SP and DP. That said, SP and DP can show correlation coefficients within a person of 0.7 to 0.8 [90,91], which could possibly be leveraged to balance accuracy and convenience. Conversely, models with more parameters can attain higher accuracy, but parameter determination may mandate many cuff BP measurements.

Examples of simple, yet physiologically-inspired, parametric models are potentially as follows:

$$P_d = K_1 \frac{l}{\tau} + K_2 \tag{6.7}$$

$$P_s = K_3 \frac{l}{\tau} + K_4 \tag{6.8}$$

$$P_s = K_1 \frac{l}{\tau} + K_2 + K_3 y, \tag{6.9}$$

where P_s and P_d are SP and DP and y is a PPG or BCG amplitude. Eqs. (6.7) and (6.8) assume that the two BP levels are linearly correlated, whereas Eqs. (6.7) and (6.9) determine the PP difference via an arterial waveform amplitude. However, more complicated models may be necessary for the camera-based approach to somehow compensate for PTT changes induced by smooth-muscle contraction and waveform-amplitude variations caused by arterial compliance and area changes. Examples of complicated models that have been employed are feedforward neural nets with a few hidden layers and many waveform features as inputs [79,92–94].

Model parameter determination:

The model parameters can be determined via a person-specific, population-based, or hybrid method (see Fig. 6.5). The three methods represent different trade-offs between accuracy and convenience. In the person-specific method, all model parameters are determined from simultaneous measurements of video camera features and cuff BP from the person during interventions that change BP. In the population-based method, all model parameters are determined from basic information about a person along with a dataset consisting of measurement pairs of the features and cuff BP from a training cohort of subjects. In the hybrid method, one parameter is determined from the features and one cuff BP measurement pair, and the remaining parameters are determined from basic person information and a similar training dataset.

While the person-specific method may be most accurate, the need for BP interventions renders it the least convenient (see Fig. 6.5). The interventions should change BP over the range of interest for accurate determination of the model parameters, while being practical enough for person compliance. A cold pressor test (hand/foot immersed in cold water) [95], mental arithmetic [96], and sustained handgrip [97] can increase BP significantly, whereas a Valsalva maneuver (exhaling with nose and mouth closed) [98] can decrease BP appreciably. However, these interventions may not be sufficiently convenient for different reasons. Exercise and hydrostatic maneuvers may be more suitable. After exercise, which can be a routine part of life, BP is markedly increased and then can fall even below pre-exercise level sometime later [99]. A hydrostatic maneuver is performed by varying the vertical height (h) of the effective BP measurement site with respect to the heart. This maneuver will cause the local BP to change by ρgh, where ρ is near water density, g is gravity, and h may be measured with the camera. A change in h of just 10 cm will cause BP to change by over seven mmHg. For the camera-based approach, the effective BP measurement site can be the forehead (for face PPG waveform features) or the midpoint between the heart and forehead (for PTT via BCG and face PPG waveforms). In this case, a change in posture from seated to supine will cause the local BP to increase by up to 17–27 mmHg (= 22 mmHg due to hydrostatic effect ± 5 mmHg due to systemic effect of the posture change). Alternatively, the effective BP measurement site could be the palm (for hand PPG waveform features) such that the hydrostatic maneuver could be performed via hand raising/lowering [100]. Natural BP variations occurring over time in a person due to stress, meals, and other factors (with standard deviation of five–ten mmHg [10]) could also be invoked, but many cuff BP measurements may be needed to capture the changes.

The number of feature-cuff BP measurement pairs must be equal to, or exceed, the number of model parameters. Hence, the person-specific method may only be applicable to simple models. Automatic rather than manual cuff BP measurements may also make more sense for this method. The model parameters may be estimated using standard methods (e.g., linear least squares estimation for the models of Eqs. (6.7)–(6.9)) or advanced methods (e.g., total least squares estimation, which would also account for the likely appreciable error in the video camera features [101]). The person-specific method has yielded low BP errors when employed in conjunction with a two-parameter model relating contact PAT measurements to BP [57], but comprehensive testing is needed.

The population-based method is most convenient because no cuff BP measurements from the person are needed, but it may be the least accurate (see Fig. 6.5). The training dataset must include video-camera feature-cuff BP measurement pairs from many subjects covering the clinical BP range. It would be best if the dataset included manual cuff BP measurements for a more accurate reference, subjects of diverse age and skin color for broad applicability, and BP interventions for attaining trending ability. Since it is possible to obtain a training dataset comprising hundreds of features-cuff BP measurement pairs [56], the population-based method is applicable to complicated models. For neural nets, training is typically achieved with back propagation via the Levenburg–Marquardt or another algorithm, and a portion of the training dataset (i.e., validation set) is utilized for hyperparameter determination [82,92–94].

The model parameters cannot be the same for everyone, so they should depend on some information about the person. Useful and readily available information are age, gender, height (e.g., to determine l in Eqs. (6.7)–(6.9)), and race. For example, P_M and P_R in Eqs. (6.3)–(6.5) decrease linearly with age (but without significant gender dependency) based on *ex vivo* aorta measurements [58,59]. Other information that could help are cardiovascular risk factors, such as cholesterol level, smoking history, and presence of diabetes. Few studies have been published on the population-based method. However, one study, which involved a two-parameter model relating PTT via contact ear and toe PPG waveforms to BP and determination of the parameters for four groups (young, male; young, female; old, male; and old, female), showed promising results [102].

The hybrid method does not involve employing a BP intervention but does require a cuff BP measurement from the person and may therefore represent a compromise between accuracy and convenience (see Fig. 6.5). The hybrid method appears promising in the sense that it may be the basis of the cuff-less BP monitoring devices that are FDA-approved. However, few studies have been published on the method. One method used a five-parameter model to relate contact finger PAT to SP and determined constant values for four of the parameters using a training dataset and the remaining intercept parameter for a new person using a cuff BP measurement [103]. The hybrid method could also be implemented in other ways. One way is to implement the population-based method, determine the model parameter that is least predictable from available person information, and then determine that parameter using a cuff BP measurement from the person. Another way is to constrain the model parameters using the single pair of waveform features and cuff BP. For example, the model parameters are determined from the measurements of comparable subjects in the training dataset in the least squares sense under the constraint that the model exactly fits the single measurement of the person.

Recalibrations:

The person-specific method and likely the hybrid method have to be applied periodically to account for aging effects on the model parameters (see Fig. 6.5). However, the required frequency for such cuff calibrations is largely unknown. We predicted the maximum time period between cuff calibrations that would not introduce significant BP error for an aortic PTT-based method using available models and data [104]. The maximum recalibration period declined significantly with age but did not depend on gender. The maximum recalibration period was at least one year for a 30-year old and six months for a 70-year old. In an unpublished preliminary study, we also found that the calibration model relating contact toe PAT to SP hardly changed over one-year of aging in seven subjects. However, these theoretical and experimental results pertain to aortic PTT, which cannot be easily measured with the camera-based approach. Note that for the population-based and some hybrid methods, the effects of aging may be

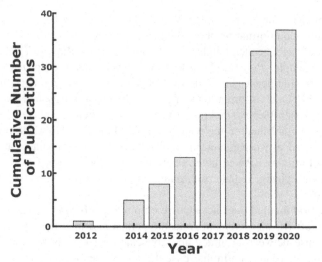

FIGURE 6.6 **Cumulative number of peer-reviewed papers on camera-based BP monitoring published up to July 2020.** The topic is receiving increasing attention.

automatically accounted for via age-dependent model parameters. Such methods may be better suited to the camera-based approach for reasons previously outlined.

6.3 Summary of previous experimental studies

As shown in Fig. 6.6, an increasing number of papers on camera-based BP monitoring are appearing in journals and conference proceedings. We examined the papers on experimental studies to understand the current evidence for putting the theory to practice. We considered all papers that provided results of BP measurements via only a video camera. Of these 23 papers, we excluded those with very few subjects or results that we could not interpret. We provide a summary of the nine remaining papers. Since this number is small and the studies are not uniform, we followed with a summary of papers on relevant contact-sensor investigations. These papers could be viewed as indicating an upper bound for the performance of camera-based BP monitoring. We specifically included those studies with results that may be most representative of the actual performance of contact sensors.

Before proceeding with the summary, we should emphasize that results reported in the field of cuff-less BP monitoring can be misleading. For example, in the case of a person-specific or hybrid calibration method, the correlation between subsequent cuff-less and cuff BP measurements over a group of subjects is sometimes presented. However, this correlation is often high due merely to the inter-subject cuff BP variations [105]. That is, because the cuff-less measurements are calibrated with baseline cuff BP, the data points from subjects with different usual BP levels will almost always fall on the identity line. Intra-subject correlation coefficients would paint a true picture of how well the cuff-less device tracks BP. As another example, in the case of a typically population-based calibration method, the BP error between cuff-less and cuff devices over a group of subjects is typically shown. However, this error can be low simply because the range of cuff BP in the group is small [56]. Consider a cuff-less device that always predicts a SP of 120 mmHg. This device will yield BP errors within the regulatory limits (i.e., five and eight mmHg bias and precision errors) [106] in subjects with uniformly distributed SP over the range of 107 to 133 mmHg. The reduction in BP error relative to the BP standard deviation of the subject cohort or, better yet, the BP error of a model that predicts BP from all other available information (e.g., age, gender, race, cardiovascular risk factors) would be far more meaningful. In addition, exercise is often used to validate a cuff-less device. However, it is already known that arterial waveforms can easily distinguish rest from exercise (via, e.g., high HR and pulse amplitude). What would be more convincing is to show that the device can track a battery of BP interventions.

6.3.1   Key camera-based BP monitoring investigations

Table 6.1 provides a summary of the key papers to date on experimental studies of camera-based BP monitoring. The studies typically include a small number of subjects and do not include subjects with diverse BP or employ a battery of BP interventions. Hence, each study by itself may not be challenging enough to truly indicate camera-based BP monitoring performance. Aggregating the results of the studies may give a better idea on the current state of camera-based BP monitoring. The studies are in fact similar in that contactless PPG waveforms are acquired via RGB cameras from non-hypertensive subjects in controlled settings, and the correlation coefficient is mainly used as an evaluation metric. Note that the signs of the correlation coefficients should be different. For example, negative coefficients between PTT and BP are expected, whereas positive coefficients between calibrated features and BP should occur. We therefore made all correlation coefficients with the physiologically correct sign to be positive and all correlation coefficients with incorrect sign (red font (gray in print version) in Table 6.1) to be negative. We then computed the unweighted average of the highest correlation coefficient reported in each study as 0.54. This level of correlation indicates that contactless PPG can indeed contain BP information. The common features are PTTs and PPG amplitude and time intervals. However, the studies have significant methodological differences, so this aggregate and best-case result should be interpreted with caution. Also, note that the studies typically employed industrial rather than consumer-grade cameras.

One study [56] stands out in Table 6.1. In this study, an iPhone front camera was used to extract contactless PPG waveforms at 17 facial locations from 1,328 normotensive, East Asian subjects. An FDA-approved finger-cuff volume-clamp device was employed for reference BP measurement [110]. A neural

TABLE 6.1 Summary of key experimental studies to date on camera-based blood pressure (BP) monitoring. PPG is photoplethysmography; PTT, pulse transit time; SP/DP, systolic/diastolic BP; HR, heart rate. Red (gray in print version) indicates physiologically incorrect sign.

Ref	Subjects (Number/Type)	BP Interventions	Reference BP Device	Camera Settings (Frames Per Sec/Lighting/Camera-Person Position)	Waveform Acquisition	Waveform Features	Calibration (Model/Parameter Determination)	Results Correlation Coefficient SP	DP	Percent Error Reduction (w.r.t. SD) SP	DP
[107]	70/Intensive Care	Noradrenalin Nitroglycerin	Radial Artery Catheter	120/Ceiling Fluorescent/Fixed	Contactless PPG	Face PPG Amplitude	None	0.54 (with pulse pressure)	—	—	—
[70]	20/Normotensive, Male, Young	Valsalva Maneuver	Finger Cuff (Volume-Clamp)	120/White LED/Fixed		Forehead to Palm PTT		0.56	—		
						Forehead PPG Time Interval	None	-0.09			
						Palm PPG Time Interval		-0.59			
						Cheek PPG Time Interval		-0.21			
[71]	7/Normotensive	Cycling	Arm Cuff (Oscillometry)	420/Extra Light Source/Fixed		Face to Palm PTT	None	-0.79	-0.36	—	
[56]	1328/Normotensive (70% training, 15% validation, 15% testing)	None	Finger Cuff (Volume-Clamp)	60/Two Extra Light Sources/Fixed		155 Facial PPG Features and Demographic Features	Multilayer Perceptron Network/Population-Based Calibration	0.67	0.47	Demographics Plus HR 8.8 / Plus Contactless PPG 25.5	8.8 / 12.0

TABLE 6.1 (continued)

Ref	Subjects (Number/Type)	BP Interventions	Reference BP Device	Camera Settings (Frames Per Sec/Lighting/Camera-Person Position)	Waveform Acquisition	Waveform Features	Calibration (Model/Parameter Determination)	Results Correlation Coefficient SP	DP	Percent Error Reduction (w.r.t SD) SP	DP
[108]	20/Normotensive	Natural (morning training, afternoon and evening testing)	Wrist Cuff (Oscillometry)	30/Ambient Light/Fixed		8 Facial PPG Features	Neural Network/Population-Based Calibration	Afternoon 0.74 Evening 0.72	0.60 0.53	–	–
[67]	20/Normotensive, Young	None	Finger Cuff (Volume Clamp)	140/White LED/Fixed		Cheek to Forehead PTT / Cheek to Palm PTT / Forehead to Palm PTT	None	−0.22 0.49 0.56	–		
[68]	20/Normotensive (for training and testing)	Breath Holding	Finger Cuff (Volume Clamp)	120/Ambient Light/Fixed		2 PPG Features Plus Cheek to Forehead PTT	Linear Model/Population-Based Calibration	0.29 (with mean BP)		–	
[72]	10/Normotensive, Young	None	Wrist Cuff (Oscillometry)	120/Ambient Light/Fixed		Face to Palm PTT	None	−0.54	0.16	–	
[109]	40/Normotensive Male (for training and testing)	None	Arm Cuff (Oscillometry)	60/Ambient Light/Fixed		PPG Amplitude	Random Forest Regression/Population-Based Calibration	–		15.0	27.0

network-population-based calibration method was developed using a portion of the data and then tested on the remaining data. The camera-based method measured SP better than a model with basic subject information and HR as input (precision error of 7.3 ± 0.02 versus 8.9 ± 0.0 mmHg). While the method hardly improved DP accuracy, SP may be more important due to the prevalence of isolated systolic hypertension [111]. The most important features appeared to be waveform shape, amplitude, and energy rather than facial PTTs. However, the subject cohort was not heterogenous, especially in terms of BP levels.

6.3.2 Relevant contact-sensor BP monitoring investigations

Table 6.2 provides a summary of germane papers on relatively challenging evaluation studies of cuff-less BP monitoring via contact sensing. These studies offer both optimism and pessimism for camera-based BP monitoring as follows.

PTT via the foot-to-foot time delay between ear and finger PPG waveforms was recently investigated as a marker of BP in a relatively diverse subject cohort under a battery of interventions [61]. This study may represent the most challenging assessment of the PTT in non-hospitalized humans to date. The intra-subject correlation coefficient between the PTT and SP/DP was no greater than 0.3. The reason may be that the time delay is actually a difference in PTT to the finger and PTT to the ear and therefore susceptible to smooth-muscle contraction. The study suggests that PTT via contactless face and hand PPG waveforms alone may be ineffective and that contactless face PTT, which is smaller and may be subject to more confounding factors, would not be able to track BP changes any better.

Complicated model-population-based calibration methods for predicting BP from a large set of finger PPG waveform features and basic subject information have recently been assessed in a pair of studies involving hundreds of subjects. One study employed a deep belief network in critically ill patients with invasive BP as reference [93]. The method yielded BP errors that were 7–10% lower than the BP standard deviation of the testing cohort of 47 subjects. The other study employed a random forest regression in various subjects from young to old with cuff BP as reference [112]. While several results were reported, the most interpretable was that the method produced BP errors that were 10–16% lower than the BP standard deviation of a testing cohort of 147 subjects. While the studies did not report the errors of models predicting BP via all available information other than the PPG waveform, they may still indicate that PPG waveforms do contain some BP information through a number of features. This pair of studies therefore supports the large camera-based study in Table 6.1.

Similar to contactless BCG for BP monitoring, which is conspicuously absent in Table 6.1, there may be no studies assessing head BCG waveforms as markers of BP in more than a few subjects. However, two studies did examine the ability of PTT as the time interval between the I- or J-wave of the wrist BCG waveform and the foot of the wrist PPG waveform and of wrist BCG wave

TABLE 6.2 Summary of experimental studies on relevant contact-sensor BP monitoring investigations. BCG is ballistocardiography; SB, slow breathing; MA, mental arithmetic; CP, cold pressor; NTG, nitroglycerin; BH, breath holding.

Ref	Subjects (Number/Type)	BP Interventions	Reference BP Device	Contact Waveform Acquisition	Waveform Features	Calibration (Model/Parameter Determination)	Results			
							Intra-Subject Correlation		Percent Error Reduction(w.r.t. SD)	
							SP	DP	SP	DP
[61]	32/50 Years Average Age, 25% Hypertensive	SB, MA, CP, NTG	Arm Cuff (Auscultation)	Ear PPG, Finger PPG	PTT	None	0.30	0.20	–	–
[93]	572/ Critically Ill (47 for testing)	Natural (30-min)	Radial or Femoral Artery Catheter	Finger PPG (Pulse Oximeter)	Various Waveform and Demographic features	Deep Belief Network/Population-Based Calibration	–	–	7.1	10.0
[112]	1249/Young to Old (147 for testing)	Natural (1 measure per week for 1 month)	Arm Cuff (Oscillometry and Auscultation)	Finger PPG (AC and DC)	19 Waveform and Demographic Features	Random Forest Regression/Population-Based Calibration	–	–	Young: 16.0 Old: 10.4	Young: 12.0 Old: 9.7
[63]	22/Normotensive, Young	CP, MA, SB, BH	Finger Cuff (Volume-Clamp)	Wrist BCG, Wrist PPG	PTT	None	0.76	0.75	–	–
[84]	23/Normotensive, Young	CP, MA, SB, BH	Finger Cuff (Volume-Clamp)	Wrist BCG	BCG G, H, I, J, K, L Timings and Amplitudes	Linear Regression/Subject-Specific Calibration	0.75	0.75	–	–

timings and amplitudes in tracking BP changes induced by a battery of interventions [63,84]. The studies showed intra-subject correlation coefficients with BP of 0.75. However, they only included healthy, young subjects. Hence, the studies suggest that contactless BCG could be useful in tracking BP changes in at least some people.

6.4 Conclusions

6.4.1 Summary

Camera-based BP monitoring may afford passive and frequent measurements during daily life with devices that are already available to many. As a result, this contactless approach could uniquely improve hypertension awareness and control rates.

Although measuring BP without touching a person may seem unrealistic, there are theoretical principles to support the feasibility of the approach. The general theory is to acquire contactless PPG or BCG waveforms, extract physiologically relevant features from the waveforms such as PTT and waveform amplitudes and time intervals, and calibrate the features to BP using a model ranging from a simple regression to a complicated neural net whose parameters are determined via previous cuff BP measurements from the person or a group of similar people. This theory is similar to most devices for cuff-less BP measurement via contact sensors, but video recordings of a person may also offer additional correlates of BP via spatial and contextual information.

Experimental studies on the camera-based approach are beginning to show that contactless PPG obtained in highly controlled settings (i.e., fixed subject-camera positioning and external lighting) can track BP with a correlation coefficient of up to 0.5. However, the studies are currently limited in terms of number and rigor of evaluation. Larger and more challenging contact-sensor studies suggest that contactless BCG waveforms can also offer BP information and confirm that single PPG waveform features can yield BP errors that may be 10–20% lower than the BP standard deviation. A number of PPG waveform features related to amplitude and shape appears to be needed. PTT may be of less value in the camera-based approach because it requires accurate measurement of two waveforms and the so-obtained time delays are confounded by physiology in contrast to aortic PTT.

6.4.2 Future research directions

The feasibility of camera-based BP monitoring is therefore unknown. Much future research is needed to reach a final verdict. Here are our recommended directions.

Innovative video processing methods should be pursued for exquisite acquisition of the arterial waveforms. In this way, for example, the smaller contactless PTTs may be detected with high fidelity. Optimizing such methods for consumer-grade cameras is essential for reaching the mass population.

Discovery of video camera features that faithfully track BP is a must. The studies to date suggest that individual features such as contactless PTT and PPG amplitude have only modest correlation with BP. Hence, combining a number of (independent) features may be a reasonable course of action. Developing a theory for the relationship between PPG waveforms and BP levels that can usually hold up under daily living could be of great value in feature identification. Complicated models would of course be needed to accommodate the large feature set.

Creation of an extensive training dataset comprising video camera recording-cuff BP measurement pairs may be the single most important direction. Such a dataset is vital for training a complicated model. It is mandatory that the dataset include diverse subjects covering the clinical range of BP and skin color. Ideally, the dataset would comprise measurements during a battery of BP interventions. In this way, BP trending in a person may be realized in addition to measuring absolute BP. However, such interventions may limit the number of subjects in the dataset. If the dataset were made freely available, the research community could work together to establish the best video processing method.

A focus should be on the development of methods invoking population-based calibration to reach everyone. However, methods based on hybrid calibration should be simultaneously pursued to improve accuracy for those people with cuff-device access.

Deep learning based on both calibration methods may be worthwhile to investigate provided that the training dataset is sufficiently large. In this approach, the features would be automatically extracted without explicitly acquiring arterial waveforms. The approach should not only afford the best possible camera-based BP measurement accuracy but may also unveil new physiology via the selected features.

Handling hydrostatic effects must eventually be addressed. That is, if a person is using a smartphone, the observed BP via the front camera may be higher by up to 25 mmHg while lying on the couch compared to sitting on a chair. The device would ideally measure the vertical height difference between the heart and local BP measurement site via the smartphone camera or only provide a measurement in the upright posture. Likewise, privacy issues would ultimately have to be eliminated for device adoption.

Lastly, camera-based BP monitoring devices must be rigorously tested, and the results should clearly show added value. For devices employing population-based calibration, the Association for Advancement of Medical Instrumentation (AAMI) protocol may be applied for evaluation [106]. This protocol, which is accepted by the FDA, involves obtaining triplicate measurements from the test device and a manual cuff device in at least 85 subjects. The test device must, for example, yield a bias error (mean of > 255 errors) of < 5 mmHg and a precision error (standard deviation of > 255 errors) of < 8 mmHg. For devices invoking person-specific/hybrid calibration, protocols that employ BP interventions must be used. An IEEE standard has been created for testing these cuff-calibrated, cuff-less devices [113], but there are no specific intervention requirements. As stated earlier, cuff-less devices can trivially track BP in exercise.

6.4.3 Outlook

We conclude with our expectations on the outcome of completing these and other research directions. Our outlook is tempered by the understanding that cuff-less BP monitoring via contact sensing has yet to be proven. For example, the two FDA-approved cuff-less devices are not backed by significant published data. This point cannot be overlooked, as results in this field can be misleading. While video camera recordings of a person do provide more information than contact sensors, it is still hard to imagine that the contactless approach could be as accurate in measuring BP than a contact approach based on the same theoretical principles. It is even harder to accept that BP could be measured with any accuracy in the wild (e.g., surveillance setting in Fig. 6.1) or under significant motion or low-signal conditions. However, people are often still and in well-lit environments, and BP measurements only in these ideal conditions would be of great value. Although camera-based BP monitoring may prove to not be useful for quantitative measurement, we think that there is a chance that it could be effective in screening people for high BP. Focus on such a screening application makes sense due to the ubiquity of the camera and would be a tremendous outcome because too many people with hypertension are unaware of their condition. Another target clinical application that may be rationale is monitoring BP in neonatal intensive care units wherein contact sensors are highly undesirable. We hope that this chapter helps facilitate research on camera-based BP monitoring to realize an important clinical application or at least improve our understanding of the underlying technology and physiology.

Acknowledgments

This work was supported in part by the US National Institutes of Health under Grant EB-018818.

References

[1] K.T. Mills, et al., Global disparities of hypertension prevalence and control: a systematic analysis of population-based studies from 90 countries, Circulation 134 (6) (2016) 441–450.

[2] Y. Ostchega, C.D. Fryar, T. Nwankwo, D.T. Nguyen, Hypertension prevalence among adults aged 18 and over: United States, 2017–2018, NCHS Data Brief 364 (2020) 1–8.

[3] S. Lewington, R. Clarke, N. Qizilbash, R. Peto, R. Collins, Age-specific relevance of usual blood pressure to vascular mortality: a meta-analysis of individual data for one million adults in 61 prospective studies, The Lancet 360 (9349) (Dec. 2002) 1903–1913, https://doi.org/10.1016/S0140-6736(02)11911-8.

[4] B.M. Psaty, et al., Health outcomes associated with antihypertensive therapies used as first-line agents: a systematic review and meta-analysis, Journal of the American Medical Association 277 (9) (1997) 739–745, https://doi.org/10.1001/jama.1997.03540330061036.

[5] C.K. Chow, et al., Prevalence, awareness, treatment, and control of hypertension in rural and urban communities in high-, middle-, and low-income countries, Journal of the American Medical Association 310 (9) (2013) 959–968.

[6] H. Wang, et al., Global, regional, and national life expectancy, all-cause mortality, and cause-specific mortality for 249 causes of death, 1980–2015: a systematic analysis for the global burden of disease study 2015, The Lancet 388 (10053) (Oct. 2016) 1459–1544, https://doi.org/10.1016/S0140-6736(16)31012-1.

[7] M. Sullivan, Hypertension is so common that almost everyone is affected at some point, Hello Heart (25 May 2020), https://www.helloheart.com/blog/hypertension-is-so-common-that-almost-everybody-is-affected-at-some-point. (Accessed 15 July 2020).

[8] T.G. Pickering, D. Shimbo, D. Haas, Ambulatory blood-pressure monitoring, The New England Journal of Medicine 354 (22) (2006) 2368–2374, https://doi.org/10.1056/NEJMra060433.

[9] US Preventive Services Task Force, Final recommendation statement: high blood pressure in adults: screening – US preventive services task force, https://www.uspreventiveservicestaskforce.org/Page/Document/RecommendationStatementFinal/high-blood-pressure-in-adults-screening, 2016. (Accessed 15 July 2020).

[10] B. Rosner, B.F. Polk, Predictive values of routine blood pressure measurements in screening for hypertension, American Journal of Epidemiology 117 (4) (Apr. 1983) 429–442, https://doi.org/10.1093/oxfordjournals.aje.a113561.

[11] R. Agarwal, J.E. Bills, T.J.W. Hecht, R.P. Light, Role of home blood pressure monitoring in overcoming therapeutic inertia and improving hypertension control: a systematic review and meta-analysis, Hypertension 57 (1) (2011) 29–38, https://doi.org/10.1161/HYPERTENSIONAHA.110.160911.

[12] Home | Sotera Wireless, https://www.soterawireless.com/. (Accessed 15 July 2020).

[13] Bio-Beat medical smart-monitoring, http://www.biobeat.cloud/. (Accessed 16 July 2020).

[14] A. Chandrasekhar, K. Natarajan, M. Yavarimanesh, R. Mukkamala, An iPhone application for blood pressure monitoring via the oscillometric finger pressing method, Scientific Reports 8 (13136) (2018) 1–6, https://doi.org/10.1038/s41598-018-31632-x.

[15] Set up and use the My BP Lab Research app, https://www.samsung.com/us/support/answer/ANS00082868/. (Accessed 15 July 2020).

[16] Home | Bold Diagnostics, https://www.bolddiagnostics.com/. (Accessed 15 July 2020).

[17] Home | CNOGA Medical, https://cnogacare.co/. (Accessed 15 July 2020).

[18] Bio-Beat medical smart-monitoring, https://www.bio-beat.com/. (Accessed 15 July 2020).

[19] FDA, Sotera Wireless ViSi FDA 510(k) substantial equivalence summary, 2013.

[20] FDA, Bio-Beat BB-613WP FDA 510(k) substantial equivalence summary, 2019.

[21] Wearable blood pressure monitor & watch | HeartGuide by OMRON, https://omronhealthcare.com/products/heartguide-wearable-blood-pressure-monitor-bp8000m/?. (Accessed 15 July 2020).

[22] How many people have smartphones worldwide | BankMyCell, BankMyCell, https://www.bankmycell.com/blog/how-many-phones-are-in-the-world. (Accessed 15 July 2020).

[23] A. Bastawrous, M.J. Armstrong, Mobile health use in low- and high-income countries: an overview of the peer-reviewed literature, Journal of the Royal Society of Medicine 106 (4) (2013) 130–142, https://doi.org/10.1177/0141076812472620.

[24] A. Chandrasekhar, C.-S.S. Kim, M. Naji, K. Natarajan, J.-O.O. Hahn, R. Mukkamala, Smartphone-based blood pressure monitoring via the oscillometric finger-pressing method, Science Translational Medicine 10 (431) (Mar. 2018), https://doi.org/10.1126/scitranslmed.aap8674.

[25] R. Mukkamala, Blood pressure with a click of a camera?, Circulation: Cardiovascular Imaging 12 (8) (Aug. 2019), https://doi.org/10.1161/CIRCIMAGING.119.009531.

Contactless Vital Signs Monitoring

[26] Time spent on smartphone everyday worldwide in 2017, Statista, May 14, 2020, https://www.statista.com/statistics/781692/worldwide-daily-time-spent-on-smartphone/, 2020. (Accessed 16 July 2020).

[27] J. Allen, Photoplethysmography and its application in clinical physiological measurement, Physiological Measurement 28 (3) (2007) 1, https://doi.org/10.1088/0967-3334/28/3/R01.

[28] A. Reisner, P.A. Shaltis, D. McCombie, H.H. Asada, Utility of the photoplethysmogram in circulatory monitoring, Anesthesiology 108 (5) (2008) 950–958, https://doi.org/10.1097/ALN.0b013e31816c89e1.

[29] W. Verkruysse, L.O. Svaasand, J.S. Nelson, Remote plethysmographic imaging using ambient light, Optics Express 16 (26) (Dec. 2008) 21434, https://doi.org/10.1364/oe.16.021434.

[30] M. Yoshioka, S. Bounyong, Regression-forests-based estimation of blood pressure using the pulse transit time obtained by facial photoplethysmogram, in: Proceedings of the International Joint Conference on Neural Networks, vol. 2017-May, Jun. 2017, pp. 3248–3253.

[31] P. Zurek, O. Krejcar, M. Penhaker, M. Cerny, R. Frischer, Continuous noninvasive blood pressure measurement by near infra red CCD camera and pulse transmit time systems, in: 2010 2nd International Conference on Computer Engineering and Applications, ICCEA 2010, vol. 2, 2010, pp. 449–453.

[32] A. Chandrasekhar, M. Yavarimanesh, K. Natarajan, J.-O. Hahn, R. Mukkamala, PPG sensor contact pressure should be taken into account for cuff-less blood pressure measurement, IEEE Transactions on Biomedical Engineering 67 (11) (2020) 3134–3140, https://doi.org/10.1109/tbme.2020.2976989.

[33] Y. Mironenko, K. Kalinin, M. Kopeliovich, M. Petrushan, Remote photoplethysmography: rarely considered factors, in: Proceedings of the IEEE/CVF Conference on Computer Vision and Pattern Recognition (CVPR) Workshops, Jun. 2020, p. 1.

[34] P. Viola, M. Jones, Rapid object detection using a boosted cascade of simple features, in: Proceedings of the IEEE Conference on Computer Vision and Pattern Recognition (CVPR), vol. 1, 2001, pp. 1511–1518.

[35] K. Zhang, Z. Zhang, Z. Li, Y. Qiao, Joint face detection and alignment using multitask cascaded convolutional networks, IEEE Signal Processing Letters 23 (10) (2016) 1499–1503.

[36] X. Xiong, F. de la Torre, Supervised descent method and its applications to face alignment, in: Proceedings of the IEEE Conference on Computer Vision and Pattern Recognition (CVPR), 2013, pp. 532–539.

[37] W. Wang, S. Stuijk, G. de Haan, Living-skin classification via remote-PPG, IEEE Transactions on Biomedical Engineering 64 (12) (2017) 2781–2792.

[38] W. Wang, S. Stuijk, G. de Haan, Unsupervised subject detection via remote PPG, IEEE Transactions on Biomedical Engineering 62 (11) (2015) 2629–2637.

[39] W. Wang, A.C. den Brinker, S. Stuijk, G. de Haan, Algorithmic principles of remote PPG, IEEE Transactions on Biomedical Engineering 64 (7) (2016) 1479–1491.

[40] M. Lewandowska, J. Rumiński, T. Kocejko, J. Nowak, Measuring pulse rate with a webcam—a non-contact method for evaluating cardiac activity, in: Proceedings of the Federated Conference on Computer Science and Information Systems (FedCSIS), 2011, pp. 405–410.

[41] M.-Z. Poh, D.J. McDuff, R.W. Picard, Advancements in noncontact, multiparameter physiological measurements using a webcam, IEEE Transactions on Biomedical Engineering 58 (1) (2010) 7–11.

[42] G. de Haan, V. Jeanne, Robust pulse rate from chrominance-based rPPG, IEEE Transactions on Biomedical Engineering 60 (10) (2013) 2878–2886.

[43] G. de Haan, A. van Leest, Improved motion robustness of remote-PPG by using the blood volume pulse signature, Physiological Measurement 35 (9) (2014) 1913.

[44] W. Wang, A.C. den Brinker, G. de Haan, Discriminative signatures for remote-PPG, IEEE Transactions on Biomedical Engineering 67 (5) (2019) 1462–1473.

[45] W. Chen, D. McDuff, Deepphys: video-based physiological measurement using convolutional attention networks, in: Proceedings of the European Conference on Computer Vision (ECCV), 2018, pp. 349–365.

[46] Q. Zhan, W. Wang, G. de Haan, Analysis of CNN-based remote-PPG to understand limitations and sensitivities, Biomedical Optics Express 11 (3) (2020) 1268–1283.

[47] F. Bousefsaf, A. Pruski, C. Maaoui, 3D convolutional neural networks for remote pulse rate measurement and mapping from facial video, Applied Sciences 9 (20) (2019) 4364.

[48] O.T. Inan, et al., Ballistocardiography and seismocardiography: a review of recent advances, IEEE Journal of Biomedical and Health Informatics 19 (4) (Jul. 2015) 1414–1427, https://doi.org/10.1109/JBHI.2014.2361732.

[49] C.S. Kim, et al., Ballistocardiogram: mechanism and potential for unobtrusive cardiovascular health monitoring, Scientific Reports 6 (1) (Sep. 2016) 1–6, https://doi.org/10.1038/srep31297.

[50] G. Balakrishnan, F. Durand, J. Guttag, Detecting pulse from head motions in video, in: Proceedings of the IEEE Conference on Computer Vision and Pattern Recognition (CVPR), 2013, pp. 3430–3437.

[51] A.V. Moco, S. Stuijk, G. de Haan, Ballistocardiographic artifacts in PPG imaging, IEEE Transactions on Biomedical Engineering 63 (9) (2015) 1804–1811.

[52] J.-Y. Bouguet, Pyramidal implementation of the affine Lucas kanade feature tracker description of the algorithm, 2001.

[53] M. Mueller, N. Smith, B. Ghanem, Context-aware correlation filter tracking, in: Proceedings of the IEEE Conference on Computer Vision and Pattern Recognition (CVPR), 2017, pp. 1396–1404.

[54] G. Farnebäck, Two-frame motion estimation based on polynomial expansion, in: Scandinavian Conference on Image Analysis, 2003, pp. 363–370.

[55] B.D. Lucas, T. Kanade, An iterative image registration technique with an application to stereo vision, in: Proceedings of the 7th International Conference on Artificial Intelligence (IJCAI), 1981, pp. 674–679.

[56] H. Luo, et al., Smartphone-based blood pressure measurement using transdermal optical imaging technology, Circulation: Cardiovascular Imaging 12 (8) (Aug. 2019) 8857, https://doi.org/10.1161/CIRCIMAGING.119.008857.

[57] R. Mukkamala, et al., Toward ubiquitous blood pressure monitoring via pulse transit time: theory and practice, IEEE Transactions on Biomedical Engineering 62 (8) (Aug. 2015) 1879–1901, https://doi.org/10.1109/TBME.2015.2441951.

[58] G.J. Langewouters, K.H. Wesseling, W.J.A. Goedhard, The static elastic properties of 45 human thoracic and 20 abdominal aortas in vitro and the parameters of a new model, Journal of Biomechanics 17 (6) (Jan. 1984) 425–435, https://doi.org/10.1016/0021-9290(84)90034-4.

[59] K.H. Wesseling, J.R.C. Jansen, J.J. Settels, J.J. Schreuder, Computation of aortic flow from pressure in humans using a nonlinear, three-element, Journal of Applied Physiology 74 (5) (1993) 2566–2573.

[60] M. Gao, H.-M.M. Cheng, S.-H.H. Sung, C.-H.H. Chen, N.B. Olivier, R. Mukkamala, Estimation of pulse transit time as a function of blood pressure using a nonlinear arterial tube-load model, IEEE Transactions on Biomedical Engineering 64 (7) (Jul. 2017) 1524–1534, https://doi.org/10.1109/TBME.2016.2612639.

[61] R.C. Block, M. Yavarimanesh, K. Natarajan, A. Carek, A. Mousavi, A. Chandrasekhar, C.S. Kim, J. Zhu, G. Schifitto, L.K. Mestha, O.T. Inan, J.O. Hahn, R. Mukkamala, Conventional pulse transit times as markers of blood pressure changes in humans, Sci. Rep. 10 (1) (2020 Oct. 2) 16373, https://doi.org/10.1038/s41598-020-73143-8.

[62] M. Gao, N.B. Olivier, R. Mukkamala, Comparison of noninvasive pulse transit time estimates as markers of blood pressure using invasive pulse transit time measurements as a reference, Physiological Reports 4 (10) (May 2016), https://doi.org/10.14814/phy2.12768.

[63] P. Yousefian, et al., The potential of wearable limb ballistocardiogram in blood pressure monitoring via pulse transit time, Scientific Reports 9 (1) (2019) 1–11.

[64] O.T. Inan, M. Etemadi, R.M. Wiard, L. Giovangrandi, G.T.A. Kovacs, Robust ballistocardiogram acquisition for home monitoring, Physiological Measurement 30 (2) (2009) 169–185, https://doi.org/10.1088/0967-3334/30/2/005.

[65] D. Shao, Y. Yang, F. Tsow, C. Liu, N. Tao, Non-contact simultaneous photoplethysmo-gram and ballistocardiogram video recording towards real-time blood pressure and abnormal heart rhythm monitoring, in: Proceedings – 12th IEEE International Conference on Auto-matic Face and Gesture Recognition, FG 2017–1st International Workshop on Adaptive Shot Learning for Gesture Understanding and Production, ASL4GUP 2017, Biometrics in the Wild, Bwild 2017, Heteroge, 2017, pp. 273–277.

[66] C. Gonzalez Viejo, S. Fuentes, D. Torrico, F. Dunshea, Non-contact heart rate and blood pressure estimations from video analysis and machine learning modelling applied to food sensory responses: a case study for chocolate, Sensors 18 (6) (Jun. 2018) 1802, https://doi.org/10.3390/s18061802.

[67] N. Sugita, K. Obara, M. Yoshizawa, M. Abe, A. Tanaka, N. Homma, Techniques for estimat-ing blood pressure variation using video images, in: Proceedings of the Annual International Conference of the IEEE Engineering in Medicine and Biology Society, EMBS, vol. 2015-November, Nov. 2015, pp. 4218–4221.

[68] M. Yoshizawa, N. Sugita, M. Abe, A. Tanaka, N. Homma, T. Yambe, Non-contact blood pressure estimation using video pulse waves for ubiquitous health monitoring, in: 2017 IEEE 6th Global Conference on Consumer Electronics, GCCE 2017, vol. 2017-January, Dec. 2017, pp. 1–3.

[69] N. Vahdani-Manaf, Biological assessments by innovative use of multi-wavelength photo-plethysmographic signals time differences, Journal of Applied Sciences 15 (11) (Nov. 2015) 1312–1317, https://doi.org/10.3923/jas.2015.1312.1317.

[70] N. Sugita, M. Yoshizawa, M. Abe, A. Tanaka, N. Homma, T. Yambe, Contactless technique for measuring blood-pressure variability from one region in video plethysmography, Journal of Medical and Biological Engineering 39 (1) (Mar. 2018) 76–85, https://doi.org/10.1007/s40846-018-0388-8.

[71] I.C. Jeong, J. Finkelstein, Introducing contactless blood pressure assessment using a high speed video camera, Journal of Medical Systems 40 (4) (2016) 1–10, https://doi.org/10.1007/s10916-016-0439-z.

[72] J.N. Jensen, M. Hannemose, Camera-based heart rate monitoring, 2014.

[73] M. Elgendi, On the analysis of fingertip photoplethysmogram signals, Current Cardiology Reviews 8 (1) (2012) 14–25.

[74] M. Hosanee, et al., Cuffless single-site photoplethysmography for blood pressure monitoring, Journal of Clinical Medicine 9 (3) (2020) 723.

[75] W. Flügge, Viscoelastic models, in: Viscoelasticity, Springer, Berlin, Heidelberg, 1975, pp. 4–33.

[76] C.L. Garrard, A.M. Weissler, H.T. Dodge, The relationship of alterations in systolic time intervals to ejection fraction in patients with cardiac disease, Circulation 42 (3) (Sep. 1970) 455–462, https://doi.org/10.1161/01.CIR.42.3.455.

[77] A.T. Reisner, D. Xu, K.L. Ryan, V.A. Convertino, C.A. Rickards, R. Mukkamala, Monitoring non-invasive cardiac output and stroke volume during experimental human hypovolaemia and resuscitation, British Journal of Anaesthesia 106 (1) (2011) 23–30, https://doi.org/10.1093/bja/aeq295.

[78] P.S. Addison, Slope transit time (STT): a pulse transit time proxy requiring only a single sig-nal fiducial point, IEEE Transactions on Biomedical Engineering 63 (11) (2016) 2441–2444.

[79] J. Dey, A. Gaurav, V.N. Tiwari, InstaBP: cuff-less blood pressure monitoring on smartphone using single ppg sensor, in: 2018 40th Annual International Conference of the IEEE Engi-neering in Medicine and Biology Society (EMBC), 2018, pp. 5002–5005.

[80] H. Shin, S.D. Min, Feasibility study for the non-invasive blood pressure estimation based on ppg morphology: normotensive subject study, Biomedical Engineering Online 16 (1) (2017) 10.

[81] Y. Liang, Z. Chen, R. Ward, M. Elgendi, Hypertension assessment using photoplethysmog-raphy: a risk stratification approach, Journal of Clinical Medicine 8 (1) (2019) 12.

[82] A. Gaurav, M. Maheedhar, V.N. Tiwari, R. Narayanan, Cuff-less PPG based continuous blood pressure monitoring—a smartphone based approach, in: 2016 38th Annual International Conference of the IEEE Engineering in Medicine and Biology Society (EMBC), 2016, pp. 607–610.

[83] M.H. Chowdhury, et al., Estimating blood pressure from the photoplethysmogram signal and demographic features using machine learning techniques, Sensors 20 (11) (2020) 3127.

[84] P. Yousefian, et al., Data mining investigation of the association between a limb ballistocardiogram and blood pressure, Physiological Measurement 39 (7) (2018) 075009, https://doi.org/10.1088/1361-6579/aacfe1.

[85] R.H. Fagard, K. Pardaens, J.A. Staessen, Relationships of heart rate and heart rate variability with conventional and ambulatory blood pressure in the population, Journal of Hypertension 19 (3) (2001) 389–397.

[86] T. F. of the E. S. of Cardiology and T. N. A. S. of Pacing-Electrophysiology Heart rate variability: standards of measurement, physiological interpretation, and clinical use, Circulation 93 (5) (Mar. 1996) 1043–1065, https://doi.org/10.1161/01.CIR.93.5.1043.

[87] B.B. Kent, J.W. Drane, B.B. Blumenstein, J.W. Manning, A mathematical model to assess changes in the baroreceptor reflex, Cardiology 57 (5) (1972) 295–310.

[88] Y. Chen, S. Shi, Y.-K. Liu, S.-L. Huang, T. Ma, Cuffless blood-pressure estimation method using a heart-rate variability-derived parameter, Physiological Measurement 39 (9) (2018) 95002.

[89] R. Mukkamala, J.-O. Hahn, Initialization of pulse transit time-based blood pressure monitors, in: J. Solà, R. Delgado-Gonzalo (Eds.), The Handbook of Cuffless Blood Pressure Monitoring, Springer International Publishing, 2019, pp. 163–190.

[90] B. Gavish, I.Z. Ben-Dov, M. Bursztyn, Linear relationship between systolic and diastolic blood pressure monitored over 24 h: assessment and correlates, Journal of Hypertension 26 (2) (2008) 199–209.

[91] A.M. Master, R.P. Lasser, The relationship of pulse pressure and diastolic pressure to systolic pressure in healthy subjects, 20–94 years of age, American Heart Journal 70 (2) (1965) 163–171.

[92] Y. Kurylyak, F. Lamonaca, D. Grimaldi, A neural network-based method for continuous blood pressure estimation from a PPG signal, in: 2013 IEEE International Instrumentation and Measurement Technology Conference (I2MTC), 2013, pp. 280–283.

[93] J.C. Ruiz-Rodríguez, et al., Innovative continuous non-invasive cuffless blood pressure monitoring based on photoplethysmography technology, Intensive Care Medicine 39 (9) (2013) 1618–1625.

[94] E. Monte-Moreno, Non-invasive estimate of blood glucose and blood pressure from a photoplethysmograph by means of machine learning techniques, Artificial Intelligence in Medicine 53 (2) (2011) 127–138.

[95] E.A. Hines, G.E. Brown, The cold pressor test for measuring the reactibility of the blood pressure: data concerning 571 normal and hypertensive subjects, American Heart Journal 11 (1) (Jan. 1936) 1–9, https://doi.org/10.1016/S0002-8703(36)90370-8.

[96] M. Al'Absi, S. Bongard, T. Buchanan, G. Pincomb, J. Lincino, W. Lovallo, Cardiovascular and neuroendocrine adjustment to public speaking and mental arithmetic stressors, Psychophysiology 34 (3) (1997) 266–275.

[97] J.S. Petrofsky, A.R. Lind, Aging, isometric strength and endurance, and cardiovascular responses to static effort, Journal of Applied Physiology 38 (1) (1975) 91–95, https://doi.org/10.1152/jappl.1975.38.1.91.

[98] G. Parati, R. Casadei, A. Groppelli, M. di Rienzo, G. Mancia, Comparison of finger and intra-arterial blood pressure monitoring at rest and during laboratory testing, Hypertension 13 (6_pt_1) (1989) 647–655.

[99] M.J. Kenney, D.R. Seals, Postexercise hypotension: key features, mechanisms, and clinical significance, Hypertension 22 (5) (1993) 653–664.

[100] D.B. McCombie, A.T. Reisner, H.H. Asada, Motion based adaptive calibration of pulse transit time measurements to arterial blood pressure for an autonomous, wearable blood pressure monitor, in: 2008 30th Annual International Conference of the IEEE Engineering in Medicine and Biology Society, 2008, pp. 989–992.

[101] S. van Huffel, J. Vandewalle, The Total Least Squares Problem: Computational Aspects and Analysis, SIAM, 1991.

[102] Y. Chen, C. Wen, G. Tao, M. Bi, G. Li, Continuous and noninvasive blood pressure measurement: a novel modeling methodology of the relationship between blood pressure and pulse wave velocity, Annals of Biomedical Engineering 37 (11) (2009) 2222–2233.

[103] H. Gesche, D. Grosskurth, G. Küchler, A. Patzak, Continuous blood pressure measurement by using the pulse transit time: comparison to a cuff-based method, European Journal of Applied Physiology 112 (1) (2012) 309–315.

[104] R. Mukkamala, J.-O.O. Hahn, Toward ubiquitous blood pressure monitoring via pulse transit time: predictions on maximum calibration period and acceptable error limits, IEEE Transactions on Biomedical Engineering 65 (6) (Jun. 2018) 1–11, https://doi.org/10.1109/TBME.2017.2756018.

[105] N. Watanabe, et al., Development and validation of a novel cuff-less blood pressure monitoring device, JACC: Basic to Translational Science 2 (6) (2017) 631–642.

[106] ISO 81060-2:2018—Non-invasive sphygmomanometers – Part 2: Clinical investigation of intermittent automated measurement type, https://www.iso.org/standard/73339.html, 2018. (Accessed 16 July 2020).

[107] A. Trumpp, et al., Relation between pulse pressure and the pulsation strength in camera-based photoplethysmograms, Current Directions in Biomedical Engineering 3 (2) (2017) 489–492.

[108] O.R. Patil, Y. Gao, B. Li, Z. Jin, CamBP: a camera-based, non-contact blood pressure monitor, in: Proceedings of the 2017 ACM International Joint Conference on Pervasive and Ubiquitous Computing and Proceedings of the 2017 ACM International Symposium on Wearable Computers, 2017, pp. 524–529.

[109] R.H. Goudarzi, S.S. Mousavi, M. Charmi, Using imaging photoplethysmography (iPPG) signal for blood pressure estimation, in: 2020 International Conference on Machine Vision and Image Processing (MVIP), 2020, pp. 1–6.

[110] FMS Finapres Medical Systems | the Finapres NOVA has received 510(k) clearance from the US FDA!, http://www.finapres.com. (Accessed 16 July 2020).

[111] X. Liu, C.J. Rodriguez, K. Wang, Prevalence and trends of isolated systolic hypertension among untreated adults in the United States, Journal of the American Society of Hypertension 9 (3) (2015) 197–205.

[112] X. Xing, Z. Ma, M. Zhang, Y. Zhou, W. Dong, M. Song, An unobtrusive and calibration-free blood pressure estimation method using photoplethysmography and biometrics, Scientific Reports 9 (1) (2019) 1–8.

[113] IEEE Standards Association, Draft standard for wearable cuffless blood pressure measuring devices, in: IEEE Standards P1708a/D3, April 2019, 2019, pp. 1708–2014.

Chapter 7

Clinical applications for imaging photoplethysmography

Sebastian Zaunseder[a] and Stefan Rasche[b]

[a]*FH Dortmund, Faculty of Information Engineering, Dortmund, Germany,* [b]*TU Dresden, Faculty of Medicine Carl Gustav Carus, Dortmund, Germany*

Contents

7.1 Overview

Imaging photoplethysmography (iPPG) has raised much interest in recent years, resulting in a large number of published works. Most works address the general feasibility and methodological aspects related to usable hardware, image processing, or signal processing, respectively. Such works typically include a small number of mostly healthy and young participants. Only a limited number of works have addressed the clinical application of iPPG. This chapter has its focus on such applications. To that end, iPPG offers variable opportunities. On the one hand, iPPG can replace current patient-monitoring equipment (to some extent). On the other hand, owing to its ability of superficial and spatial perfusion analyses, iPPG has diagnostic relevance in variable clinical contexts. This chapter details both areas. First, we highlight iPPG's usage for patient monitoring purposes. After that, we summarize further diagnostic applications of iPPG. In either case, we present the clinical view on iPPG and application scenarios rather than focusing on methodological details. Finally, we provide a summary and an outlook regarding future developments and requirements.

7.2 Patient monitoring and risk assessment

7.2.1 Current monitoring—target groups and technology

Patient monitoring denotes the (continuous) measurement of physiologic parameters over time. It aims at instantaneous estimates of a patient's condition, eventually generates alarms, and provides a base to take further actions (interventions, treatments). According to [42], at least three categories of patients need monitoring.

1. Patients with compromised physiological regulation.
2. Stable patients with a condition that could suddenly change to become life threatening.
3. Patients in a critical state.

Generally, patients at intensive care units (ICUs), as well as patients during and after surgery or certain interventions, meet these requirements and are regularly monitored. According to the clinical field, patient conditions, and local practice, the extent of monitoring varies considerably. It ranges from intermittent non-invasive measurements up to invasive continuous measurements. Table 7.1 provides a comprehensive overview of clinically used vital parameters and common ways for their assessment during monitoring.

7.2.2 Patient monitoring by iPPG—measures of relevance

Motivation: Although clinically indispensable, current monitoring devices or equipment, respectively, pose inherent limitations and disadvantages. Invasive monitoring carries a particular risk of infection, vascular lesions, and thrombosis. Even non-invasive contact probes are a significant source of contamination and tissue damage. Contact-based monitoring further impedes the mobility of patients and causes discomfort. Depending on its complexity, the monitoring setup is time consuming and resource intensive. Finally, it can hamper the access to the patient for interventions and nursing. iPPG has significant advantages over current monitoring equipment and can overcome such limitations. At the same time, according to Table 7.1, it can yield multiple parameters that have fundamental importance for monitoring.

Monitoring of heart rate: Heart-rate monitoring is the genuine application of iPPG. It is the most fundamental and most widespread task. Consequently, there are a large number of works related to heart-rate monitoring [3,56]. Most works exploit the variation of light absorption associated with the blood volume pulse. However, the cardiac ejection and the traveling pulse wave also generate movements that can be analyzed to yield a pulse signal (global or local ballistocardiographic effects [56]). Recent developments have ensured that iPPG can

TABLE 7.1 Overview on clinically used vital parameters and common ways for their assessment during monitoring.

Vital parameter	Information content	Typical assessment
General monitoring [8,12]		
Body temperature	Hypothermia, fever	Thermistors, thermocouple
Heart rate	Arrhythmia, abnormal heart rate	Electrocardiography, photoplethysmography
Blood pressure	Global hemodynamics	Sphygmomanometer, arterial catheter
Respiratory rate	Respiratory drive	Impedance pneumography, oronasal flow sensor, respiration belt
Oxygen saturation	Respiratory function, vascular function	Pulse oximetry
Extended general monitoring [8,12]		
Pain	Current discomfort	Multivariate personal assessment
Level of consciousness	Cerebral processes	Multivariate personal assessment (e.g. AVPU, Glasgow coma scale)
Urine output	Renal function, vascular function	Urinary catheter
Hemodynamic monitoring [10,11,20,52]		
Electrical heart activity	Electrical conduction, repolarization, ischemia	Electrocardiography
Cardiac output	Hemodynamic state	Indicator dilution (catheter), CO rebreathing
Central venous pressure	Cardiac function	Venous catheter
Pulmonary arterial pressure	Cardiac function	Pulmonal artery catheter
Heart/valve mechanics	Intracardiac flow and filling, myocardial movement	Cardiac ultrasound
Microcirculation	Local perfusion	Clinical signs, Laser-Doppler/Speckle, dark field imaging, tissue PCO2
Neurological Monitoring [34,37]		
Brain activity	Coma, epileptic seizures	Electroencephalogram
Intracranial pressure	Fluid accumulation	Intracranial catheter
Tissue oxygenation	Cerebral perfusion	NIRS, intracranial catheter

continued on next page

TABLE 7.1 (*continued*)

Vital parameter	Information content	Typical assessment
Respiratory/metabolic monitoring [11]		
Breath carbon dioxide	Lung function	Capnography
Blood gas	Lung function, gas exchange	Laboratory, multispectral PPG
Blood sugar	Hypo-/hyperglycemia	Laboratory, catheter

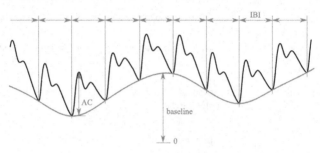

FIGURE 7.1 Exemplary illustration of respiratory effects in a photoplethysmogram. Respiration modulates the interbeat intervals (IBI) via the parasympathetic branch of the autonomous nervous system. Variations in blood filling due to respiration additionally cause baseline variations and beat-to-beat variations in the AC component.

be used for heart-rate monitoring even under difficult recording conditions in view of movements and light variations. Image processing has developed from simple static regions of interest to highly flexible time-dependent regions of interest (ROI). Signal processing's development has led to powerful methods of color-channel combinations, which can efficiently augment pulse even under challenging conditions. Though most works focus on mean heart rate, e.g., the heart rate averaged over some seconds, it is possible to detect single heart beats and reveal heart rate variability, as well [3,56,57].

Monitoring of respiratory rate: Similar to heart rates, monitoring of respiratory rates from videos has been under investigation for a long period. Even the respiration affects videos in different ways. It modulates the periodicity of heart beats, effects the vascular filling, and causes movements (see Fig. 7.1) [14,40]. Either effect, alone or in combination, can be used to derive respiration but the analysis of movements is more common.

Monitoring of oxygen saturation: Clinically, the monitoring of oxygen saturation is the most important application of contact photoplethysmography. It is thus natural to employ iPPG in a similar way. Wieringa et al. published early attempts of camera-based pulse oximetry [53]. Compared to hear-rate and respiratory monitoring, pulse oximetry by iPPG is more complex. On the one hand, RGB cameras do not provide an optimal base to determine the oxygen saturation because the wavelengths are below the isosbestic point. The combination of cameras operating at suitable wavelengths (e.g., red and near-infrared) overcomes this limitation. On the other hand, distortions heavily affect the signal morphology and can easily lead to erroneous readings. The progress in video processing, particularly the identification and detailed tracking of suitable ROIs, allows improved arterial oxygen-saturation (SpO2) monitoring [15]. Besides, simulation studies and advances in signal processing have improved SpO2 assessment techniques and finally proved its applicability [49,50].

However, compared to monitoring heart rate and respiratory rate, monitoring of oxygen saturation is technically much more demanding. Amongst others, the need for at least two wavelengths—preferably one of them in near-infrared range, which shows poor signal quality—and respective illumination concepts, as well as the need for proper calibration, are troublesome.

Monitoring of blood pressure: (Arterial) blood pressure monitoring by cameras has become a focus of iPPG in recent years. Blood pressure and its variations affect variable video characteristics. Morphological signal characteristics, in the simplest case the amplitude of iPPG signals, show variation with respect to the blood pressure or pulse pressure, respectively [36,47]. Other studies relate to the ideas of pulse-wave velocity and exploit temporal characteristics. Such approaches use phase shifts between multiple measurement sites [21] or combine ballistocardiographic signals and blood-volume signals [41]. Driven by the progress in the field of machine learning, even approaches that integrate various signal characteristics, eventually in a black box fashion, are in use to estimate the blood pressure [27].

Further available information: The aforementioned measures are most central to the research activities with respect to iPPG and fundamental to today's routine clinical monitoring (compare to Table 7.1). However, videos provide even more information with relevance for clinical monitoring.

- A couple of works detail methods for visual pain assessment exploiting facial expressions [16]. Even heart-rate variability, which can be derived from videos as well, was associated with pain and awareness [19]. In the same way, iPPG might add relevant information to general monitoring.
- Some works have highlighted the possibility of pulse-wave analysis (PWA) and pulse-wave decomposition from iPPG recordings [13]. Such analysis techniques might yield hemodynamic function beyond the blood pressure and thus be of importance for hemodynamic monitoring. In this context, even the analysis of the jugular venous pulse should be mentioned. The JVP is closely related to central venous pressure and right atrial pressure, respectively, and thus of major interest. Amelard et al. [2] and Gracia-Lopez et al. both show that the camera-based acquisition of JVP and detection of its fiducial points are feasible. However, future studies will have to investigate its usability.
- Rasche et al. show that iPPG can replace, to some extent, Laser Speckle Contrast Analysis (LASCA) and thus allow hemodynamic analysis with respect to the microcirculation [35].

- Monitoring of respiratory rate is common. Interestingly, current works try to reveal information about the respiration other than respiratory rate. For example, Liu et al. detail a method that aims to capture spirometric information like forced expiratory volume and peak expiratory flow [26], which has importance for general respiratory monitoring.
- Recently, Pilz et al. show the relationship of iPPG to brain waves. This interaction might gain importance for neurological monitoring [33].
- Nishidate et al. detail a method to access tissue metabolic measures (melanin, oxygenated blood, and deoxygenated blood) [32]. The approach yields information beyond pulse oximetry. By using multispectral camera systems, similar approaches might be expandable to monitor multiple substances and contribute to contactless metabolic monitoring.

Certainly, these approaches are not readily applicable. Particularly, the high number of confounding factors is likely to affect the results under real-world conditions. However, they indicate, according to Table 7.1, the high potential of iPPG for patient monitoring.

General validation: As discussed before, classical monitoring has some limitations. The consideration on accessible information underlines the high potential of cameras for monitoring purposes. Thereby, cameras can be used alone or combined with conventional sensors or other non-contact modalities. A striking example is a combination with infrared cameras, i.e., thermography [40]. By this combination, one can acquire the five most important vital parameters for monitoring without any contact. However, a couple of aspects have to be investigated before camera-based patient monitoring becomes a reality. Generally, many studies in the field focus on the principal feasibility. Elaborated tests according to guidelines or normatives are not common. Even worse, evidence on the effectiveness with the critically ill is widely lacking. Few studies considering patients, e.g., show heart-rate assessment to be possible in cardiovascular impaired patients and during arrhythmia like atrial fibrillation. However, other arrythmic events, like ventricular arrhythmia, are likely to heavily affect iPPG and probably hinder a proper analysis. Obviously, much more clinical evidence is needed before iPPG can be part of routine monitoring or replace it (in certain areas).

7.2.3 Patient monitoring by iPPG—realistic usage scenarios

iPPG features highly relevant information for clinical monitoring. It allows a much more convenient assessment, but at the expense of information content and reliability. The management of the critically ill requires highly accurate physiological measurements. In such situations, some inconvenience is preferred for more reliable and informative signals. Consequently, in our opinion, iPPG is not likely to replace current equipment to monitor standard vital parameters of the critically ill on a larger scale (e.g., if a patient is at high risk for cardiac arrhythmia, the electrocardiogram is indispensable because it provides much more information than the PPG ever could). However, there are specific clinical applications related to patient monitoring where iPPG might gain importance.

Neonatal monitoring is one example. Particularly, premature neonates are at risk of severe complications (including sudden infant death). Monitoring of heart rate and respiratory in neonates is therefore common but the available equipment is troublesome. Today's contact sensors are difficult to apply, generate stress to neonates, and can eventually cause lesions or ischemia, respectively. Various works thus have addressed non-contact monitoring of neonates and preterm infants, respectively, by various techniques. Owing to their wide possibilities, cameras are an interesting choice [1,4,7]. Besides monitoring in clinics, ambulatory neonatal monitoring might be an interesting application for the future.

Apart from neonatal monitoring, situations in intensive care are rare, where monitoring standard vital parameters is beneficial but contact-based sensors cannot be applied for objective medical reasons. However, the ease of use and higher convenience of a non-contact monitoring to capture usual vital signals should not be underestimated. Together with a ubiquitous availability of the technically easy to implement iPPG, it is conceivable to monitor the vital signs of almost every patient in a hospital. Based on current data, a significant reduction of critical incidents and even cardiac arrests in hospitals could be assumed thereof. According to an expert opinion, two of three cardiac arrests in hospitals are avoidable [17]. The most crucial points to prevent these incidents are an early detection of deterioration and a sufficient sensitivity of diagnostic tools. The delayed diagnosis of deterioration is the most common reason for a cardiac arrest, though abnormal vitals are already present hours before the event [17]. Several tools have been implemented to aid identifying patients at risk (e.g., Modified early warning score MEWS, Cardiac arrest risk triage CART; see Table 7.2 for an example). However, contemporary scores prevent barely more than every other critical incident [6]. Key to a higher success rate is the consistent recording of the score items. Respiratory rate, heart rate, and diastolic blood pressure were identified as the most predictive parameters in general-ward patients to predict cardiac arrest or the transfer to the intensive care unit [6,9]. Since (at least) the first two of these are easily derived by iPPG, an "always-on" monitoring could improve the feasibility of risk scores and eventually the safety of patients in hospitals. Further available information, e.g., derived by the analysis of facial expressions, might feature the usage of videos for risk assessment in clinical settings. A closely related area is (intermittent) monitoring in nursing homes. Today, automated monitoring is not common. However, as in hospitals, a user-friendly technique such as iPPG can be used (at least) for intermittent measurements and help to recognize deterioration early. Similarily, iPPG offers wide opportunities for home care in telemedical scenarios. Even here, a high level of availability, easy applicability, and versatile information make the use of cameras a promising approach.

TABLE 7.2 Details on the modified early warning score MEWS. The basic idea of a clinical risk score consists of summing up univariate scores in a multivariate measure. For MEWS, an increasing value is associated with an increased risk of admission to an intensive care unit or death. Note that used parameters and thresholds vary and risk scores, even MEWS, are not without controversy. However, the example serves to demonstrate how simple parameters and their (coarse) assessment can contribute to risk assessment.

Score	3	2	1	0	1	2	3
RR	>35	31–35	21–30	9–20			<7
SpO2	<85	85–89	90–92	>92			
T		>38.9	38–38.9	36–37.9	35–35.9	34–34.9	<34
SBP		>=200		101–199	81–100	71–80	<=70
HR	>129	110–129	100–109	50–99	40–49	30–39	<30
AVPU				A	V	P	U

RR – Respiratory rate in breaths per minute, SpO2 – Arterial oxygen saturation in %
T – Body temperature in °C, SBP – Systolic blood pressure, HR – Heart rate in beats per minute
AVPU – Alert, Verbal, Unresponsive, Pain

The aforementioned applications focus on common vital signals like heart rate and respiratory rate. These specific applications impose special requirements regarding a user-friendly signal acquisition rendering iPPG's usage advantageous compared to using contact-based sensors. Clinical routine monitoring, in turn, is likely to adhere to contact-based sensors because they have a higher reliability. iPPG still might find application in routine monitoring because it features complementary information: iPPG allows a specific assessment of skin microcirculation. Skin microcirculation is known to carry prognostic information and can be used to guide therapy. Monitoring skin microcirculation is not common in today's routine monitoring. Alternative techniques like laser speckle provide information about the skin microcirculation, as well, but they have specific limitations, e.g., regarding their application. In this regard, iPPG has high potential to find application in routine monitoring and add information about the microcirculation in the future (see Sects. 7.3.2 and 7.3.3 for some more detailed information on the diagnostic potential). However, more research including prospective patient studies are required to prove iPPG's clinical value in this regard.

7.3 Application beyond patient monitoring

7.3.1 Sleep medicine

One possible application, which is closely related to the aforementioned principles of patient monitoring, is sleep analysis. Objective sleep analysis relies on the polysomnogram or the polygram. Both techniques commonly use sensors attached to the skin (see Table 7.3). As the measurement equipment interferes with sleep, sleep medicine is an interesting application for non-contact measurement techniques [58]. The most important information of the polysomnogram relates to brain, eye, and muscle activities, which is difficult to gain from videos (though it even might be impossible, see e.g., [33]). The polygram, in turn, essentially features information on the body position, pulse, and respiration. Such measures are accessible from videos making it an interesting application using cameras. Some studies have shown iPPG's applicability during sleep to derive the heart rate and respiratory rate and, by multiwavelength recordings, oxygen saturation [24,51]. Importantly, only near-infrared wavelengths can be used, which renders sleep monitoring difficult owing to a reduced signal quality in the near-infrared range. A combination with an infrared camera can help to improve the results [40]. Future studies will have to show the benefit of non-contact sleep monitoring. However, regarding its application during sleep, the camera-based approach has to compete with alternative solutions, e.g., user-friendly wearables or pressure-sensitive foils, which can be placed under the mattress [58]. A certain advantage is iPPG's versatile applicability concerning the accessible information (beyond heart rate, respiratory rate, and oxygen saturation, which are fundamental to sleep medicine, recent works showed iPPG's correlation to blood pressure and via the autonomous function to EEG parameters; additionally, videos allow movement analysis, which is highly relevant for sleep analysis, as well).

7.3.2 Local perfusion analysis

iPPG selectively measures the peripheral cutaneous circulation owing to the circumscribed propagation of light through the skin. The two-dimensional data acquisition provides maps that allow the analysis of spatial perfusion properties (see Fig. 7.2).

Early investigations on iPPG have already focused on the analysis of skin lesions like port-wine stain and allergic reactions, respectively [5,18]. Such works proved a characteristic behavior of iPPG measures, most importantly the strength of perfusion determined by pulsation amplitude. Thatcher et al. showed that burned tissue can be distinguished from healthy tissue and wound bed tissue in a porcine model [46]. Recently, Mamontov et al. showed variations in iPPG measures owing to systemic scleroderma [28]. Kukel et al. [23] show iPPG to capture local effects of cardiac surgery on chest-wall perfusion and enable assessment of recovery. Though not directly related to lesions, but even exploiting local effects on the microcirculation, Rubins et al. propose to use iPPG to monitor the effect of local anesthesia by iPPG [38] and relate local iPPG characteristics to neuropathy [39].

Such applications hold strong potential for clinical usage of iPPG. While absolute values are and will be difficult to use, in particular, the analysis of the local pulsation compared to reference sites/surrounding tissue or its temporal changes

TABLE 7.3 Signals that are typically recorded during polysomnography and polygraphy, together with common techniques for their assessment. Note that varying setups can be used (particularly in polygraphy). Information according to [58].

Signal/contained information	Recording technique	Typically done by Polysomnography	Typically done by Polygraphy
Brain function	Electroencephalogram	Yes, 3–6 channels	No
Eye movements	Electrooculogram	Yes, 2 channels	No
Chin muscle tone	Electromyogram	Yes, 2–3 channels	No
Leg muscle tone	Electromyogram	Yes, 2 channels	optional
Behavior	Infrared video	Yes, 1 channel	No
Body movements	3D accelerometer	Yes, 1 channel	optional
Body position	3D accelerometer	Yes, 3 channel*	Yes, 3 channel*
Heart function	Electrocardiogram	Yes, 1–3 channel	optional
Pulse rate	Finger photoplethysmogram	Yes, 1 channel	Yes, 1 channel
Oxygen saturation	Finger oxymetry	Yes, 1 channel	Yes, 1 channel
Oronasal air flow	Pressure transducers/thermocouples/thermistors	Yes, 1–2 channels	Yes, 1–2 channels
Thoracic/abdominal respiratory effort	Piezoelectric/inductive plethysmographic belts	Yes, 2 channels	Yes, 2 channels
Snoring	Microphone/from air flow sensor	Yes, 1 channel	Yes, 1 channel

* A single sensor provides three-3 dimensional information, i.e., three channels.

should yield meaningful results. Possible applications include the assessment of allergic reactions and all kinds of skin lesions like wounds and burns. In today's practice, the assessment of such phenomena often invokes subjective procedures and measures. Using iPPG might help to make the assessments more comparable.

However, even here, other techniques, particularly other optical ones, are applicable, e.g., coherence tomography and hyperspectral imaging, just to name two of them [29,45]. Though they are not the most sensitive technologies, the striking advantages of simplicity and ease of use might help iPPG to find wider application and become part of routine care in said applications. In this regard, widening the focus of assessing the perfusion to a more detailed metabolic analysis, by combining multiple wavelengths, should be highly beneficial.

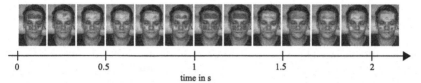

0   0.5   1   1.5   2

time in s

FIGURE 7.2 Exemplary visualization of the perfusion in selected regions of interest over time. Red (mid gray in print version) color indicates high blood volume and blue (dark gray in print version).color indicates low blood volume. Besides a cyclic variation due to the heart rate, one can observe a spatio-temporal pattern of the perfusion. The temporal evolution of blood filling makes possible mapping the amplitude and phase, respectively, of the perfusion. Analysis of such maps can direct at absolute values, spatial variability, or its symmetry.

7.3.3 Skin microcirculation as diagnostic proxy

The previous section addressed situations in which local factors locally affect the perfusion. iPPG can capture the perfusion and is, consequently, indicative for the underlying local factor. Even systemic or central processes and pathologies affect the dermal perfusion (typically via the autonomous nervous system). Microcirculatory features have been recognized for their clinical value, particularly in the critically ill, but are not covered yet by contemporary monitoring equipment. The dermal microcirculation and its assessment by iPPG thus can serve as diagnostic proxy, which opens up another field of clinical applications for iPPG.

A couple of works have shown that iPPG is generally sensitive to distant stimulation and autonomous activity [48,54]. Zaproudina et al. characterize migraine by iPPG and propose it as a possible biomarker [55]. They report a lateralization of blood perfusion and asynchronous pulsation due to autonomic dysfunction. Future applications might exploit the significant clinical merits of peripheral perfusion in the assessment of organ perfusion, organ failure, and shock conditions [25].

These approaches are still in their beginning, but they might gain importance in future clinical care. Using iPPG may allow a more objective assessment of skin perfusion than contemporary methods, while keeping the effort manageable. To exploit the full potential of iPPG in this regard, novel metrics to analyze two-dimensional perfusion pattern might be beneficial.

7.3.4 Further applications

Beyond the aforementioned applications, there are further clinical applications that bear some potential for iPPG's future clinical use. Venous occlusion plethysmography is one of them. The technique typically uses contact sensors to study the peripheral perfusion after mechanical occlusion of a measurement site (forearm or leg). Venous occlusion plethysmography allows variable statements on the hemodynamic and vascular state, particularly those also related to the venous system. Recent works show that iPPG can capture the reaction to occlusion and thus replace contact measurement equipment [22,30]. iPPG's ease of use and the advantage of non-interference with vascular refilling are advantages compared to common contact techniques. By suitable analysis of the pulse-wave shape, iPPG can yield even more information on the cardiovascular state with clinical relevance, e.g., the aging index [13,43]. iPPG and the usage of cameras, in general, further might gain importance in pandemic screening at clinic admission. Particularly in combination with infrared cameras, iPPG yields valuable information. Negishi et al. employed such a system for influenza screening [31]. Against the current situation of a pandemic, a wider usage seems to be possible. A more convenient spirometric test, eventually more often repeated, [26] and generating triggers for magnet resonance imaging are other potential clinical applications [44]. However, clinical use would require further improvements and validation.

7.4 Summary and outlook

iPPG is a promising diagnostic approach that addresses various clinical needs. Its ease of use and its information content make it interesting for multiple clinical applications. Essentially, four fields of application seem to be feasible:

1. iPPG might replace current (contact-based) patient-monitoring equipment where monitoring is currently done but particularly troublesome, e.g., neonatal monitoring.
2. iPPG might open up patient monitoring to areas in which the limitations of current equipment have hampered a wider monitoring so far, e.g., monitoring in the general ward or geriatric monitoring.
3. iPPG might replace current (contact-based) diagnostic procedures where non-contact sensing has advantages over contact sensors (more convenient, more objective), e.g., sleep medicine or skin perfusion as a diagnostic proxy.
4. iPPG might open up novel diagnostic procedures, e.g., an objective assessment of scleroderma or wound healing.

Current limitations regarding iPPG's clinical use mostly relate to the very restricted number of clinical trials so far. The questions to answer in this regard are twofold. First, controlled recording conditions and selected patient groups might gloss over the real performance of iPPG. More realistic environments, including artifacts and noise from multiple sources, and multi-morbidity in patients are likely to affect iPPG's performance. Also skin color is an important factor that must be considered properly. Maybe such factors are not an obstacle, but this has to be proven. Secondly, even if iPPG is applicable under real-world conditions, its clinical benefit has to be confirmed. This typically requires larger (epidemiological) studies.

Additionally, the usage of videos raises privacy issues (videos reveal a lot of information). It further goes along with high demands on data storage, particularly as some applications are known to suffer heavily from video compression. At the same time, depending on the specific application, iPPG not only has to be validated, but it competes with alternative innovative techniques, e.g., other non-contact measurement techniques like radar and novel imaging modalities like optical coherence tomography. The aforementioned limitations are serious but (in part) common to novel diagnostic techniques and must not hinder iPPG's clinical use. In fact, we believe that, due to its versatility in combination with its ease of use, iPPG will enter clinical applications. In this regard, future developments will foster additional applications. In particular, the combination of wavelengths and more sophisticated two-dimensional analysis approaches hold high potential for future refinements and extended use.

Acknowledgments

We want to thank Alexander Woyczyk and Vincent Fleischhauer for their support regarding literature review. We further want to thank the Deutsche Forschungsgemeinschaft (DFG, German Research Foundation) for funding our work on iPPG (project 401786308).

References

[1] Lonneke A.M. Aarts, et al., Non-contact heart rate monitoring utilizing camera photoplethysmography in the neonatal intensive care unit? A pilot study, Early Human Development 89 (12) (Dec. 2013) 943–9484, https://doi.org/10.1016/j.earlhumdev.2013.09.016, issn: 03783782.

[2] Robert Amelard, et al., Non-contact hemodynamic imaging reveals the jugular venous pulse waveform, Scientific Reports 7 (Jan. 2017) 40150, https://doi.org/10.1038/srep40150, issn: 2045-2322 arXiv:1604.05213.

[3] Christoph Hoog Antink, et al., A broader look: camera-based vital sign estimation across the spectrum, Yearbook of Medical Informatics 28 (1) (2019) 102–114, https://doi.org/10.1055/s-0039-1677914, issn: 23640502.

[4] Nikolai Blanik, et al., Remote vital parameter monitoring in neonatology – robust, unobtrusive heart rate detection in a realistic clinical scenario, Biomedizinische Technik. Biomedical Engineering 61 (6) (2016) 631–643, https://doi.org/10.1515/bmt-2016-0025, issn: 1862278X.

[5] C.R. Blazeket, et al., Assessment of allergic skin reactions and their inhibition by antihistamines using photoplethysmography imaging (PPGI), Journal of Allergy and Clinical Immunology 117 (2) (Feb. 2006) S226, https://doi.org/10.1016/j.jaci.2005.12.894.

[6] Matthew M. Churpek, et al., Derivation of a cardiac arrest prediction model using ward vital signs*, Critical Care Medicine 40 (7) (July 2012) 2102–2108, https://doi.org/10.1097/CCM.0b013e318250aa5a, issn: 1530-0293.

[7] Juan Carlos Cobos-Torres, Mohamed Abderrahim, José Martinez Orgado, Non-contact, simple neonatal monitoring by photoplethysmography, Sensors (Switzerland) 18 (12) (2018) 1–14, https://doi.org/10.3390/s18124362, issn: 14248220.

[8] Cristiano André da Costa, et al., Internet of health things: toward intelligent vital signs monitoring in hospital wards, Artificial Intelligence in Medicine 89 (May) (July 2018) 61–69, https://doi.org/10.1016/j.artmed.2018.05.005, issn: 09333657.

[9] Brian H. Cuthbertson, et al., Can physiological variables and early warning scoring systems allow early recognition of the deteriorating surgical patient?, Critical Care Medicine 35 (2) (Feb. 2007) 402–409, https://doi.org/10.1097/01.CCM.0000254826.10520.87, issn: 0090-3493.

[10] Philipe Franco Do Amaral Tafner, et al., Recent advances in bedside microcirculation assessment in critically ill patients, Revista Brasileira de Terapia Intensiva 29 (2) (2017) 238–247, https://doi.org/10.5935/0103-507X.20170033, issn: 19824335.

[11] Jesse M. Ehrenfeld, Maxime Cannesson (Eds.), Monitoring Technologies in Acute Care Environments, Springer, New York, 2014, isbn: 978-1-4614-8556-8.

[12] Malcolm Elliott, Alysia Coventry, Critical care: the eight vital signs of patient monitoring, British Journal of Nursing 21 (10) (May 2012) 621–625, https://doi.org/10.12968/bjon.2012.21.10.621, issn: 0966-0461.

[13] Vincent Fleischhauer, et al., Pulse decomposition analysis in photoplethysmography imaging, Physiological Measurement 41 (9) (Oct. 2020) 095009, https://doi.org/10.1088/1361-6579/abb005, issn: 1361-6579.

[14] Mark van Gastel, Sander Stuijk, Gerard de Haan, Robust respiration detection from remote photoplethysmography, Biomedical Optics Express 7 (12) (Dec. 2016) 4941–4957, https://doi.org/10.1364/BOE.7.004941, issn: 2156-7085.

[15] Alessandro R. Guazzi, et al., Non-contact measurement of oxygen saturation with an RGB camera, Biomedical Optics Express 6 (9) (Sept. 2015) 3320, https://doi.org/10.1364/BOE.6.003320, issn: 2156-7085.

[16] T. Hassan, D. Seus, J. Wollenberg, K. Weitz, M. Kunz, S. Lautenbacher, J.-U. Garbas, U. Schmid, Automatic detection of pain from facial expressions: a survey, IEEE Transactions on Pattern Analysis and Machine Intelligence 43 (6) (Jun. 2021) 1815–1831, https://doi.org/10.1109/TPAMI.2019.2958341.

[17] Timothy J. Hodgetts, et al., Incidence, location and reasons for avoidable in-hospital cardiac arrest in a district general hospital, Resuscitation 54 (2) (Aug. 2002) 115–123, https://doi.org/10.1016/s0300-9572(02)00098-9, issn: 0300–9572.

[18] Markus Huelsbusch, Vladimir Blazek, Contactless mapping of rhythmical phenomena in tissue perfusion using PPGI, in: Anne V. Clough, Chin-Tu Chen (Eds.), Proceedings of SPIE, vol. 4683, Apr. 2002, p. 110.

[19] R. Huhle, et al., Effects of awareness and nociception on heart rate variability during general anaesthesia, Physiological Measurement 33 (2) (Feb. 2012) 207–217, https://doi.org/10.1088/0967-3334/33/2/207, issn: 1361-6579.

[20] Johan Huygh, et al., Hemodynamic monitoring in the critically ill: an overview of current cardiac output monitoring methods, F1000Research 5 (2016) 2855, https://doi.org/10.12688/f1000research.8991.1, issn: 2046-1402.

[21] In Cheol Jeong, Joseph Finkelstein, Introducing contactless blood pressure assessment using a high speed video camera, Journal of Medical Systems 40 (4) (Apr. 2016) 77, https://doi.org/10.1007/s10916-016-0439-z, issn: 1573-689X.

[22] Alexei A. Kamshilin, Valeriy V. Zaytsev, Oleg V. Mamontov, Novel contactless approach for assessment of venous occlusion plethysmography by video recordings at the green illumination, Scientific Reports 7 (1) (2017) 464, https://doi.org/10.1038/s41598-017-00552-7, issn: 2045-2322.

[23] Kukel Imre, et al., Contact-free optical assessment of changes in the chest wall perfusion after coronary artery bypass grafting by imaging photoplethysmography, Applied Sciences (ISSN 2076-3417) 10 (18) (Sept. 2020) 6537, https://doi.org/10.3390/app10186537].

[24] Michael H. Li, Azadeh Yadollahi, Babak Taati, Noncontact vision-based cardiopulmonary monitoring in different sleeping positions, IEEE Journal of Biomedical and Health Informatics (ISSN 2168-2194) 21 (5) (Sept. 2017) 1367–1375, https://doi.org/10.1109/JBHI.2016.2567298.

[25] A. Lima, J. Takala, Clinical significance of monitoring perfusion in non-vital organs, Intensive Care Medicine 40 (7) (Jul. 2014) 1052–1054, https://doi.org/10.1007/s00134-014-3345-1.

[26] Chenbin Liu, et al., Noncontact spirometry with a webcam, Journal of Biomedical Optics (ISSN 1560-2281) 22 (5) (May 2017) 57002, https://doi.org/10.1117/1.JBO.22.5.057002.

[27] Hong Luo, et al., Smartphone-based blood pressure measurement using transdermal optical imaging technology, Circulation: Cardiovascular Imaging 12 (8) (2019) 1–10, https://doi.org/10.1161/CIRCIMAGING.119.008857, issn: 19420080.

[28] Oleg V. Mamontov, et al., Novel instrumental markers of proximal scleroderma provided by imaging photoplethysmography, Physiological Measurement 41 (4) (2020) 044004, https://doi.org/10.1088/1361-6579/ab807c, issn: 13616579.

[29] Jörg Marotz, et al., Extended perfusion parameter estimation from hyper-spectral imaging data for bedside diagnostic in medicine, Molecules 24 (22) (2019) 1–16, https://doi.org/10.3390/molecules24224164, issn: 14203049.

[30] Kazuya Nakano, et al., Non-contact imaging of venous compliance in humans using an RGB camera, Optical Review 22 (2) (2015) 335–341, https://doi.org/10.1007/s10043-015-0041-5, issn: 13499432.

[31] Toshiaki Negishi, et al., Contactless vital signs measurement system using RGB-thermal image sensors and its clinical screening test on patients with seasonal influenza, Sensors (Switzerland) 20 (8) (2020), https://doi.org/10.3390/s20082171, issn: 14248220.

[32] Nishidate Izumi, et al., Noninvasive imaging of human skin hemodynamics using a digital red-green-blue camera, Journal of Biomedical Optics 16 (8) (2011) 086012, https://doi.org/10.1117/1.3613929, issn: 10833668.

[33] Christian S. Pilz, et al., Predicting brainwaves from face videos, in: International Workshop and Challenge on Computer Vision for Physiological Measurement, June. 2020.

[34] James Ralph, Naginder Singh, Advanced neurological monitoring, Surgery (United Kingdom) 34 (2) (2016) 94–96, https://doi.org/10.1016/j.mpsur.2015.11.006, issn: 18781764.

[35] Stefan Rasche, et al., Association of remote imaging photoplethysmography and cutaneous perfusion in volunteers, Scientific Reports (ISSN 2045-2322) 10 (1) (Dec. 2020) 16464, https://doi.org/10.1038/s41598-020-73531-0.

[36] Stefan Rasche, et al., Remote photoplethysmographic assessment of the peripheral circulation in critical care patients recovering from cardiac surgery, Shock (Augusta, Ga.) (ISSN 1540-0514) 52 (2) (Aug. 2019) 174–182, https://doi.org/10.1097/SHK.0000000000001249.

[37] Anette Ristic, Raoul Sutter, LuziusA Steiner, Current neuromonitoring techniques in critical care, Journal of Neuroanaesthesiology and Critical Care (ISSN 2348-0548) 02 (02) (Aug. 2015) 97–103, https://doi.org/10.4103/2348-0548.154234.

[38] Uldis Rubins, Aleksejs Miscuks, Marta Lange, Simple and convenient remote photoplethysmography system for monitoring regional anesthesia effectiveness, in: EMBEC & NBC 2017, Tampere, Finland, 2018, pp. 378–381.

[39] U. Rubins, Z. Marcinkevics, I. Logina, A. Grabovskis, E. Kviesis-Kipge, Imaging photoplethysmography for assessment of chronic pain patients, in: Optical Diagnostics and Sensing XIX: Toward Point-of-Care Diagnostics, vol. 1088508, February 2019, p. 8, https://doi.org/10.1117/12.2508393, 2019.

[40] Gaetano Scebba, Giulia Da Poian, Walter Karlen, Multispectral video fusion for non-contact monitoring of respiratory rate and apnea, IEEE Transactions on Biomedical Engineering 68 (1) (Jan. 2021) 350–359, https://doi.org/10.1109/TBME.2020.2993649.

[41] Dangdang Shao, et al., Simultaneous monitoring of ballistocardiogram and photoplethysmogram using a camera, IEEE Transactions on Biomedical Engineering (ISSN 1558-2531) 64 (5) (2017) 1003–1010, https://doi.org/10.1109/TBME.2016.2585109.

[42] Edward H. Shortliffe, James J. Cimino (Eds.), Biomedical Informatics, Springer, London, 2014, isbn: 978-1-4471-4473-1.

[43] Michele Sorelli, et al., Pulse decomposition analysis in camera-based photoplethysmography, in: 2019 41st Annual International Conference of the IEEE Engineering in Medicine and Biology Society (EMBC), IEEE, ISBN 978-1-5386-1311-5, July 2019, pp. 3179–3182.

[44] Nicolai Spicher, et al., Initial evaluation of prospective cardiac triggering using photoplethysmography signals recorded with a video camera compared to pulse oximetry and electrocardiography at 7T MRI, Biomedical Engineering Online (ISSN 1475-925X) 15 (1) (Nov. 2016) 126, https://doi.org/10.1186/s12938-016-0245-3.

[45] Jeffrey E. Thatcher, et al., Imaging techniques for clinical burn assessment with a focus on multispectral imaging, Advances in Wound Care 5 (8) (2016) 360–3782, https://doi.org/10.1089/wound.2015.0684, issn: 21621934.

[46] Jeffrey E. Thatcher, et al., Multispectral and photoplethysmography optical imaging techniques identify important tissue characteristics in an animal model of tangential burn excision, Journal of Burn Care & Research (ISSN 1559-047X) 37 (1) (2016) 38–52, https://doi.org/10.1097/BCR.0000000000000317.

[47] Alexander Trumpp, et al., Relation between pulse pressure and the pulsation strength in camera-based photoplethysmograms, Current Directions in Biomedical Engineering (ISSN 2364-5504) 3 (2) (2017) 489–492, https://doi.org/10.1515/cdbme-2017-0184.

[48] Alexander Trumpp, et al., Vasomotor assessment by camera-based photoplethysmography, Current Directions in Biomedical Engineering (ISSN 2364-5504) 2 (1) (Jan. 2016) 199–202, https://doi.org/10.1515/cdbme-2016-0045.

[49] Mark Van Gastel, Sander Stuijk, Gerard De Haan, Camera-based pulse-oximetry – validated risks and opportunities from theoretical analysis, Biomedical Optics Express (ISSN 2156-7085) 9 (1) (Jan. 2018) 102–119, https://doi.org/10.1364/BOE.9.000102.

[50] Wim Verkruysse, et al., Calibration of contactless pulse oximetry, Anesthesia and Analgesia (ISSN 1526-7598) 124 (1) (Jan. 2017) 136–145, https://doi.org/10.1213/ANE.0000000000001381.

[51] Tom Vogels, et al., Fully-automatic camera-based pulse-oximetry during sleep, in: 2018 IEEE/CVF Conference on Computer Vision and Pattern Recognition Workshops (CVPRW), IEEE, ISBN 978-1-5386-6100-0, June 2018, pp. 1430–14308.

[52] X. Watson, M. Cecconi, Haemodynamic monitoring in the perioperative period: the past, the present and the future, Anaesthesia 72 (Jan. 2017) 7–15, https://doi.org/10.1111/anae.13737, issn: 00032409.

[53] F.P. Wieringa, F. Mastik, A.F.W. van der Steen, Contactless multiple wavelength photoplethysmographic imaging: a first step toward "SpO2 camera" technology, Annals of Biomedical Engineering (ISSN 0090-6964) 33 (8) (Aug. 2005) 1034–1041, https://doi.org/10.1007/s10439-005-5763-2.

[54] Alexander Woyczyk, Stefan Rasche, Sebastian Zaunseder, Impact of sympathetic activation in imaging photoplethysmography, in: 2019 IEEE/CVF International Conference on Computer Vision Workshop (IC-CVW), IEEE, ISBN 978-1-7281-5023-9, Oct. 2019, pp. 1697–1705.

[55] Nina Zaproudina, et al., Asynchronicity of facial blood perfusion in migraine, PLoS ONE (ISSN 1932-6203) 8 (12) (Jan. 2013) e80189, https://doi.org/10.1371/journal.pone.0080189.

[56] Sebastian Zaunseder, et al., Cardiovascular assessment by imaging photoplethysmography a review, Biomedical Engineering/Biomedizinische Technik (ISSN 1862-278X) 63 (5) (Oct. 2018) 617–634, https://doi.org/10.1515/bmt-2017-0119.

[57] Sebastian Zaunseder, et al., Heart beat detection and analysis from videos, in: 2014 IEEE 34th International Scientific Conference on Electronics and Nanotechnology (ELNANO), IEEE, ISBN 978-14799-4580-1, Apr. 2014, pp. 286–290.

[58] Sebastian Zaunseder, et al., Unobtrusive acquisition of cardiorespiratory signals, Somnologie 21 (2) (June 2017) 93–100, https://doi.org/10.1007/s11818-017-0112-x, issn: 1432–9123.

Chapter 8

Applications of camera-based physiological measurement beyond healthcare

Daniel McDuff

Microsoft Research, Redmond, WA, United States

Contents

8.1 The evolution from the lab to the real world

It can take many years for new technologies to transition from laboratory discoveries to mature applications. The field of computer vision (CV) is six decades old, yet it is only in the past ten years that CV has started to be used in many commercial systems. Research into camera-based vital signs monitoring began much more recently, within the past 20 years, and therefore we should expect that more time might be required for the technology to mature. Physiological measurement via cameras involves capturing and processing subtle changes in the appearance of the body, in many cases imperceptible to the unaided human eye, to recover underlying signals about how the body is functioning. Early digital imaging hardware often had insufficient resolution or too much sensor noise to recover many of these changes. Thanks to large investments in the development of imaging hardware, in large part due to the popularity of consumer photography and "camera phones", hardware is no longer a significant limitation for many applications. Still, given the relatively recent emergence of this field, we are only beginning to see the deployment of real-world applications, and there is a long way to go to realize their full potential. When discussing current and future applications of camera-based physiological measurement, it is helpful first to consider how far the technology has come and what the limitations and remaining challenges are.

In this chapter, I will primarily consider vital signs measurement via motion (e.g., respiration, ballistocardiography) and color or absorption changes (e.g., photoplethysmography) as captured by digital cameras covering the visible and infrared ranges and, to a lesser extent, thermal cameras. There is certainly work beyond this scope using novel devices that can image further beneath the skin and capture additional metrics. However, I will not have the space here to consider all imaging work and will focus in particular of measurement using ubiquitous sensors (e.g., webcams) and the types of applications that these enable.

In 2000, Wu et al. [56] presented preliminary results from a non-contact photoplethysmographic imager demonstrating that a charge-coupled device (CCD) camera could be used to recover the the cardiac pulse. Following this, Takano and Ohta [50] published a study using a grayscale CCD camera to non-invasively measure respiration and pulse rates. At a similar time, Garbey et al. [17] showed that a thermal camera could also be used to detect a cardiac pulse signal. An important step in these early methods was spatially averaging pixel information to reduce the camera quantization error. From these measurements, it was possible to capture the photoplethysmogram (PPG). Leveraging a color imaging device, it was observed that signals in the green channel often provided the highest signal-to-noise ratio (SNR) of the typical RGB channels [52] with a maximum SNR at approximately 570 nm [7]. Subsequent work extended these findings to show that heart rate variability [7] and subtle PPG waveform dynamics, such as the timing diastolic peak/infections and the diachrotic notch, could also be detected [28]. Using PPG signals [41], via respiratory sinus arrhythmia, it is also possible to capture respiration. However, in real-world applications, simply spatially averaging individual color channels or aggregating optical flow across a region of interest will lead to signals easily corrupted by sources other than the underlying physiological process. They are highly susceptible to noise from motion, ambient lighting, and sensor artifacts. Addressing practical limitations such as these has been one of the main focuses of the research community in recent years. Most early algorithms proposed unsupervised signal separation techniques [14,22,42,55]. More recently, the supervised learning has out-paced these methods on benchmark datasets [12,26,59].

The ballistocardiography (BCG) signal can be measured using optical flow to capture subtle motions of the body resulting from the cardiac pulse [6,48]. The BCG provides rich and complementary information to the PPG signal. However, camera-based BCG measurement methods are highly sensitive to other body motions, making it difficult to design practical applications. It is perhaps more reasonable to think that BCG measurement from inertial measurement units (IMU) would be a more practical companion to camera-based PPG measurement [18]. Combining PPG and BCG is attractive for calculating pulse transit time (PTT) [47] and deriving measurements that correlate with blood pressure.

While motions due to BCG are very subtle, respiration or breathing motions are often much more obvious in videos. A number of approaches to measuring respiration motions from video have been proposed. Tarassenko et al. [51] used a similar pipeline to PPG measurement, with spatial averaging and filtering to extract the breathing rate from a video. End-to-end methods can also work effectively for this task [12], treating pulse and breathing measurement as a multitask problem can help reduce the computational burden of calculating multiple metrics from video [26].

Supervised neural models can help improve the accuracy of physiological measurements from video [11,12,21,26,59]. These models, which often feature millions of parameters, can capture more complex relationships between raw pixel intensities and physiological signals. An example of this is the case of dealing with video compression [27,59], where machine learning algorithms can successfully learn to recover signals even in the presence of video compression artifacts. Given the large number of parameters, these models typically require sizable training datasets, and, as with all supervised learning techniques, these examples need to match the distribution of examples seen in the application domain. These large models are often non-trival to deploy due to their computational requirements. Efficient methods that can run on devices are needed for many applications [26].

Personalization is a way in which machine learning methods can help improve the state of the art in physiological measurement [21,25]. Meta-learning is a quickly growing field of machine learning in which the goal is to produce models that can successfully adapt to new domains (e.g., individuals not seen in the training set). Given the large variability in appearance and physiology between people, these methods are particularly attractive for camera-based sensing of vital signs. Early research has shown promising results, even when unsupervised methods are used for adaptation [25]. Effective personalization is likely to be necessary for reaching the required accuracy needed for many applications of camera-based vital signs measurement. More research is needed in this domain.

Thermal cameras can be (one or two orders of magnitude) more expensive than RGB devices and are currently far less ubiquitous. However, thermal camera can be used to capture similar physiological information [17,34], including the cardiac pulse and breathing rate. Thermal cameras have several other advantages over cameras in the visible or NIR ranges. For example, they can operate in low-light conditions and image deeper into the skin. These devices can be used to measure body temperature [45] and sweat (e.g., perinasal perspiration) [5]. It is unlikely that RGB cameras could be used to capture the same information due to the penetration of visual light below the skin being too small. Therefore, thermal cameras are a natural complement to RGB imaging devices and combined could increase the power of camera-based sensing.

8.2 The promise for ubiquitous computing

One of the visions of ubiquitous computing is the ability for people to interact with computing using any device, in any location, and in any format. A motivation for designing methods for measuring physiological signals using CV is the ubiquitous nature of cameras. The popularity of digital photography has led to a huge investment into the miniaturization of imaging devices, which are now standard hardware on almost all personal electronics—from phones to laptops to smart speakers and even some fridges. Sensing is a tool that can enable new applications; however, these applications will typically only flourish if there is a need that can be satisfied and a path to productization.

8.2.1 Fitness and wellness

For many people, the first applications of camera-based vital signs measurement that come to mind are in fitness and wellness. While there are many potential applications of camera-based sensing, some of the first that have evolved are in clinical contexts, such as in-patient care. In Neonatal Intensive Care Units (NICUs), the measurement of heart rate and other physiological parameters is essential; however, newborn infants are vulnerable and have delicate skin. Attaching electrodes to their bodies (the current standard of care) can lead to skin damage that could increase the chance of infections or, at very least, cause discomfort to the baby. This is perhaps the first application of camera-based measurement that is undergoing serious real-world validation [1,53]. There are several reasons that this kind of deployment is tractable. First, the illumination in a hospital ward can be controlled to some extent. Second, the image sensor and position of the camera and the patient can be also be controlled (a baby cannot typically move outside of an incubator and thus remains within the field of view of the camera). Third, contact sensors necessary for validation of camera-based measurement can easily be collected as that is the current standard of care. Patients with severely damaged skin, for example, burn victims, also experience discomfort from wearing electrodes for extended periods of time, and monitoring of adults in intensive care units (ICUs) is a another application of camera-based sensing in in-patient settings.

However, there are many applications beyond critical care in which camera-based measurement could offer benefits. One that comes immediately to mind is telehealth. One effect of the 2020 COVID-19 pandemic is the sharp increase (by more than a factor of ten) in the number of medical appointments held via telehealth platforms because of the increased pressures on healthcare systems, the desire to protect healthcare workers, and restrictions on travel [49]. These are longstanding arguments for telehealth and will still be valid after the end of the current pandemic. Healthcare systems are likely to maintain a high number of telehealth appointments beyond the current pandemic [44]. In telehealth contexts, objective measurement of vital signs is often challenging or impossible. Most patients may not have access to traditional medical devices for measuring vital signs, and/or their technological literacy might make it difficult for them to learn to use new devices.

Physiological measurement using cameras is attractive for telehealth applications because it essentially relies only on the hardware already being used for video conferencing and no additional contact devices need to be synchronized or charged. A software upgrade could allow physicians to gather objective measurements of a patient during a call or outside of an appointment window. Video conferencing platforms such as Microsoft Teams, Google Meet, or Zoom could easily integrate these algorithms. Heart rate, respiration rate, and blood oxygenation alone could be useful to clinicians, but there are a number of other metrics that might make the tools more powerful. The subtle dynamics, or morphology, of the blood-volume pulse (BVP) signal can be measured using non-contact camera-based approaches [28]. Derivative metrics calculated from the waveform have been shown to be indicators of blood pressure [3,15] and could be helpful in assessing the impact of chronic conditions, such as hypertension and the associated risk of mortality. Atrial fibrillation (AF) is another condition that is associated with significant morbidity, i.e., increased mortality. As with PRV, the measurement to detecting AF requires recovering the pulse waveform with a high signal-to-noise ratio. Several validation studies have shown promising results on AF detection from non-contact recordings [43,57]. Such a tool might be particularly helpful for heart-failure patients or others at high risk. The software may not need to entirely replace a blood pressure cuff or other contact sensors but could supplement these devices to help collect complementary longitudinal data.

Beyond these applications, there are several other promising opportunities for camera-based measurement. The comfort of sleep studies [58] could be improved substantially using non-contact measurement, and it may even be possible to conduct such studies at a patient's home. The assessment of Jugular Venous Pressure (JVP) [2] could be made more accurate using video measurement and magnification techniques. Over the next decade, we are likely to see numerous commercial applications of camera-based health sensing systems that promote wellness and help people manage their health remotely with less dependence on traditional hospital visits.

In consumer fitness, there could be a big market for contact-less vital signs monitoring on equipment, whether in the home or in gyms. High-end home-fitness equipment, such as Pelaton[1] and Mirror,[2] have become increasingly popular, especially during the 2020 COVID-19 pandemic. Many pieces of electronic equipment could easily have image sensors integrated into them. These sensors do not require being touched and therefore could be made more sanitary and comfortable than contact sensors, such as hand grips on treadmills that measure pulse information. Smartphone applications already exist that leverage

---

[1] https://www.onepeloton.com.

[2] https://www.mirror.co/.

the device camera to measure heart rate[3] from either the finger tip and/or the face. However, to our knowledge, these applications are mostly only robust to small motions and are not validated for applications in which the subject may be running on a treadmill or cycling on a static bike. Many of these problems will be solved in the near future thanks to advances in deep learning and greater volumes of data (both real and synthetic).

Non-contact measurement enables new forms of interfaces with health data. Digital mirrors that integrate cameras and display the calculated health metrics could rise in popularity building on the success of high-tech fitness systems [40]. While the necessary hardware makes these much more expensive than a typical bathroom mirror, they could be attractive if they can offer accurate and useful health tracking. More and more devices are now available with cameras. This creates new opportunities for combining camera-based measurement with other technologies.

Baby monitors are another technology that can leverage camera-based tracking. Similar arguments for camera-based tracking apply in consumer baby monitoring as in the NICU application. Baby monitors that offer optical respiration measurement are already on the market (e.g., MikuCare[4]); however, it is likely that the next generation of these devices will try to integrate heart/rate tracking and other metrics. As with all health monitoring applications, the design of all these systems needs to be thought through carefully. Simply monitoring more and more information about oneself or another person (e.g., a child) could lead to increased anxiety about what these data mean.[5] In all applications of camera-based measurement, we should strongly promote a user-centered design philosophy, in which the sensing and user interface solve a clear need for the consumer and minimize potential harms.

8.2.2 Affective computing

Beyond health and well-being applications, there are many uses for camera-based physiological measurement. Affective computing encompasses computing systems that sense, interpret, and respond to human emotions [39]. Changes in cardiopulmonary parameters with cognitive and psychological stress can be captured using camera-based methods [8,9,28,31]. While average heart rate is often not very indicative of stress, heart-rate variability (HRV) and other peripheral signals can be quite highly correlated with sympathetic nervous system changes. From camera-based measurement via PPG, we can capture pulse-rate variability (PRV) which is highly correlated with HRV.

Measurement of PRV has been applied to tracking cognitive loads during computer tasks [8,31]. The same study found that both PRV and respiration were

---

[3] https://www.azumio.com/s/instantheartrate/index.html, https://apps.apple.com/us/app/cardiio-heart-rate-monitor/id542891434.

[4] https://mikucare.com/.

[5] https://mashable.com/2017/02/18/raybaby-baby-breathing-monitor/.

significantly different when participants were under stress compared to at rest. Quantitative evaluation of peripheral hemodynamics provides a way of monitoring tissue metabolism and can also be used to infer autonomic nervous-system activity. Peripheral hemodynamics can be characterized via the plethysmograph and vasomotion signals. These capture changes in blood volume in the surface of the skin and spontaneous oscillations in the tone of blood vessel walls, independent of heart beat, innervation, or respiration. Vasomotion and peripheral pulse amplitude change with nervous system activity. It have been found that significant decreases in peripheral pulse signal power and vasomotion signal power occur during periods of cognitive and psychological stress [32,35]. Thermal cameras have been used to measure stress in workplace contexts via changes in body temperature and sweat [5,45].

While these signals (HRV/PRV, vasomotion and pulse amplitude) are indicators of arousal, they are unspecific. Therefore, it is necessary to combine these measurements with other sensor data or contextual cues for practical applications. In the case of affective computer applications, this may include the context of the applications being used, their location, or activity data, other nonverbal cures (such as facial expressions, voice tone), and/or movement of peripheral devices (such as a mouse or keyboard typing patterns).

There are many applications of affective state measurement. In education, such techniques could help with building online learning systems and MOOCs that adapt to the state of the learner [38]. In the automotive industry, there is acute interest in the measurement of vital signs and affect because of the implications this could have for road safety and driver experience. Driving is an application in which ambient illumination levels and dynamic lighting changes are very significant. Thus infrared cameras, despite their having a lower signal-to-noise ratio, are employed [37]. In-cabin driver monitoring could be used to measure the stress of drivers [19]. This is another application of long-term unobtrusive physiological measurement where wearing contact devices for extended periods may be infeasible or simply disrupt the experience of a users.

One of the first industries to leverage affective computing technologies at scale was media measurement and testing; increasingly, this type of testing is happening online so that participants do not need to leave their homes. Physiological measurement via cameras can be used to help infer changes in emotional states in response to content [10], and camera-based methods are the only practical way to do this given the ubiquity of the sensors.

8.2.3 Biometric recognition and liveness detection

One application for which camera-based physiological measurement seems particularly suited is liveness detection for visually-based biometric recognition systems [20,23,24]. In facial authentication systems, such as Apple Face ID or Microsoft Hello, there is a risk that a photograph, video, or mask might be used to unlock a device. However, these do not have all the properties of a real human face, in part due to the lack of subtle physiological markers (e.g., changes in light absorption due to PPG). It is possible that physiological features (e.g., BCG waveform morphology) might even be used directly for identification purposes and not only in liveness detection [18]. Another related application is the detection of fake media content. Sophisticated computer vision technology can be used to create fake videos of people that are difficult to distinguish from real footage. These have been employed in harmful ways to generate videos of a politician's appearing to say things that they did not and pornographic content featuring celebrities. Recently, the rise of "deep fakes" has led to increased efforts to find algorithms that can be used to detect them.[6] Several attempts have been made to use the presence of a PPG signal, and therefore a heart beat, within a video to detect if it is fake or not [13,46]. While it is possible that a PPG signal could also be "faked" in a video, this will at least make deep-fake creation more challenging.

8.2.4 Avatars, remote communication, and mixed reality

There are many scenarios in which people could interact with embodied virtual avatars and holograms. The quality of these interactions depends, in part, on the physical representation of the agent or avatar. There are many properties that create believable embodied characters. In Masahiro Mori's [33] graphical representation of the uncanny valley, zombies and corpses fall at the very lowest point, which is sometimes known as the "death mask effect". There is a great deal of overlap in the techniques used for measuring physiological signals in video and those used to magnify them. These method could be used to augment the appearance of avatars to create more lifelike representations.

Augmented reality could even allow physiological signals to be superimposed on the appearance of human beings from the perspective of the viewer [29]. This could be used to help improve people's interoceptive abilities and enable the study of how knowledge of an interlocutor's physiological state could help increase empathy. For example, superimposing physiological signals in this way can help people interpret the arousal of another subject [30].

8.3 Challenges

There remains a considerable gap between the performance of contact sensors and camera-based methods when it comes to measuring vital signs in everyday contexts. Most camera-based measurement solutions are attractive because camera hardware is ubiquitous. However, as a result, there is typically no dedicated light source and the applications rely on ambient illumination. In low-light conditions, the changes in pixel intensities due to physiological processes can be very small, and in brightly lit environments, skin pixels can become close to saturation, similarly meaning the variance in pixel intensity from physiological changes may be very small.

---

[6] https://www.kaggle.com/c/deepfake-detection-challenge.

Measurement on mobile devices [26] is possible at high frame rates of up to 150 FPS. However, in the mobile context, there may be greater levels of noise due to participants holding the imaging devices in their hands, rather than having them fixed as would often be the case with a webcam attached to a computer.

The design of sensors and algorithms for physiological measurement are subject to biases. Skin type influences the signal-to-noise ratio (SNR), and indeed several studies have found this to be the case [4,16,36,48,54,55]. The larger melanin concentration in people with darker skin absorbs more light, making the intensity of light returning to the camera lower and thus the iPPG signal weaker.

8.4 Ethics and privacy implications

The measurement of physiological signals using imaging devices has great potential but also raises serious ethical concerns. The unobtrusive and ubiquitous nature of cameras means that theoretically surveillance could be performed at scale and without the subject's knowledge. This may be justified by claims that such monitoring could be used to screen for people with symptoms of a virus during a pandemic or to try to detect a nervous individual. However, there is very little evidence that camera-based measurement alone would be accurate enough and effective for these applications, and there are many possible negative outcomes that might result even if the technology were accurate. As just described, there is considerable evidence that camera-based methods currently do not perform uniformly across people of all skin types, and the populations that are subject to the worst performance might also be those that are already subject to disproportionate targeting and systematic negative biases.

As with any new technology, it is important to consider how it could be used by "bad actors" or applied in a negligent or irresponsible manner. Application without sufficient forethought concerning the implications could undermine the positive applications of these methods. Non-contact sensing could be used to measure personal physiological information without the knowledge of the subject. Military or law enforcement organizations may try to apply this in an attempt to detect individuals who appear "nervous" via signals such as an elevated heart rate or irregular breathing. An employer may surreptitiously screen prospective employees for health conditions without their knowledge during an interview. Criminals may try to collect physiological signals for the purposes of biometric identification. These applications would set a very dangerous precedent and would probably be illegal. Just as is the case with traditional contact sensors, it must be made very transparent when camera-based physiological measurement is being used and subjects should be required to consent before physiological data is measured or recorded. There should be no penalty for individuals who decline to be measured. Ubiquitous sensing offers the ability to measure signals in more contexts, but that does not mean that this should necessarily be acceptable. Just because cameras may be able to measure these signals in new context, or with less effort, it does not mean they should be subject to any less regulation than existing sensors.

8.5 Regulation

The need to consider validation and certification: The US Federal Drug Administration (FDA) mandates that testing of a new device for cardiac monitoring should show "substantial equivalence" in accuracy with a legal predicate device,[7] which means a contact sensor. This standard has not been achieved. The attraction of camera-based measurement is that ubiquitous hardware can be leveraged. However, this means that the cameras on many devices may need to be tested even if the image sensors are quite similar. To my knowledge, there are no camera-based physiological measurement systems that have achieved clearance (results of a successfully reviewed and accepted 501(k) submission) or approval (rigorous review and approval process required for Class III medical devices). However, I expect that there are a number of companies who have started, or are planning to start, such a process.

8.6 Summary

Camera-based measurement of physiology has the potential to enable ubiquitous, unobtrusive health and affect measurement. While the field is still in an early stage, there have been considerable advancements in recent years, in particular, towards the robust measurement of vital signs in specific contexts. However, to realize many of the applications that have been touted, the field should focus more efforts on validating longitudinal measurement in-situ, e.g., measurement in hospitals, the home, workplaces or gyms. The technology raises important ethical questions, and the research community should also contribute to addressing these, in addition to developing hardware and algorithms.

References

[1] Lonneke A.M. Aarts, et al., Non-contact heart rate monitoring utilizing camera photoplethysmography in the neonatal intensive care unit—a pilot study, Early Human Development 89 (12) (2013) 943–948.

[2] Freddy Abnousi, et al., A novel noninvasive method for remote heart failure monitoring: the EuleriAn video Magnification apPLications In heart Failure studY (AMPLIFY), NPJ Digital Medicine 2 (1) (2019) 1–6.

[3] Paul S. Addison, Slope transit time (STT): a pulse transit time proxy requiring only a single signal fiducial point, IEEE Transactions on Biomedical Engineering 63 (11) (2016) 2441–2444.

[4] Paul S. Addison, et al., Video-based heart rate monitoring across a range of skin pigmentations during an acute hypoxic challenge, Journal of Clinical Monitoring and Computing 32 (5) (2018) 871–880.

[5] Fatema Akbar, et al., Email makes you sweat: examining email interruptions and stress using thermal imaging, in: Proceedings of the 2019 CHI Conference on Human Factors in Computing Systems, 2019, pp. 1–14.

---

[7] https://www.fda.gov/regulatory-information/search-fda-guidance-documents/cardiac-monitor-guidance-including-cardiotachometer-and-rate-alarm-guidance-industry.

[6] Guha Balakrishnan, Fredo Durand, John Guttag, Detecting pulse from head motions in video, in: Proceedings of the IEEE Conference on Computer Vision and Pattern Recognition, 2013, pp. 3430–3437.

[7] Ethan B. Blackford, Justin R. Estepp, Daniel McDuff, Remote spectral measurements of the blood volume pulse with applications for imaging photoplethysmography, in: Optical Diagnostics and Sensing XVIII: Toward Point-of-Care Diagnostics, vol. 10501, International Society for Optics and Photonics, 2018, p. 105010Z.

[8] Frédéric Bousefsaf, Choubeila Maaoui, Alain Pruski, Remote assessment of the heart rate variability to detect mental stress, in: 2013 7th International Conference on Pervasive Computing Technologies for Healthcare and Workshops, IEEE, 2013, pp. 348–351.

[9] Frédéric Bousefsaf, Choubeila Maaoui, Alain Pruski, Remote detection of mental workload changes using cardiac parameters assessed with a low-cost webcam, Computers in Biology and Medicine 53 (2014) 154–163.

[10] Mihai Burzo, et al., Towards sensing the influence of visual narratives on human affect, in: Proceedings of the 14th ACM International Conference on Multimodal Interaction, 2012, pp. 153–160.

[11] Sitthichok Chaichulee, et al., Cardio-respiratory signal extraction from video camera data for continuous non-contact vital sign monitoring using deep learning, Physiological Measurement 40 (11) (2019) 115001.

[12] Weixuan Chen, Daniel McDuff, Deepphys: video-based physiological measurement using convolutional attention networks, in: Proceedings of the European Conference on Computer Vision (ECCV), 2018, pp. 349–365.

[13] Umur Aybars Ciftci, Ilke Demir, Lijun Yin, Fakecatcher: detection of synthetic portrait videos using biological signals, IEEE Transactions on Pattern Analysis and Machine Intelligence (2020).

[14] Gerard De Haan, Vincent Jeanne, Robust pulse rate from chrominance-based rPPG, IEEE Transactions on Biomedical Engineering 60 (10) (2013) 2878–2886.

[15] Mohamed Elgendi, et al., The use of photoplethysmography for assessing hypertension, NPJ Digital Medicine 2 (1) (2019) 1–11.

[16] Bennett A. Fallow, Takashi Tarumi, Hirofumi Tanaka, Influence of skin type and wavelength on light wave reflectance, Journal of Clinical Monitoring and Computing 27 (3) (2013) 313–317.

[17] Marc Garbey, et al., Contact-free measurement of cardiac pulse based on the analysis of thermal imagery, IEEE Transactions on Biomedical Engineering 54 (8) (2007) 1418–1426.

[18] Javier Hernandez, Daniel McDuff, Rosalind W. Picard, Bioinsights: extracting personal data from "still" wearable motion sensors, in: 2015 IEEE 12th International Conference on Wearable and Implantable Body Sensor Networks (BSN), IEEE, 2015, pp. 1–6.

[19] Javier Hernandez, et al., AutoEmotive: bringing empathy to the driving experience to manage stress, in: Proceedings of the 2014 Companion Publication on Designing Interactive Systems, 2014, pp. 53–56.

[20] Javier Hernandez-Ortega, et al., Time analysis of pulse-based face antispoofing in visible and NIR, in: Proceedings of the IEEE Conference on Computer Vision and Pattern Recognition Workshops, 2018, pp. 544–552.

[21] Eugene Lee, Evan Chen, Chen-Yi Lee, Meta-rPPG: remote heart rate estimation using a transductive meta-learner, arXiv preprint, arXiv:2007.06786.

[22] Magdalena Lewandowska, et al., Measuring pulse rate with a webcam—a non-contact method for evaluating cardiac activity, in: Computer Science and Information Systems (FedCSIS), 2011 Federated Conference on., IEEE, 2011, pp. 405–410.

[23] Si-Qi Liu, Xiangyuan Lan, Pong C. Yuen, Remote photoplethysmography correspondence feature for 3D mask face presentation attack detection, in: Proceedings of the European Conference on Computer Vision (ECCV), 2018, pp. 558–573.

[24] Si-Qi Liu, et al., 3D mask face anti-spoofing with remote photoplethysmography, in: European Conference on Computer Vision, Springer, 2016, pp. 85–100.

[25] Xin Liu, et al., MetaPhys: unsupervised few-shot adaptation for NonContact physiological measurement, in: Proceedings of the ACM Conference on Health, Inference, and Learning, 2021.

[26] Xin Liu, et al., Multi-task temporal shift attention networks for on-device contactless vitals measurement, in: NeurIPS, 2020.

[27] Daniel McDuff, Ethan B. Blackford, Justin R. Estepp, The impact of video compression on remote cardiac pulse measurement using imaging photoplethysmography, in: 2017 12th IEEE International Conference on Automatic Face & Gesture Recognition (FG 2017), IEEE, 2017, pp. 63–70.

[28] Daniel McDuff, Sarah Gontarek, Rosalind W. Picard, Remote detection of photoplethysmographic systolic and diastolic peaks using a digital camera, IEEE Transactions on Biomedical Engineering 61 (12) (2014) 2948–2954.

[29] Daniel McDuff, Christophe Hurter, Mar Gonzalez-Franco, Pulse and vital sign measurement in mixed reality using a HoloLens, in: Proceedings of the 23rd ACM Symposium on Virtual Reality Software and Technology, 2017, pp. 1–9.

[30] Daniel McDuff, Ewa Nowara, "Warm Bodies": a post-processing technique for animating dynamic blood flow on photos and avatars, in: Proceedings of the 2021 CHI Conference on Human Factors in Computing Systems, ACM, 2021.

[31] Daniel McDuff, et al., Cogcam: contact-free measurement of cognitive stress during computer tasks with a digital camera, in: Proceedings of the 2016 CHI Conference on Human Factors in Computing Systems, ACM, 2016, pp. 4000–4004.

[32] Daniel McDuff, et al., Non-contact imaging of peripheral hemodynamics during cognitive and psychological stressors, Scientific Reports 10 (1) (2020) 1–13.

[33] M. Mori, The uncanny valley, Energy 7 (4) (1970) 33–35.

[34] Ramya Murthy, Ioannis Pavlidis, Panagiotis Tsiamyrtzis, Touchless monitoring of breathing function, in: The 26th Annual International Conference of the IEEE Engineering in Medicine and Biology Society, vol. 1, IEEE, 2004, pp. 1196–1199.

[35] Nishidate Izumi, et al., RGB camera-based noncontact imaging of plethys-mogram and spontaneous low-frequency oscillation in skin perfusion before and during psychological stress, Optical Diagnostics and Sensing XIX: TowardPoint-of-Care Diagnostics, vol. 10885, International Society for Optics and Photonics, 2019, p. 1088507.

[36] Ewa M. Nowara, Daniel McDuff, Ashok Veeraraghavan, A MetaAnalysis of the impact of skin tone and gender on non-contact photoplethysmography measurements, in: Proceedings of the IEEE/CVF Conference on Computer Vision and Pattern Recognition Workshops, 2020, pp. 284–285.

[37] Ewa Magdalena Nowara, et al., SparsePPG: towards driver monitoring using camera-based vital signs estimation in near-infrared, in: 2018 IEEE/CVF Conference on Computer Vision and Pattern Recognition Workshops (CVPRW), IEEE, 2018, pp. 1353–135309.

[38] Phuong Pham, Jingtao Wang, AttentiveLearner: improving mobile MOOC learning via implicit heart rate tracking, in: International Conference on Artificial Intelligence in Education, Springer, 2015, pp. 367–376.

[39] Rosalind W. Picard, Affective Computing, MIT Press, 2000.

[40] Ming-Zher Poh, Daniel McDuff, Rosalind Picard, A medical mirror for non-contact health monitoring, in: ACM SIGGRAPH 2011 Emerging Technologies, 2011, p. 1.

[41] Ming-Zher Poh, Daniel McDuff, Rosalind W. Picard, Advancements in noncontact, multi-parameter physiological measurements using a webcam, IEEE Transactions on Biomedical Engineering 58 (1) (2010) 7–11.

[42] Ming-Zher Poh, Daniel McDuff, Rosalind W. Picard, Non-contact, automated cardiac pulse measurements using video imaging and blind source separation, Optics Express 18 (10) (2010) 10762–10774.

[43] Ming-Zher Poh, et al., Diagnostic assessment of a deep learning system for detecting atrial fibrillation in pulse waveforms, Heart 104 (23) (2018) 1921–1928.

[44] Kim Pollock, Michael Setzen, Peter F. Svider, Embracing telemedicine into your otolaryngology practice amid the Covid-19 crisis: an invited commentary, American Journal of Otolaryngology (2020) 102490.

[45] Colin Puri, et al., StressCam: non-contact measurement of users' emotional states through thermal imaging, in: CHI'05 Extended Abstracts on Human Factors in Computing Systems, 2005, pp. 1725–1728.

[46] Hua Qi, et al., DeepRhythm: exposing deepfakes with attentional visual heartbeat rhythms, in: Proceedings of the 28th ACM International Conference on Multimedia, 2020, pp. 4318–4327.

[47] Dangdang Shao, et al., Non-contact simultaneous photoplethysmogram and ballistocardiogram video recording towards real-time blood pressure and abnormal heart rhythm monitoring, in: 2017 12th IEEE International Conference on Automatic Face & Gesture Recognition (FG 2017), IEEE, 2017, pp. 273–277.

[48] Dangdang Shao, et al., Simultaneous monitoring of ballistocardiogram and photoplethysmogram using a camera, IEEE Transactions on Biomedical Engineering 64 (5) (2016) 1003–1010.

[49] Anthony C. Smith, et al., Telehealth for global emergencies: implications for coronavirus disease 2019 (Covid-19), Journal of Telemedicine and Telecare (2020), 1357633X20916567.

[50] Chihiro Takano, Yuji Ohta, Heart rate measurement based on a timelapse image, Medical Engineering & Physics 29 (8) (2007) 853–857.

[51] L. Tarassenko, et al., Non-contact video-based vital sign monitoring using ambient light and auto-regressive models, Physiological Measurement 35 (5) (2014) 807.

[52] Wim Verkruysse, Lars O. Svaasand, J. Stuart Nelson, Remote plethys-mographic imaging using ambient light, Optics Express 16 (26) (2008) 21434–21445.

[53] Mauricio Villarroel, et al., Non-contact physiological monitoring of preterm infants in the neonatal intensive care unit, npj Digital Medicine 2 (1) (2019) 1–18.

[54] Wenjin Wang, Sander Stuijk, Gerard De Haan, A novel algorithm for remote photoplethysmography: spatial subspace rotation, IEEE Transactions on Biomedical Engineering 63 (9) (2015) 1974–1984.

[55] Wenjin Wang, et al., Algorithmic principles of remote PPG, IEEE Transactions on Biomedical Engineering 64 (7) (2017) 1479–1491.

[56] Ting Wu, et al., Photoplethysmography imaging: a new noninvasive and noncontact method for mapping of the dermal perfusion changes, Optical Techniques and Instrumentation for the Measurement of Blood Composition, Structure, and Dynamics 4163 (2000) 62–70.

[57] Bryan P. Yan, et al., Contact-free screening of atrial fibrillation by a smartphone using facial pulsatile photoplethysmographic signals, Journal of the American Heart Association 7 (8) (2018) e008585.

[58] Cheryl C.H. Yang, et al., Relationship between electroencephalogram slow-wave magnitude and heart rate variability during sleep in humans, Neuroscience Letters 329 (2) (2002) 213–216.

[59] Zitong Yu, et al., Remote heart rate measurement from highly compressed facial videos: an end-to-end deep learning solution with video enhancement, in: Proceedings of the IEEE International Conference on Computer Vision, 2019, pp. 151–160.

Part II

Wireless sensor-based vital signs monitoring

Chapter 9

Radar-based vital signs monitoring

Jingtao Liu[a,b], Yuchen Li[a,b], and Changzhan Gu[a,b]

[a]*MoE Key Lab of Artificial Intelligence, AI Institute, Shanghai Jiao Tong University, Shanghai, China,* [b]*MoE Key Lab of Design and EMC of High-Speed Electronic Systems, Department of Electronic Engineering, Shanghai Jiao Tong University, Shanghai, China*

Contents

9.1 Introduction

Vital signs monitoring has become a hot research topic in the fields of consumer electronics, biomedical applications, and through-wall detections. In these application scenarios, mainstream solutions include contact-based sensors, camera-based solutions, and microwave radar systems.

The advantages and disadvantages of contact-based sensors come from their characteristics: they need to be in contact to take effect. Therefore, wearing one for a long time or requiring frequent contact will cause the user an uncomfortable feeling, which greatly reduce the convenience of using the sensors. For a camera-based solution, the advantage is that it has a wide range of uses, and tracking and detecting objects is more intuitive. But, it also has several disadvantages. First, it is extremely dependent on light. When the light conditions are not good or there is interference from other light sources, the imaging ability drops rapidly. Second, it cannot penetrate objects. When there is an obstacle between the camera and the target, the camera-based solution is completely ineffective. Third, most camera-based solutions can only provide two-dimensional image information. Even if some cameras can provide depth information [1,2], there is still no method that can better distinguish between people and other static objects.

Compared with conventional contact-based sensors and camera-based solutions, microwave radar systems have their own natural advantages. First, the microwave radar system is a contactless solution, and therefore is convenient to use. Second, microwave radar systems do not depend on light and can penetrate obstacles well [3–5]. More importantly, radar sensors are more sensitive to small movements. Based on Doppler and micro-Doppler characteristics, radar sensors can detect very small movements of target objects such as human respiration and heartbeats [6–8].

Continuous-wave (CW) radar has the advantages of low transmit power, high sensitivity, and simple structure. Therefore, it is widely used in various fields. Typically, CW radars are divided into two categories: 1) unmodulated CW systems and 2) modulated CW solutions. A typical example of an unmodulated continuous-wave system is an interferometer (Doppler) radar [9,10], which operates on a single-tone CW to obtain the target's phase history. In addition, this kind of radar has high accuracy in displacement and velocity measurements [4,6,11]. However, it is difficult for CW radar to obtain the absolute range information of the target. Modulated continuous-wave radar includes frequency-shift keyed radar [12], stepped frequency continuous-wave radar [13], and frequency modulated continuous-wave (FMCW) radar [14–20]. As one of the most popular types, FMCW radar can easily and accurately obtain the accurate distance of the target. In addition, if the coherence of the system is achieved, FMCW radar can extract Doppler information related to the radial velocity of the target and measure the displacement of the target. However, the hardware and signal processing of the FMCW system to measure the range profile is much more complicated than the hardware and signal processing of the unmodulated CW system. Moreover, the displacement accuracy of the FMCW radar may not be as good as that of the unmodulated CW system, which can easily achieve sub-mm accuracy [14]. By making the radar coherent, the phase history of the target can be preserved during the coherent processing interval (CPI), so that Doppler information can be derived, which provides two dimensions: distance and Doppler. This two-dimensional information can help isolate the required moving target from the surrounding static clutter, thereby achieving the detection of vital signals.

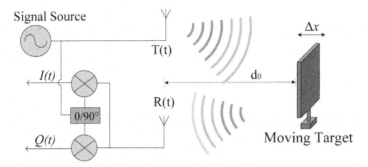

FIGURE 9.1 Common configuration of CW radar.

9.2 Vital signs monitoring through continuous-wave radar

9.2.1 Theory

9.2.1.1 Basic theory

Fig. 9.1 shows the block diagram of a common CW radar. As shown in Fig. 9.1, CW radar detects the movements of the target by transmitting a single-tone electromagnetic wave to the target and receiving the reflected signal, which modulates the motion information of the target. The theory is detailed as follows.

The transmitted signal $T(t)$ is modeled as [21]:

$$T(t) = \exp\left(j\left(2\pi f_c t + \varphi_0\right)\right),\tag{9.1}$$

where f_c is the frequency of $T(t)$ and φ_0 is the initial phase.

The reflected signal $R(t)$ is:

$$R(t) = \rho T(t - \Delta t),\tag{9.2}$$

where ρ is the propagation loss and Δt is the time interval of the signal from the transmitting (TX) antenna to the receiving (RX) antenna. $\Delta t = \frac{2x(t)}{c}$, where $x(t)$ is the distance from the radar to the target.

After mixing, the intermediate frequency (IF) signal is obtained as:

$$S(t) = T(t)R^*(t) = \exp\left(j\left(2\pi f_c \Delta t\right)\right) = \rho \exp j\left(2\pi f_c \frac{2x(t)}{c}\right)$$

$$= \exp j\left(\frac{4\pi \Delta x(t)}{\lambda_c} + \frac{4\pi d_0}{\lambda_c}\right)$$

$$= I(t) + jQ(t) \stackrel{\text{def}}{=} e^{j\Phi(t)},\tag{9.3}$$

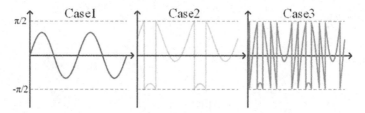

FIGURE 9.2 Three cases of arctangent demodulation results.

where $\lambda_c = c/f_c$ is the wavelength of the electromagnetic wave, $x(t) = \Delta x(t) + d_0$, and $\Delta x(t)$ is the motion of the target at around distance d_0.

The motion of the target can be extracted from the phase $\Phi(t)$:

$$x(t) = \frac{c\Phi(t)}{4\pi f_c} = \frac{\lambda_c \Phi(t)}{4\pi}. \qquad (9.4)$$

Note that $\Phi(t)$ is a periodic function whose cycle is 2π. Therefore, only the vibration of $x(t)$, i.e., $\Delta x(t)$, can be extracted. The speed of the target is:

$$v = \frac{dx(t)}{dt} = \frac{\frac{c}{4\pi f_c}d\Phi(t)}{dt} = \frac{cf_d}{2f_c}, v \ll c, \qquad (9.5)$$

where $f_d = \frac{1}{2\pi}\frac{d\Phi(t)}{dt}$ is the Doppler shift.

9.2.1.2   Phase demodulation algorithm

As discussed in Sect. 9.2.1.1, the motion of the target can be extracted from the phase $\Phi(t)$. Thus, the key to extracting the motion is to demodulate $\Phi(t)$. Many demodulation algorithms have been proposed. The classic algorithm is arctangent demodulation algorithm [21–23], as shown in Eq. (9.6).

$$\Phi_{\text{atan}}(t) = arctan\left(\frac{Q(t)}{I(t)}\right) \qquad (9.6)$$

However, the value of the arctangent function is restricted to $(-\pi/2, \pi/2)$. Fig. 9.2 shows three cases of arctangent demodulation. The vibration amplitude of $\Phi(t)$ is smaller than π in cases I and II. The transition occurs when the DC component is close to $-\pi/2$ or $\pi/2$. When the vibration amplitude of $\Phi(t)$ is larger than π, as shown in case III, the transitions are inevitable. Compensations of π or $-\pi$ need to be done at the transition point. However, the automatic compensation is difficult to implement when the vibration amplitude becomes larger and the noise exists [24].

Another phase demodulation algorithm is the differential and cross-multiplication (DACM) algorithm [25]. Unlike the arctangent demodulation algorithm, the DACM algorithm obtains the result by the derivation of the arctangent function [25]:

$$\omega(t) = \frac{d\Phi_{atan}(t)}{dt} = \frac{I(t)\,Q'(t) - I'(t)\,Q(t)}{I(t)^2 + Q(t)^2},$$ (9.7)

where $Q'(t)$ and $I'(t)$ are the time derivatives of $Q(t)$ and $I(t)$, respectively. $\omega(t)$ is also the differential of $\Phi(t)$. Therefore, $\Phi(t)$ can be reconstructed by the integration of $\omega(t)$. In the digital domain, the $\Phi[n]$ could be reconstructed by [25]:

$$\Phi[n] = \sum_{k=2}^{n} \frac{I[k]\{Q[k] - Q[k-1]\} - \{I[k] - I[k-1]\}\,Q[k]}{I^2[k] + Q^2[k]},$$ (9.8)

where the differentiation is approximated by a forward difference and the integration is replaced with an accumulation. Note that $\omega(t)$ is a single-valued function, so the codomain restriction does not exist in this algorithm.

Recently, a modified DACM algorithm has been reported [26], which has a simplified expression but much improved performance for high-linear motion detection. Different from the traditional DACM algorithm shown in (Eq. (9.7)), the modified DACM algorithm does not include arctangent function. The theory is as follows. The baseband I/Q signal shown in Fig. 9.1 could be modeled as:

$$I(t) = \cos\left[\frac{4\pi\,\Delta x(t)}{\lambda_c} + \frac{4\pi\,d_0}{\lambda_c}\right]$$ (9.9)

$$Q(t) = \sin\left[\frac{4\pi\,\Delta x(t)}{\lambda_c} + \frac{4\pi\,d_0}{\lambda_c}\right].$$ (9.10)

First, apply differentiation to Eqs. (9.9) and (9.10):

$$I'(t) = -\sin\left[\frac{4\pi\,\Delta x(t)}{\lambda_c} + \frac{4\pi\,d_0}{\lambda_c}\right] \cdot 4\pi\,\Delta x'(t)/\lambda_c,$$ (9.11)

$$Q'(t) = \cos\left[\frac{4\pi\,\Delta x(t)}{\lambda_c} + \frac{4\pi\,d_0}{\lambda_c}\right] \cdot 4\pi\,\Delta x'(t)/\lambda_c.$$ (9.12)

Second, since $\cos^2\Phi + \sin^2\Phi = 1$, $\Delta x'(t)$ can be extracted from Eqs. (9.11) and (9.12):

$$\Delta x'(t) = \frac{\lambda}{4\pi}\left[I(t)\,Q'(t) - I'(t)\,Q(t)\right]$$ (9.13)

As shown, the differential form of the target motion does not include the arctangent function. The motion trajectory $\Delta x(t)$ can be obtained by applying an integration function to $\Delta x'(t)$. In digital domain, the motion trajectory can be obtained by:

$$x[n] = \frac{\lambda}{4\pi} \sum_{k=2}^{n} I[k-1]Q[k] - I[k]Q[k-1]. \qquad (9.14)$$

It should be noted that the relationships between $\omega(t)$, $\Phi[n]$ and $\Delta x'(t)$, $x[n]$ are $\Delta x'(t) = \omega(t)\lambda/4\pi$ and $x[n] = \Phi[n]\lambda/4\pi$, respectively. Comparing Eqs. (9.8) and (9.14), the proposed algorithm does not include $I^2[k] + Q^2[k]$, which decreases the signal-to-noise ratio requirements of the signal and thereby this improves the accuracy of the demodulation.

9.2.1.3 Advancements in CW radar

In the past few decades, researchers around the world have greatly promoted the development of Doppler radar. Many techniques have been proposed in signal processing and hardware that greatly improve the robustness and accuracy of radar motion detection. [27–30]. On the hardware side, many architectures for CW radar have been explored. A typical example is self-injection-locked (SIL) CW radar [31]. In signal processing, many algorithms that can significantly improve the performance of CW radar are proposed. To deal with the I/Q imbalance caused by the deficiency of hardware, algorithms based on radar-measured data and ellipse fitting are proposed. [32,33]. To deal with AC coupling in low-complexity Doppler radar, a digital post-distortion (DPoD) technique [34] is proposed to compensate for the signal distortions in the digital baseband domain.

In total, the advancements have given CW radar great application prospects. One typical example is contactless vital signs monitoring.

9.2.2 Vital signs monitoring

9.2.2.1 Cardiopulmonary monitoring

As early as in 1975, CW radar has been used in respiration monitoring by Professor James C. Lin [35]. Since then, the application of CW radar in cardiopulmonary monitoring has drawn much attention [29,36].

Fig. 9.3 shows an experimental setup of an instrument-based CW radar for cardiopulmonary monitoring [37]. The instrument-based CW radar system includes Agilent spectrum analyzer E4407B, Agilent vector signal generator E8267C, and Agilent vector signal analyzer 89600S. E8267C is utilized as a local oscillator (LO) to generate stable RF signal. E4407B is utilized as a mixer to convert the motion modulated RF signal to IF signal. 89600S is utilized as an analog-to-digital converter (ADC) and signal processor. The subject person sits around 1 m from the radar and breaths normally. A pulse sensor (HK-2000B) is

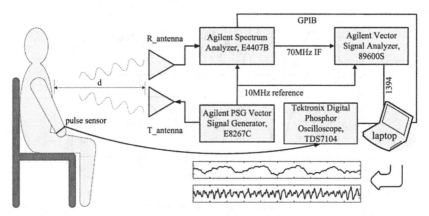

FIGURE 9.3 Experimental setup of an instrument-based CW radar [37].

wrapped around the wrist of the subject to detect the heartbeat signal. The signal is digitized by TDS7104 and used as a reference signal for the heartbeat signal detected by the CW radar. A 2.4-GHz printed patch antenna is used to validate the cardiopulmonary monitoring performance of the system, which means the radar system works at 2.4 GHz.

The captured respiration and heartbeat curve of the experiment are shown in Figs. 9.4 and 9.5, respectively. The upper curve in Fig. 9.4 is the baseband output signal, which is also known as raw data. The lower curve in Fig. 9.4 is the filtered respiration signal. The filtered respiration signal is the same as the raw data in periodicity but becomes smoother after being filtered. The upper curve of Fig. 9.5 represents the filtered beat signal, and the lower curve represents the reference heartbeat signal detected by the pulse sensor HK-2000B. As shown, the heartbeat cycles of the two curves in Fig. 9.5 match 100%, indicating that, under these experimental conditions, the heartbeat-detection accuracy of the CW radar system is the same as that of HK-2000B.

In addition to the time-domain form of the signal, the frequency-domain form of the signal can more intuitively express the respiratory and heart rates. By performing FFT on the data used in Figs. 9.4 and 9.5, the frequency-domain spectra of the signals are obtained and shown in Fig. 9.6. The solid line is obtained from the raw data, and the dashed line is obtained from the filtered heartbeat data. The abscissa is converted from Hz to beat/min, indicating the number of breaths or heartbeats in one minute. The signal amplitude is normalized. As shown, the harmonics of the heartbeat signal and the respiration signal can be easily identified, and the breath and heartbeat rates can be easily obtained by finding the peak values of the spectra. In this case, the heartbeat is around 17 beats/min, and the breath rate is around 78 beats/min.

Microwaves are penetrating. Therefore, CW radar can penetrate obstacles for cardiopulmonary monitoring. The experiment shown in Fig. 9.7 is used as

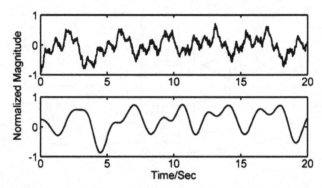

FIGURE 9.4 Detection results. The upper trace represents the baseband output recorded by VSA software. The lower trace represents the filtered respiration signal [37].

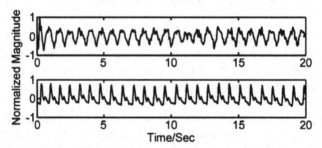

FIGURE 9.5 Detection results. The upper trace represents the filtered heartbeat signal. The lower trace represents the reference signal [37].

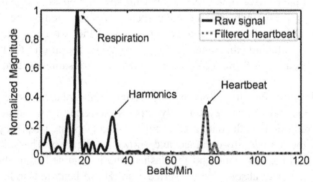

FIGURE 9.6 Frequency spectrum of the raw signal and filtered heartbeat signal. The breathing rate is about 17 beats/min, and the heartbeat rate is about 78 beats/min [37].

an example to reveal the penetration ability of the CW radar [37]. The radar system used in Fig. 9.7 is the same CW radar system shown in Fig. 9.3. As shown, the subject person faces the obstructions from a distance of d_1, and the

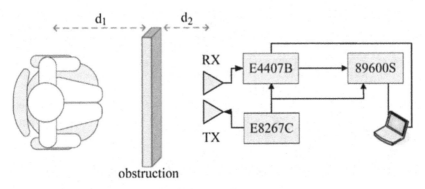

FIGURE 9.7 Measurement setup with obstructions [37].

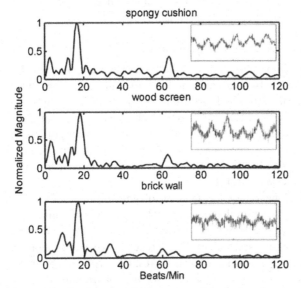

FIGURE 9.8 Measured baseband spectra with three types of obstructions (spongy cushion, wood screen, and brick wall). The insets indicate the corresponding raw time-domain signals [37].

obstruction is a distance d_2 from the radar. The obstructions include a spongy cushion (19-cm thick), a wood screen (3.7-cm thick), and a brick wall (30-cm thick). The measured results are shown in Fig. 9.8. As shown, radar can still accurately perform cardiopulmonary monitoring with three different obstacles in between, which means the radar system is good at penetrating obstructions. The insets of Fig. 9.8 are the raw signals.

There are also many other CW radar systems designed for cardiopulmonary monitoring [38–40]. In [39], a compact CW radar system for cardiopulmonary monitoring was designed. The radar system is smaller than a palm and can send the sampled baseband data to a PC remotely by Bluetooth [39]. With the devel-

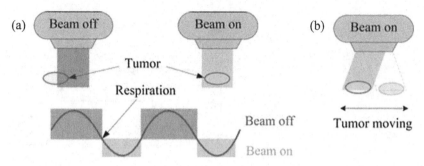

FIGURE 9.9 Mechanism of (a) respiratory-gated radiotherapy and (b) tumor tracking [6].

opment of the integrated circuit, the radar system can be made even smaller and has prospects for wide commercial application

9.2.2.2 Cancer medical application

Cancer is an important chronic disease that has always troubled people. Radiation therapy has been a major modality for cancer treating. Studies show that improved local control and survival rates can be achieved through increased radiation. However, the tumors in many anatomic sites (e.g., lungs and liver) move significantly (~2–3 cm) with respiration. The movement of the tumor with the breath makes it a significant challenge to provide a sufficient radiation dose without causing secondary cancer or severe radiation damage to the surrounding healthy tissue [6,41,42].

Motion-adaptive radiotherapy can solve the problem of tumor movement in the process of radiation-dose delivery, where respiratory gating and tumor tracking are two promising methods. However, an accurately measured respiration pattern is necessary for both of the approaches to either generate gating signals or derive the real-time tumor locations [6]. CW radar is a promising method to accurately monitor the respiration of the patient without additional equipment having to be attached.

The mechanism of respiratory-gated radiotherapy is shown in Fig. 9.9(a). Only when the tumor moves into the radiation coverage is the radiation dose delivered to the tumor. The radiation is turned off in the other situations to protect the healthy tissue. Gating of the radiation beam is based on either the amplitude or phase of the respiration signal. Two essential parameters, i.e., duty cycle and residual motion, characterize a gated treatment. Residual motion is the amount of tumor motion during the incidence of the radiation beam, and duty cycle is the percentage of time that the radiation is turned on during a breathing cycle. In respiratory gating, a trade-off must be made between the duty cycle and residual motion to decrease the side effects of radiotherapy. A longer duty cycle makes the overall treatment shorter but also means longer radiation exposure and larger residual motion of the tumor, leading to more exposure of healthy

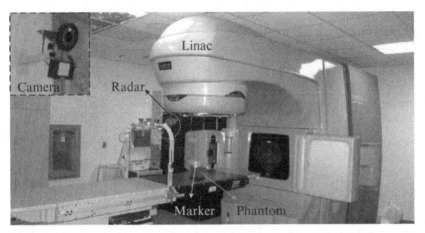

FIGURE 9.10 An actual motion-adaptive radiotherapy system based on radar respiration sensing [6].

tissues to the radiation. For a particular patient, the duty cycle must be determined by the doctor in the treatment-preparation stage according to the specific treatment strategies.

Real-time tumor tracking for motion-adaptive radiotherapy can also be realized by the CW radar [43]. In respiratory gating, a trade-off between the duty cycle and residual tumor motion must be leveraged to achieve better treatment. However, in this treatment regimen, the radiation beam tracks the moving tumor dynamically in real time, as shown in Fig. 9.9(b). Therefore, the demerit is eliminated by this type of real-time tumor-tracking treatment. An actual motion-adaptive radiotherapy system based on radar respiration sensing is shown in Fig. 9.10.

9.3 Vital signs monitoring using FMCW radar

9.3.1 Composition of an FMCW radar system

The composition of an FMCW radar system primarily includes: a sawtooth-wave signal generator, a transceiver module, a digital-to-analog converter, and a digital signal processor and controller. The signal is generated by the sawtooth-wave signal generator, passed through the frequency multiplier and the power amplifier (PA) in turn, and transmitted by the transmitting antenna. The electromagnetic wave is reflected by the object and received by the receiving antenna. The received signal is mixed with the transmitted signal after passing through a low-noise amplifier (LNA) to obtain an intermediate frequency (IF) signal. Finally, the intermediate frequency signal is filtered and ADC sampled sequentially, and then sent to the back end for further signal processing. The block diagram of a typical FMCW radar system is shown in Fig. 9.11.

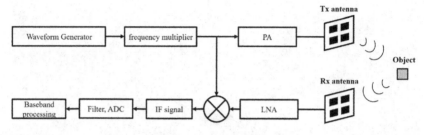

FIGURE 9.11 Block diagram of a typical FMCW radar system.

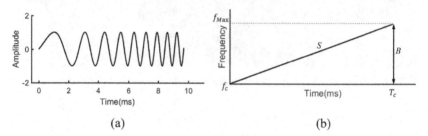

FIGURE 9.12 Block diagram of a typical FMCW radar system.

9.3.2 Analysis of an FMCW radar IF signal

For FMCW radar, a variety of modulations is possible. The transmitter frequency can vary up and down as follows: sine wave, sawtooth wave, triangle wave, square wave, etc.

Sawtooth modulation is the most used in FMCW radars, so the following analysis is based on the sawtooth wave.

The sawtooth wave of an FMCW radar is a frequency linear-modulation method. The frequency of electromagnetic waves changes linearly with time. The schematic diagram is shown in Fig. 9.12. Fig. 9.12(a) shows the time-domain representation. The transmitted signal in a cycle is usually called a chirp. Fig. 9.12(b) shows the frequency-time diagram of a chirp.

Denoting T_c as the chirp repetition period, B is the bandwidth of the chirp, and $S = B/T_c$ is the slope of the frequency modulation. The mathematical expression for the transmitted signal within one frequency ramp interval is:

$$s_{\text{Tx}}(t) = \exp\left(\text{j}\left(2\pi f_c t + \pi S t^2 + \phi\right)\right), \qquad (9.15)$$

where f_c is the center frequency of the frequency ramp, ϕ is the initial phase residual, and $t \in [-T_c/2, T_c/2]$. Suppose that there is a reflection point, and its distance from the radar as a function of time is $R(\tau)$. Assuming that the movement of the scattering point is relatively slow, $R(\tau)$ can thus be regarded as a constant within a certain period. This is a "stop-and-go" hypothesis, which is extremely common when dealing with slow-moving targets. Therefore, for the scattering point located at $R(\tau)$, the echo signal received by the FMCW radar is a function of the time delay Δt and a certain amplitude attenuation σ of the transmitted signal. Among them, the time delay is:

$$\Delta t = \frac{2R(\tau)}{c} \tag{9.16}$$

Therefore, the echo signal can be expressed as:

$$s_{Rx}(t) = \sigma s_{Tx}(t - \Delta t)$$
$$= \sigma \cdot \exp\left(j\left(2\pi f_c (t - \Delta t) + \pi S (t - \Delta t)^2 + \phi\right)\right). \tag{9.17}$$

According to Fig. 9.11, the received signal is mixed with the transmitted signal. After that, the resulting mixed signal is low-pass filtered and the intermediate frequency signal is obtained:

$$s_{IF}(t) = f_{LPF}\{s_{Tx}(t) \cdot s_{Rx}(t)\}$$
$$= \sigma \cdot \exp\left(j\left(\frac{4\pi S R(\tau) t}{c} + \frac{4\pi f_c R(\tau)}{c} - \frac{4\pi S R^2(\tau)}{c^2}\right)\right). \tag{9.18}$$

It can be seen from Eqs. (9.16)–(9.18) that the obtained intermediate frequency signal is a sinusoidal motion with frequency f_{IF}:

$$f_{IF} = \frac{2S R(\tau)}{c}. \tag{9.19}$$

9.3.3 FMCW radar parameter estimation

Based on the intermediate frequency signal obtained in Sect. 9.2.2, parameters such as the position, velocity, and angle of the target object can be estimated. The parameter estimation methods are discussed separately next.

9.3.3.1 FMCW radar range estimation

Since the frequency of the EM wave emitted by the FMCW radar system changes linearly with time, the frequency of the received EM wave also changes linearly with time. Therefore, when the frequency of the IF signal obtained by mixing the transceiver signals is determined, the distance of the object can be obtained according to the known slope S of the sawtooth-wave modulation.

As shown in Fig. 9.13, the transmitted signal is received by the radar after time τ. Therefore, the transmitted Tx chirp and the received Rx chirp are two parallel lines with the same slope on the frequency-time graph. Note that the slope of the transmitted sawtooth wave is S, and the duration of the transmitted signal is t_2. The frequency-time diagram of the IF signal is shown in Fig. 9.13(b). The frequency of the intermediate frequency signal $f_{IF} = S\tau$.

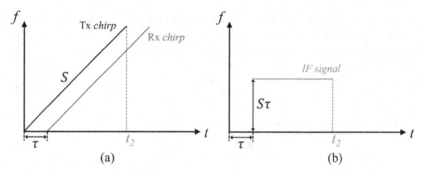

FIGURE 9.13 FMCW intermediate frequency signal diagram.

Therefore, when there is a single object in front of the radar, the frequency-time diagram of the generated intermediate frequency signal is a single-valued line, and the frequency is proportional to the EM wave-propagation time. From Eq. (9.16), we know that the propagation time is proportional to the distance from the object to the radar, that is $\tau = 2d/c$, where d is the distance from the object to the radar.

In summary, the distance from the object to the radar can be obtained from the frequency f_{IF} of the intermediate frequency signal obtained in Eq. (9.18):

$$d = \frac{cf_{IF}}{2S} \tag{9.20}$$

It should be noted that, when using ADC for sampling, the time-point of interest must be within the time window of $[\tau, t_2]$.

9.3.3.2 FMCW radar velocity estimation

The FMCW radar speed estimation requires the use of multiple periods of Tx chirps. By extracting the changes of the phase information of multiple chirps, the displacement of the object within the two sawtooth-wave periods is obtained, and then the estimation of the object's velocity is completed.

The first step is the phase analysis of the FMCW radar IF signal. The top graph in Fig. 9.14 is the time-domain waveform of the transmitted signal. Assuming that two chirps are continuously transmitted, the repetitive time of the transmitted signal waveform is T_c. Since the repetitive time T_c is generally in the μs level, which is very small relative to the time axis, the image of the two consecutive transmitted signal waveforms on the time axis can be considered to be overlapping. That is, the top image is actually a superimposed image of two transmitted waveforms with very short intervals on the time axis.

Suppose the first Tx chirp emitted returns after time τ, forming the light blue (light gray in print version)waveform in the middle image in Fig. 9.14, and the second Tx chirp returns after time $(\tau + \Delta t)$ as a dark blue (dark gray in print version) waveform. The Δt delay is caused by a slight displacement of

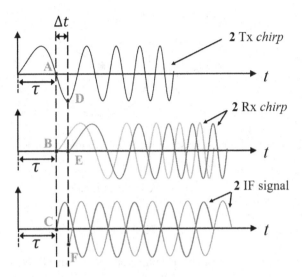

FIGURE 9.14 Phase analysis of FMCW radar IF signal.

the object within two repetitive periods of emitted Tx signal waves. It should be noted that τ and Δt are also very short periods of time, so they are enlarged for display here.

The bottom figure of Fig. 9.14, is the IF signal formed by the two received signals and the transmitted signal, respectively. By analyzing the phases of the two IF signals, the moving distance Δd of the object during time T_c can be obtained.

It can be obtained from Fig. 9.14 that the phase difference between point A and point D is $\Delta\Phi = 2\pi f_c \Delta t = 4\pi \Delta d/\lambda$. At the same time, we get the relationship in the waveform: point B and point E are 0 phase; the phase of point C is the negative value of phase A; and the phase of point F is the negative value of phase D. Therefore, the phase difference $\Delta\Phi$ between point A and point D is also the phase difference between point C and point F.

According to the conclusion in Sect. 9.3.2, the expression of sinusoidal IF signal can be rewritten as follows:

$$S_{\text{IF}}(t) = A \sin(2\pi f t + \Phi), \tag{9.21}$$

where $f = S \cdot 2d/c$ is the frequency of the signal, c is the speed of light.

When there are two IF signals, the phase difference $\Delta\Phi$ is:

$$\Delta\Phi = \frac{4\pi \Delta d}{\lambda}. \tag{9.22}$$

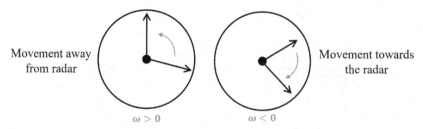

FIGURE 9.15 Relationship between phase sign and object-movement direction.

Note $\omega = \Delta\Phi = 4\pi vTc/\lambda$, so the velocity estimation is:

$$v = \frac{\lambda\omega}{4\pi T_c}. \qquad (9.23)$$

Taking the example that has been mentioned before, for a FMCW radar operating at 77 GHz, the phase change $\Delta\Phi = 4\pi \Delta d/\lambda = 180°$. It can be concluded that the frequency is almost negligible for small movement changes, but the phase information is very sensitive to small distance changes. Therefore, the phase information is used to complete the velocity estimation with higher resolution.

It should be noted that, because the speed is detected using changes in phase information, it must be effective within the range of $|\omega| < \pi$, which limits the maximum upper limit of speed estimation v_{Max}:

$$v_{Max} < \frac{\lambda}{4T_c}. \qquad (9.24)$$

In addition, the direction of the object's movement relative to the radar can be distinguished by the sign of ω.

As shown in Fig. 9.15, when $\omega > 0$, the object moves in the direction away from the radar. When $\omega < 0$, the object moves in the direction of the radar.

9.3.3.3 FMCW radar angle estimation

The angle estimation of the FMCW radar requires at least two receiving antennas. The schematic diagram is shown in Fig. 9.16.

Assuming that the distance between the two receiving antennas is d, since d is relatively small compared to the distance from the object to the radar, the reflected signals reaching the two receiving antennas can be regarded as having parallel incidence. Assuming the normal angle between the object and the receiving antenna is θ, the reflected signals received by the receiving antennas 1 and 2 have a wave-path difference $\Delta d = d \sin\theta$. Therefore, the phase difference of the reflected signals received by the two receiving antennas contains the angle information of the target.

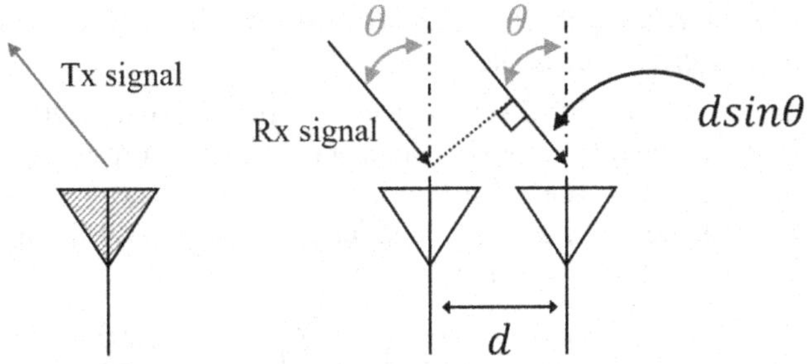

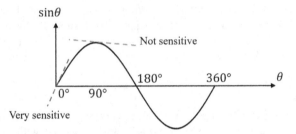

FIGURE 9.16   Schematic diagram of FMCW radar-angle estimation.

FIGURE 9.17   FMCW radar angle-estimation sensitivity with angle.

The phase difference ω between Rx antennas can be expressed as:

$$\omega = \frac{2\pi \Delta d}{\lambda} = \frac{2\pi d \sin\theta}{\lambda}.$$ (9.25)

Inversely, transform Eq, (9.13) to obtain the angle information of the target:

$$\theta = \sin^{-1}\left(\frac{\lambda\omega}{2\pi d}\right).$$

It should be noted that for Eqs. (9.14), this is the first time that nonlinearity occurs in FMCW radar parameter estimation. The nonlinearity of the arc sine function makes the sensitivity differ at different angles.

As shown in Fig. 9.17, when the angle is close to 0°, since the slope of the sine function is large, the change of the arc sine function to the angle is obvious, but when it reaches 90°, the slope of the sine function gradually approaches 0, therefore, the accuracy of angle estimation decreases rapidly. That is, when the object is directly in front of the radar, the angle estimation is the most accurate. The closer the target is to the sides of the radar, the worse the angle estimation accuracy will be.

9.3.3.4   FMCW radar phase-based range-tracking algorithm for vital signs monitoring

Since the displacement of vital signs, such as respiration and heartbeat, is very small, the phase-based range-tracking algorithm has a better result for obtaining accurate vital signs detection. If a Fourier transform is performed over each period of the IF signal (2–4), its associated range profile is derived. The IF signal after Fourier transform can be expressed as:

$$s_{\mathrm{IF}}(f) = \sigma \cdot T_c \cdot \exp\left(j\frac{4\pi f_c R(\tau)}{c}\right) \mathrm{sinc}\left(T_c(f - \frac{2SR(\tau)}{c})\right), \quad (9.26)$$

where $\mathrm{sinc}(x) = \sin(\pi x)/(\pi x)$. After a simple scaling process of the frequency axis, the corresponding range profile can be extracted.

For an LFMCW radar intended to monitor vital signs, a close look must be given to the exponential factor in Eq. (9.26). Denote the phase history in Eq. (9.27) as ϕ_d, then the phase history ϕ_d is simply related to the range evolution of the target $R(\tau)$ by

$$\phi_d = \frac{4\pi f_c R(\tau)}{c}. \quad (9.27)$$

Hence, a proper range tracking of the target requires the preservation of the phase history. Assume that the signal samples associated with distinct chirp intervals are stacked in rows. This constitutes the raw-data matrix, which is denominated M[n, m] ($n = 1, 2, \ldots, N$; $m = 1, 2, \ldots, M$, N being the number of transmitted ramps and M being the number of samples per chirp). The corresponding signal processing to derive the range evolution $R(\tau)$ is divided into four steps:

Step 1 Perform a fast Fourier transform of each row of the raw-data matrix M[n, m]. Denote the resulting range-profile matrix as R[n, m].
Step 2 Choose the range bin $m*$ in which the target is found. Synthesize the signal s[n]=R[n, m^*], which is a column of the range-profile matrix R[n, m].
Step 3 Exact the phase of the signal s[n] and unwrap it. Denote the phase of the signal s[n] as $\psi[n]$.
Step 4 From Eq. (9.27), calculate the range estimation as $\hat{R}[n] = c\psi[n]/(4\pi f_c)$.

9.3.4   Examples of FMCW radar on contactless vital signs monitoring

In recent years, FMCW radar has been widely used in non-contact range tracking of vital signs, e.g., respiration. Next, some examples are given to analyze several typical FMCW radars for vital signs monitoring.

9.3.4.1   Respiration monitoring

The body-surface movements due to physiological motions modulate the phase of the received radar signal and can be further processed to extract the breathing

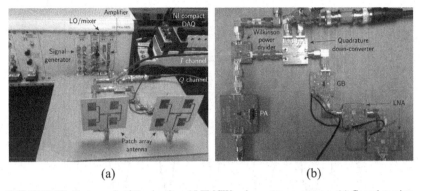

(a)   (b)

FIGURE 9.18 Photograph of the developed LFMCW radar-system prototype. (a) Complete view. (b) Detail [14].

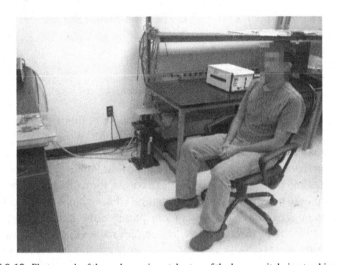

FIGURE 9.19 Photograph of the real experimental setup of the human vital-sign tracking test [14].

and heart rates. A deramping-based LFMCW radar scheme has been proposed [14]. The described LFMCW radar architecture is conceptually simple, and the deramping process greatly simplifies its hardware implementation mainly in terms of sampling speed for the Rx ADC.

The photograph of the constructed LFMCW radar system is depicted in Fig. 9.18. It consists of Tx, Rx, and signal-acquisition modules. A photograph of the experiment setup for vital-sign sensing from human target is depicted in Fig. 9.19.

By using the phase-based range-tracking algorithm mentioned in Sect. 9.3.3.4, minor movements of the human body can be detected. Fig. 9.20 plots the measured human respiration detection result. The respiration rate of the subject was about 13 cycles/min, and it almost coincides with the data measured by radar.

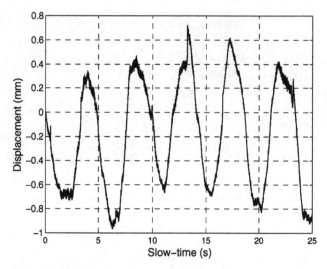

FIGURE 9.20   Detected human respiration pattern [14].

9.3.4.2   Indoor human tracking

Radar-based human tracking can be applied to indoor healthcare scenarios, such as fall detection of elderly people. Some portable FMCW radar prototypes for indoor human tracking are presented [15,16]. The RX part proceeds with the mixing of a replica of the transmitted signal, and the simultaneous sampling of the reference and baseband signals enables a correct formatting of the baseband-signal samples to construct the raw-data matrix, which guarantees the coherence of the system. Therefore, the system can preserve the phase history of the targets, and videos of inverse synthetic aperture radar (ISAR) images can be reconstructed.

9.3.4.3   Hybrid radar systems for human tracking and identification

Several hybrid radar systems that integrate the FMCW mode and interferometry mode have been published [17,18]. The FMCW mode is responsible for absolute range detection, and the interferometry mode takes care of weak physiological-movement monitoring. In [17], two radar modes share most of the RF components, and signal paths and analog switches configured by an on-board microcontroller are used to select the operational modes.

9.3.4.4   Other types of FMCW radar vital signs detection applications

A PCB realization of a K-band portable FMCW radar with beamforming array is presented [19]. It demonstrates an alternative approach to achieve portable and low-cost beamforming array radar systems with vector controllers and a six-port circuit. Range-gating and beamforming techniques enable the signal of interest to be isolated from surrounding clutter [20].

In addition, some companies such as Texas Instruments, Infineon, and Calterah have developed a series of FMCW radar products [44], and these products have greatly contributed to the development of vital signs monitoring using FMCW radars. TI's AWR6843 is an integrated single chip mmWave sensor based on FMCW radar technology capable of operation in the 60-GHz to 64-GHz band. It is built with TI's low power 45-nm RFCMOS processor and enables unprecedented levels of integration in an extremely small form. Infineon's BGT60TR13C Radar Sensor is a short range 60-GHz radar chirp utilized in smartphone Pxiel4 for gesture recognition. The radar chirp is about the size of a match head. CALTERAH's CAL60S244-IB radar sensor is a 4T4R 60 GHz-FMCW Radar SoC with an antenna array embedded in the package.

9.4   Conclusion

This chapter summarizes the basic theory and the signal processing algorithms of FMCW/CW radars. On this basis, several application examples of FMCW/CW radars in contactless vital signs monitoring are introduced, including cardiopulmonary monitoring, human-gait recognition, cancer medical application, and indoor human tracking. With the advantage of high integrability, strong environmental adaptability, low-power consumption, and penetrability, FMCW/CW radars have wide application prospects in our lives, and contactless vital signs monitoring is one of the most important scenarios.

References

[1] S. Zhang, P.S. Huang, High-resolution, real-time three-dimensional shape measurement, Opt. Eng. 45 (12) (Dec. 2006) 123601-1–123601-8.

[2] Y. Wang, K. Liu, Q. Hao, D.L. Lau, L.G. Hassebrook, Period coded phase shifting strategy for real–time 3-D structured light illumination, IEEE Trans. Image Process. 20 (11) (Nov. 2011) 3001–3013.

[3] F.-K. Wang, T.-S. Horng, K.-C. Peng, J.-K. Jau, J.-Y. Li, C.-C. Chen, Seeing through walls with a self-injection-locked radar to detect hidden people, in: IEEE MTT-S Int. Microw. Symp. Dig, Jun. 2012, pp. 1–3.

[4] F.K. Wang, T.S. Horng, K.C. Peng, J.K. Jau, J.Y. Li, C.C. Chen, Detection of concealed individuals based on their vital signs by using a see-through-wall imaging system with a self-injectionlocked radar, IEEE Trans. Microw. Theory Tech. 61 (1) (Jan. 2013) 696–704.

[5] X. Liu, H. Leung, G.A. Lampropoulos, Effect of wall parameters on ultra-wideband synthetic aperture through-the-wall radar imaging, IEEE Trans. Aerosp. Electron. Syst. 48 (4) (Oct. 2012) 3435–3449.

[6] C. Gu, et al., Accurate respiration measurement using DC-coupled continuous-wave radar sensor for motion-adaptive cancer radiotherapy, IEEE Trans. Biomed. Eng. 59 (11) (Nov. 2012) 3117–3123.

[7] Y. Kim, H. Ling, Through-wall human tracking with multiple Doppler sensors using an artificial neural network, IEEE Trans. Antennas Propag. 57 (7) (Jul. 2009) 2116–2122.

[8] H. Gao, L. Xie, S. Wen, Y. Kuang, Micro-Doppler signature extraction from ballistic target with micro-motions, IEEE Trans. Aerosp. Electron. Syst. 46 (4) (Oct. 2010) 1969–1982.

[9] C. Li, X. Yu, C.-M. Lee, D. Li, L. Ran, J. Lin, High-sensitivity software-configurable 5.8-GHz radar sensor receiver chip in 0.13-μm CMOS for noncontact vital sign detection, IEEE Trans. Microw. Theory Tech. 58 (5) (May 2010) 1410–1419.

[10] C. Gu, G. Wang, Y. Li, T. Inoue, C. Li, A hybrid radar-camera sensing system with phase compensation for random body movement cancellation in Doppler vital sign detection, IEEE Trans. Microw. Theory Tech. 61 (12) (Dec. 2013) 4678–4688.

[11] C. Li, J. Ling, J. Li, J. Lin, Accurate Doppler radar noncontact vital sign detection using the RELAX algorithm, IEEE Trans. Instrum. Meas. 59 (3) (Mar. 2010) 687–695.

[12] H. Rohling, C. Moller, Radar waveform for automotive radar systems and applications, in: Proc. IEEE Radar Conf, May 2008, pp. 1–4.

[13] M. Mercuri, D. Schreurs, P. Leroux, SFCW microwave radar for in-door fall detection, in: Proc. IEEE Topical Conf. Biomed. Wireless Technol., Netw., Sens. Syst. (BioWireleSS), Santa Clara, CA, USA, Jan. 2012, pp. 53–56.

[14] G. Wang, J. Muñoz-Ferreras, C. Gu, C. Li, R. Gómez-García, Application of linear-frequency-modulated continuous-wave (LFMCW) radars for tracking of vital signs, IEEE Trans. Microw. Theory Tech. 62 (6) (June 2014) 1387–1399.

[15] Z. Peng, J. Muñoz-Ferreras, Y. Tang, R. Gómez-García, C. Li, Portable coherent frequency-modulated continuous-wave radar for indoor human tracking, in: 2016 IEEE Topical Conference on Biomedical Wireless Technologies, Networks, and Sensing Systems (BioWireleSS), Austin, TX, 2016, pp. 36–38.

[16] Z. Peng, J. Muñoz-Ferreras, R. Gómez-García, L. Ran, C. Li, 24-GHz biomedical radar on flexible substrate for ISAR imaging, in: 2016 IEEE MTT-S International Wireless Symposium (IWS), Shanghai, 2016, pp. 1–4.

[17] Z. Peng, et al., A portable FMCW interferometry radar with programmable low-IF architecture for localization, ISAR imaging, and vital sign tracking, IEEE Trans. Microw. Theory Tech. 65 (4) (April 2017) 1334–1344.

[18] G. Wang, C. Gu, T. Inoue, C. Li, A hybrid FMCW-interferometry radar for indoor precise positioning and versatile life activity monitoring, IEEE Trans. Microw. Theory Tech. 62 (11) (Nov. 2014) 2812–2822.

[19] Z. Peng, L. Ran, C. Li, A K-band portable FMCW radar with beamforming array for short-range localization and vital-Doppler targets discrimination, IEEE Trans. Microw. Theory Tech. 65 (9) (Sept. 2017) 3443–3452.

[20] A. Ahmad, J.C. Roh, D. Wang, A. Dubey, Vital signs monitoring of multiple people using a FMCW millimeter-wave sensor, in: 2018 IEEE Radar Conference (RadarConf18), 2018, pp. 1450–1455.

[21] X. Wang, et al., Noncontact distance and amplitude-independent vibration measurement based on an extended DACM algorithm, IEEE Trans. Instrum. Meas. 63 (1) (2014).

[22] S. Kim, C. Nguyen, On the development of a multifunction millimeter-wave sensor for displacement sensing and low-velocity measurement, IEEE Trans. Microw. Theory Tech. 52 (11) (Nov. 2004) 2503–2512.

[23] B. Park, O. Boric-Lubecke, V.M. Lubecke, Arctangent demodulation with DC offset compensation in quadrature Doppler radar receiver systems, IEEE Trans. Microw. Theory Tech. 55 (5) (May 2007) 1073–1079.

[24] K. Itoh, Analysis of the phase unwrapping problem, Appl. Opt. 21 (14) (Jul. 1982) 2470.

[25] F. Schadt, F. Mohr, M. Holzer, Application of Kalman filters as a tool for phase and frequency demodulation of IQ signals, in: Proc. IEEE Int. Conf. Comput. Technol. Electr. Electron. Eng., Jul. 2008, pp. 421–424.

[26] Wei Xu, Changzhan Gu, Jun-da Mao, Noncontact high-linear motion sensing based on a modified differentiate and cross-multiply algorithm, in: IEEE MTT-S International Microwave Symposium, 2020.

[27] F.-K. Wang, T.-S. Horng, K.-C. Peng, J.-K. Jau, J.-Y. Li, C.-C. Chen, Single-antenna Doppler radars using self and mutual injection locking for vital sign detection with random body movement cancellation, IEEE Trans. Microw. Theory Tech. 59 (12) (Dec. 2011) 3577–3587.

[28] T. Kao, Y. Yan, T. Shen, A. Chen, J. Lin, Design and analysis of a 60-GHz CMOS Doppler micro-radar system-in-package for vital-sign and vibration detection, IEEE Trans. Microw. Theory Tech. 61 (4) (Apr. 2013) 1649–1659.

[29] A.D. Droitcour, O. Boric-Lubecke, V.M. Lubecke, J. Lin, G.T.A. Kovacs, Range correlation and I/Q performance benefits in single-chip silicon Doppler radars for noncontact cardiopulmonary monitoring, IEEE Trans. Microw. Theory Tech. 52 (3) (Mar. 2004) 838–848.

[30] W. Xu, C. Gu, C. Li, M. Sarrafzadeh, Robust Doppler radar demodulation via compressed sensing, Electron. Lett. 48 (22) (Oct. 2012).

[31] Chao-Hsiung Tseng, Yi-Hua Lin, 24-GHz self-injection-locked vital-sign radar sensor with CMOS injection-locked frequency divider based on push–push oscillator topology, IEEE Microw. Wirel. Compon. Lett. 28 (11) (2018).

[32] A. Singh, et al., Data-based quadrature imbalance compensation for a CW Doppler radar system, IEEE Trans. Microw. Theory Tech. 61 (4) (Apr. 2013) 1718–1724.

[33] M. Zakrzewski, et al., Quadrature imbalance compensation with ellipse-fitting methods for microwave radar physiological sensing, IEEE Trans. Microw. Theory Tech. 62 (6) (Jun. 2013) 1400–1408.

[34] C. Gu, Z. Peng, C. Li, High-precision motion detection using low-complexity Doppler radar with digital post-distortion technique, IEEE Trans. Microw. Theory Tech. 64 (3) (March 2016).

[35] J.C. Lin, Noninvasive microwave measurement of respiration, Proc. IEEE 63 (10) (Oct. 1975).

[36] C. Li, J. Lin, Random body movement cancellation in Doppler radar vital sign detection, IEEE Trans. Microw. Theory Tech. 56 (12) (Dec. 2008).

[37] C. Gu, C. Li, J. Lin, J. Long, J. Huangfu, L. Ran, Instrument-based noncontact Doppler radar vital sign detection system using heterodyne digital quadrature demodulation architecture, IEEE Trans. Instrum. Meas. 59 (6) (June 2010).

[38] C. Gu, C. Li, Distortion analysis of continuous-wave radar sensor for complete respiration pattern monitoring, in: IEEE Topical Conference on Biomedical Wireless Technologies, Networks, and Sensing Systems, Jan. 2013.

[39] W. Hu, Z. Zhao, Y. Wang, H. Zhang, F. Lin, Noncontact accurate measurement of cardiopulmonary activity using a compact quadrature Doppler radar sensor, IEEE Trans. Biomed. Eng. 61 (3) (March 2014).

[40] J. Tu, T. Hwang, J. Lin, Respiration rate measurement under 1-D body motion using single continuous-wave Doppler radar vital sign detection system, IEEE Trans. Microw. Theory Tech. 64 (6) (June 2016).

[41] S.B. Jiang, Radiotherapy of mobile tumors, Semin. Radiat. Oncol. 16 (4) (Oct. 2006) 239–244.

[42] S.B. Jiang, Technical aspects of image-guided respiration gated radiation therapy, Med. Dosim. 31 (2) (2006) 141–151.

[43] C. Gu, R. Li, C. Li, S.B. Jiang, A multi-radar wireless system for respiratory gating and accurate tumor tracking in lung cancer radiotherapy, in: Proc. 33rd Annu. Int. Conf. IEEE Eng. Med. Biol. Soc., Boston, MA, Aug. 2011, pp. 417–420.

[44] A. Santra, et al., Short-range multi-mode continuous-wave radar for vital sign measurement and imaging, in: 2018 IEEE Radar Conference (RadarConf18), Oklahoma City, OK, 2018, pp. 0946–0950.

Chapter 10

Received power-based vital signs monitoring

Jie Wang[a], Alemayehu Solomon Abrar[b], and Neal Patwari[a]

[a]*Washington University in St. Louis, St. Louis, MO, United States,* [b]*Microsoft Azure, Redmond, WA, United States*

Contents

10.1 Introduction

Vital signs, including respiration rate and pulse rate, are critical monitoring tools that help physicians make decisions about diagnosis and treatment. Non-contact radio-frequency (RF) human sensing techniques exploit radio signals to detect human vital signs without requiring a user to carry or wear any sensors. In this chapter, we explore the capability to monitor the vital signs of respiration and pulse rates using changes in received power. Such received power-based vital signs monitoring is a promising potential candidate in smart-health and smart-home applications.

Six in ten adults in the US have a chronic disease, and four in ten adults in the US have two or more [1]. Home health monitoring systems for people with chronic medical conditions are a means to continually monitor their conditions without hospitalization. A traditional way of monitoring respiration and heart activity is to make use of electrocardiography (ECG) or phonocardiogram (PCG) sensors that require electrodes attached to the body. However, the expense and discomfort of the sensors and wires make these methods a poor fit for continual home health monitoring. Contact-free sensing methods are promising in terms of cost and comfort.

Various non-contact vital signs monitoring techniques have been developed in recent years, many of which are the subjects of this book. These systems may be based on video sensing or radio-frequency (RF) sensing. Video-based non-contact vital sign systems monitor cardiac activities using remote photoplethysmography (rPPG), in which the color variations of the human skin induced by blood-volume changes can be captured by a RGB camera [2,3]. This method is less sensitive to the surroundings than RF-based systems and suitable for continuous monitoring scenarios [4]. Unfortunately, the average performance of rPPG is significantly lower for people with darker skin color, lower for women compared to men, and lower under low-light conditions or with motion [5]. It would also suffer from body motion due to dramatic variations of light reflection [6]. Finally, video-based sensing raises privacy issues.

Another RF-based vital signs monitoring approach is to use radar. Radar-based vital signs monitoring systems apply time-of-flight (TOF) or channel impulse response (CIR) measurements to detect certain human activities [7,8]. Such systems have been shown to be accurate for estimation even at long distances [9]. RF sensing generally has minimal privacy concerns, however, radar-based monitors use a very large bandwidth, which then must compete with wireless communications for use of the scarce spectrum. Radar hardware can be significantly more expensive and require higher transmit powers to combat $1/d^4$ scattering losses. Another RF-based monitoring method uses WiFi channel state information (CSI). Some commercial off-the-shelf (COTS) WiFi devices enable measurement of CSI, which is altered by breathing and pulse-induced vibrations. Similar problems exist in CSI-based monitoring due to its use of a wide bandwidth within the WiFi bands, however, its transmit power can be relatively low. We note that received power measurements may similarly use low transmit power, but, in contrast to WiFi CSI, can be made from narrowband signals.

Although vital signs monitoring techniques that use received power show some advantages over other methods, the following challenges have to be addressed first to detect vital signs using these measurements:

- Low-amplitude signal: The amplitude of the pulse-induced and breathing-induced received power signal is very small. Furthermore, variations in received power due to pulse (typically less than 0.01 dB) are about an order of magnitude smaller than those due to breathing.
- Quantization: Most commercial transceiver ICs report received power, which quantizes received power with a 1-dB or larger step size. As we describe in this chapter, this quantization is the fundamental limitation of past work in received power-based vitals monitoring.
- Noise: The noise power in the received power signal can be significantly larger than the pulse signal power. In addition, the noise is heavy-tailed, prone to large impulses. In other words, the low signal-to-noise (SNR) of pulses in the power measurements may degrade estimation performance.

- Non-sinusoidal waveform: The movement of the skin due to the pulse more closely resembles a repeating impulse than a sinusoid, so the spectral analysis is suboptimal.

In this chapter, we describe a hardware and software RF-sensing system that utilizes the narrowband received power measurements to handle these issues and obtain accurate pulse-rate and respiration-rate estimates, simultaneously. The first idea is to develop a fine-resolution received power measurement system based on low-cost narrowband transceivers. Our intuition is that received power is primarily noisy due to quantization. Therefore, we use choose a narrowband transceiver SoC that enables us to obtain raw signal samples, which we use to calculate unquantized received power. Second, because we have a single-dimensional signal, which precludes PCA-based denoising methods as used in CSI-based monitoring systems, temporal features are exploited to remove noise. We use a combination of Hampel and bandpass filters to separate breathing patterns from pulse features and also cancel out heavy-tailed, high-frequency noise. Finally, we introduce an estimator to combine the harmonics in the magnitude spectrum to improve estimation performance. Our experimental results indicate that the proposed received power-based vital signs monitoring system can estimate respiration rates with a root mean squared error (RMSE) typically less than 1.0 breaths per minute (bpm) and monitor pulse rates within 1.6 bpm of RMSE in various experimental settings.

The rest of the chapter is organized as follows. Section 10.2 introduces related work in received power-based vital signs monitoring techniques. In Sect. 10.3, we present the physical model for vital signs monitoring and the proposed estimation method. Detailed implementation of the received power based vital signs monitoring system is described in Sect. 10.4. In Sect. 10.5, we present our experimental results and estimation accuracy. Section 10.6 summarizes the work and draws conclusions.

10.2 Related work

Respiration and pulse rates are important vital signs that can help monitor people's health state, indicate sleep quality, and reveal anomalies. In this section, we describe non-invasive technologies for vital signs monitoring, focusing on received power-based methods.

In recent years, research scholars have put forward various approaches to realize reliable and accurate vital signs monitoring using RF sensing. Approaches can generally be classified into two types:

1. Radar approaches: These transmit and receive from the same location (and thus are monostatic) and record either the time delays or the Doppler shift of the scattered signal [10,11].
2. Repurposing wireless transceivers: These approaches use WiFi or other transceivers to record channel measurements on the link between a separate transmitter and receiver [12].

10.2.1 Radar approaches

One category for respiration-rate monitoring is radar-based sensing, which includes three different types of radar measurements: (1) Continuous-wave (CW) Doppler radar [13–17]; (2) impulse radio ultra-wideband (IR-UWB) radar [18–21]; and (3) frequency-modulated continuous-wave (FMCW) radar [22–24]. With the advantages of a non-contact nature and easy implementation, numerous novel applications have been proposed according to the radar-based vital signs monitoring techniques. The CW Doppler radar system have been shown to be a potential alternative in sudden infant death syndrome (SIDS) detection [25] and long-term sleep apnea monitoring [26]. The IR-UWB radar for vital signs monitoring in [21] also displays its ability to monitor driving activity for car-crash prevention.

CW Doppler radar estimates vital signs through the phase or frequency shifts in the reflected signal as compared to the transmitted signal [13,17]. An IR-UWB radar system directly measures the time-of-arrival (ToA) of the multipath components that reflect from the body and looks for changes that can be attributed to breathing and pulse [18,19]. By separating multipath components at different time delays, it is possible to realize multiple person monitoring. The third radar system, FMCW, measures the same time delays in the channel in the frequency domain, by linearly changing the frequency [22] of a transmitted sinusoidal signal, which can simplify transceiver design compared to impulse radar. Among all the FMCW radar systems, the wide ranges of bandwidth available at the millimeter wave (mmWave) radar sensor [27], along with the emergence of low-cost radar hardware, make it an attractive band for vital signs monitoring [28,29]. The wide bandwidth at mmWave frequencies enables fine-time resolution, and the small wavelength permit small antenna arrays, and thus beamforming, for the purpose of vital signs monitoring [30]. However, all radar systems measure scattering from the body, which is subject to the radar scattering model, which states that the signal power is proportional to $1/d^4$ in free space [31]. SNR issues are one of the main challenges for radar systems [32] that can be compensated for by increasing transmit power.

10.2.2 Repurposing wireless transceivers

In this category, the methods use low-cost COTS wireless devices for the purpose of making measurements of the radio channel that are changed by a person's breathing or pulse. While such systems are unable to measure the phase and frequency of a received signal to the same degree of accuracy as a radar system because the transmitter and receiver are separate devices, they benefit from being bistatic because they don't have a strong self-interference component (in which the transmitted signal leaks directly into the receiver chain). Furthermore, the multipath components that are affected by the person are often of the same order of magnitude as those unaffected by the person's motion.

Repurposed transceivers vary by orders of magnitude in the bandwidth they require, based on the protocol they implement. Most RF sensors in the recent literature use WiFi CSI, which leverages a 20-MHz channel [33]; others use Zigbee signals that occupy a two-MHz frequency band [34], and our proposal is to use low-rate low bandwidth protocols on the order of ten kHz for RF sensing [35]. As wireless communications continues to expand its spectral footprint, we expect that minimizing RF-sensing bandwidth will be of increasing importance.

Further, repurposed transceivers vary by what channel information is available for RF sensing. OFDM transceivers, like WiFi, measure CSI, and recent hacks of the Intel 5300 [33] and Atheros [36] WiFi chipsets have enabled using CSI for RF-sensing applications. Historically, received power (also called received signal strength, RSS) has been the primary channel measurement available from commercial wireless devices, and thus for RF sensing [37–39].

Received power measurements were used in the first systems that demonstrated breathing monitoring on repurposed wireless devices [40]. Retrospectively, it took three years to publish [41] because those who worked in wireless were trained and conditioned to understand the channel as a highly time-varying medium due to the motion of people or the motion of the wireless devices, and thus that breathing-induced changes would be insignificant compared to other time variations. However, when people and endpoints are not in motion, the channel is typically quite static.

Received power has been implemented for breathing monitoring using Zigbee [42], WiFi [43,44]. The general framework of received-power-based breathing monitoring is to filter the received power measurements according to the normal breathing-rate range and then estimate the breathing rate as the frequency that corresponds to the peak value in the spectrum. The key problem with received power is that it is typically quantized with a 1-dB step size or higher when it is provided by the transceiver IC as a received signal strength indicator (RSSI). In general, RSSI refers to the quantized received power in the entire received signal. In WiFi, this is the power inclusive of all subchannels, while in Zigbee or BLE there is only one channel. One may use a one-channel transceiver to hop across frequency channels to sequentially measure received power vs. frequency. Since breathing may have an amplitude on the order of 0.1 dB and pulse rate may induce an amplitude on the order of 0.01 dB, it is unlikely that breathing may be observed from RSSI quantized with a 1-dB step size [45], and even more unlikely that the pulse-induced changes appear [35]. The reported systems that have used received power for vital signs monitoring have done so using one or more of five methods to overcome this fundamental problem:

- *Link diversity*: By measuring hundreds of links, one can ensure that *some* links measure the signal [40]. As a downside, this generally requires many nodes, but on the upside, this also enables localization of a breathing person [46]. Antenna diversity from MIMO devices provides this diversity with fewer nodes [43].

- *Frequency diversity*: By measuring multiple frequency channels, one can observe the RSSI at various average values due to the frequency selectivity of multipath channels, and again, ensure that some channels observe the signal [47]. Note that CSI measurements also use frequency diversity.
- *Time diversity*: This refers to oversampling the signal in time, $\gg 2\times$ the maximum vital-sign frequency. By then averaging or low-pass filtering, one can observe changes in the probability that the RSSI value crosses a threshold, even if the mean never does [47].
- *Increasing the noise*: In the time diversity method, low standard deviation of the measurement can, counter-intuitively, be a problem. If there is very low noise, the RSSI value will rarely cross a threshold due to breathing. A higher noise standard deviation, optimally $1/4$ of the quantization step size, improves the ability to measure vital signs from an oversampled RSSI stream [35,48].
- *Reduce or remove quantization*: If it is possible to obtain from standard wireless transceivers, we can observe the received power itself, not a quantized version. For example, transceivers that provide access to samples can be used to calculate a floating-point received power with 0.013 dB median error [45]. CSI from WiFi is given with an eight-bit quantized value, but as the gain control is separate, this resolution is quite sufficient [33].

Methods that use CSI are successful both because it is not coarsely quantized and because it provides frequency and MIMO antenna-pair diversity. It also provides the phase, in addition to the amplitude, of the channel. Research has shown that WiFi CSI-based systems enable concurrent pulse and respiration rates estimation. For instance, the WiFi network system in [49,50] leverages CSI to monitor pulse and respiration rates and the ability to monitor up to two people during sleep. The "ResBeat" module in [51] and the "PhaseBeat" system in [52] rely on CSI phase differences, including its stability and periodicity, for long-term pulse and respiration-rate monitoring. A separate chapter in this book is devoted strictly to CSI-based vital signs monitoring, so we refer the reader to it for a more detailed discussion.

However, we note that the problems of RF-based vital signs monitoring are not yet solved. We experimentally compared IR-UWB, WiFi CSI, and two different narrowband received power methods in [53]. The results show that all four methods can be very accurate, but that all four have a percentage of time during which the patient's breathing is not observable, and the rate estimates are essentially random guesses.

In the following sections, we will present a vital signs monitoring system that uses a low-power narrowband receiver and measures received power with little quantization error. Compared to our past work [45,53], it measures received power measured on the links between multiple antennas of the transmitter and receiver, and it uses both frequency diversity and time diversity. We demonstrate how such a system can provide reliable vital signs monitoring despite its very low bandwidth and its use of a low-cost commercial IoT transceiver.

10.3 Received power-based vital signs monitoring

In this section, we will provide the received power model for vital signs monitoring and estimate rates from received power. Again, received power is the squared magnitude of the received signal. The model is first developed for linear received power, but we then extend the model to logarithmic (dB) received power. Note that RSSI is a quantized version of dB received power. Any of these (linear or dB received power or RSSI) can be measured in various frequency channels by changing the channel of operation of the transmitter and receiver, so we write it as a function of f, frequency.

10.3.1 Received power model

We consider a pair of narrowband radio transceivers used as transmitter and receiver as shown in Fig. 10.1. We assume that a transmitter and receiver are separated by a distance d. The transmitted RF signal propagates along multiple paths, including the line-of-sight (LOS) path and those reflected by nearby objects such as walls, furniture, and the human body. Note that we quantify the effects of the variation in signal strength caused by the chest movements and skin vibrations of a human subject near the wireless link.

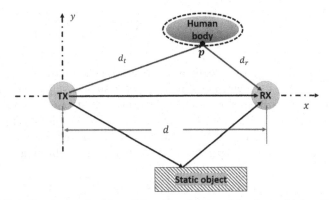

FIGURE 10.1 Signal transmitted from TX positioned at the origin takes multiple paths before it arrives at an RX node at $(d, 0)$. The signal can take time-invariant paths including line-of-sight path and reflections from stationary objects and can also be reflected off the human body positioned d_t and d_r away from TX and RX.

At a given carrier frequency f and time t, the frequency domain representations of transmitted and received signals can be denoted by $X(f, t)$ and $Y(f, t)$, respectively. Then, $Y(f, t) = H(f, t) \times X(f, t)$, where $H(f, t)$ represents a complex-valued channel frequency response (CFR).

The transmitted wireless signal can take multiple paths and be reflected from nearby objects before it reaches the receiver antenna. Assuming that there are N

multipath components, $H(f, t)$ becomes:

$$H(f, t) = e^{-j2\pi \Delta f t} \sum_{k=1}^{N} a_k(t) e^{-j2\pi f \tau_k(t)}, \qquad (10.1)$$

where $a_k(t)$ is the complex-valued attenuation coefficient, $\tau_k(t)$ is the propagation delay for the kth multipath component, and Δf is the carrier frequency offset between the transmitter and receiver. We also define $\eta_k(t)$ to be the path length for k^{th} dynamic path, which would be $d_t + d_r$ in Fig. 10.1 for the path reflecting from the human body. Note that the phase

$$2\pi f \tau_k(t) = \frac{2\pi}{\lambda} \eta_k(t).$$

We show that, when the lengths of different paths $\{\eta_k(t)\}_k$ change due to slight periodic motion, the CFR power $\|H(f, t)\|^2$ also varies periodically. The CFR can be represented as the sum of dynamic and static components. The static component $H_s(f)$ includes CFR from all static paths where the t dependence is removed since it does not change with time. The dynamic components, $k \in \mathcal{D}$, include the paths with periodic changes in path length due to breathing and skin vibrations due to a pulse [41]. Thus

$$H(f, t) = e^{-j2\pi \Delta f t} \left[H_s(f) + \sum_{k \in \mathcal{D}} a_k(f) e^{-j\frac{2\pi}{\lambda} \eta_k(t)} \right]. \qquad (10.2)$$

We consider an otherwise stationary person with chest movements and skin vibrations due to vital signs. For simplicity, we assume a single dynamic path with a complex attenuation coefficient a which is constant over time and has path length $\eta(t)$. If there are multiple dynamic paths and the kth path has $\eta_k(t) = \kappa \eta(t)$ for some constant κ, we could simplify (10.2) to the same form. At any single frequency, we can drop f from the notation for simplification:

$$H(t) = e^{-j2\pi \Delta f t} \left[H_s + a e^{-\frac{j2\pi \eta(t)}{\lambda}} \right]. \qquad (10.3)$$

The dynamic path length $\eta(t)$ is considered to change as the person's body moves with inhalation, exhalation, and the vibration caused by blood flow. We diagram the phasor sum in (10.3) in Fig. 10.2.

We focus on (10.3) because it helps us understand two critical observations we often observe in measured data, first regarding the amplitude of the vital sign–induced signal and second regarding the frequency harmonics that are observed.

First, for any measured link, there can be a *multipath invisibility problem* when θ, the angle between H_s and the affected component a, is close to 0° or 180°. Further, the amplitude of the induced signal varies as a function of θ. In

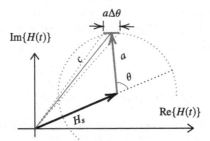

FIGURE 10.2 In (10.3), unaffected waves H_s and affected waves $ae^{-\frac{j2\pi\eta(t)}{\lambda}}$ add up in a phasor sum. The phase $\theta = 2\pi\eta(t)/\lambda$ changes with t, tracing an arc length $a\,\Delta\theta$ and changing the power of the combined RF signal $c = \|H(t)\|^2$.

[48], we show that this model leads to a dB received power change of

$$\Delta P \approx \frac{80\pi\,\Delta z}{(\ln 10)\lambda}\left(\frac{\beta\sin\theta}{1+\beta^2+2\beta\cos\theta}\right), \tag{10.4}$$

where Δz is the physical skin displacement due to breathing or pulse and $\beta = |H_s/a|$. More discussion and numerical examples are given in [48].

Second, we can use (10.3) to understand the harmonics that we observe in the frequency domain. In the Fourier representation of $\eta(t)$, we would expect to see components at the breathing and pulse rates, in addition to other frequencies. For simplicity, we consider the simplest case in which there is one frequency component at f_0, in which case the dynamic path length is approximated as $\eta(t) = \eta(0) + b\sin 2\pi f_0 t$. Then, the CFR power is calculated as follows:

$$\|H(t)\|^2 = 2\|H_s a\|\cos\left(\frac{2\pi b}{\lambda}\sin 2\pi f_0 t + \phi_s\right) + \|H_s\|^2 + \|a\|^2, \tag{10.5}$$

where $\phi_s = \frac{2\pi\eta(0)}{\lambda}$. The cosine term in (10.5) can be can be expanded using Fourier series [54]:

$$\cos\left(\frac{2\pi b}{\lambda}\sin 2\pi f_0 t + \phi_s\right) = \sum_{m=-\infty}^{\infty} J_m\left(\frac{2\pi b}{\lambda}\right)\cos(2\pi m f_0 t + \phi_s). \tag{10.6}$$

Therefore, the CFR power has the following form:

$$\|H(t)\|^2 = \mathcal{A}\sum_{m=-\infty}^{\infty} J_m\left(\frac{2\pi b}{\lambda}\right)\cos(2\pi m f_0 t + \phi_s) + \mathcal{B}, \tag{10.7}$$

where $\mathcal{A} = 2\|H_s a\|$ and $\mathcal{B} = \|H_s\|^2 + \|a\|^2$.

Eq. (10.7) shows that, even if the movement of the body is a single sinusoid, the CFR power is a periodic function with harmonics of the frequency of that sinusoid. The received power $r(t)$ is typically measured in logarithmic scale, $r(t) = 10 \log_{10}(\|Y(t)\|^2)$. For low-amplitude changes, the received power can be approximated as the weighted sum of cosine functions, each with a frequency of integer multiple of f_0 [55].

10.3.2 Estimating rates from received power

There are various methods reported in the literature that can be used to estimate respiration rate and/or pulse rate from received power measurements. Power measurements have noise sometimes larger than the vital sign–induced signal. In addition to high-frequency noise, the measurements may have a DC component, heavy-tailed noise, and interference from other motion. The first step in estimating respiration and pulse rates is to apply appropriate filtering on the received power samples to reduce out-of-band noise. We first apply a Hampel filter [56] in both respiration and pulse rate estimation. However, two different band-pass filters are utilized, respectively. A fourth-order Butterworth band-pass filter is used for respiration-rate detection, while a eighth-order Chebyshev filter is employed to estimate pulse rate due to the difficulty of pulse-induced skin-motion detection than breathing-pattern extraction. We consider the frequency ranges for normal breathing and pulse rates to be $[0.1, 0.4]$ Hz and $[0.85, 1.7]$ Hz respectively. For breathing-rate estimation, the cutoff frequencies of the band-pass filter are the boundary values of the breathing frequency range. To make use of the higher harmonics for pulse-rate estimation, we apply a band-pass filter with cutoff frequencies $(0.85, 4)$ Hz.

Once the DC component and high-frequency noise are filtered out, the resulting waveform is a periodic signal at a frequency of the vital sign of interest of the subject. If this periodic signal was a single-tone sinusoid in additive uncorrelated Gaussian noise (AUGN), the maximum likelihood estimator (MLE) is [41]:

$$\hat{f} = \underset{f}{\arg\max} \left\| \sum_{i=0}^{N-1} r(i) e^{2\pi f T_s i} \right\|^2, \tag{10.8}$$

where $f_{min} \leq f \leq f_{max}$ is the frequency variable, N is the number of samples, and T_s is the sampling period. The proposed MLE approach essentially matches with FFT-based breathing-rate estimation in [42–44] yet has two advantages: (1) higher efficiency: It can apply FFT directly within the desired breathing rate range rather than the whole frequency band, i.e., $[f_{min}, f_{max}] = [0, \frac{1}{2T_s}]$; and (2) finer resolution: The frequency bin can be flexibly determined by MLE, which leads to better frequency resolution and more accurate estimation of the frequency $\hat{f}$.

Since the changes in received power due to heart beat are significantly smaller and the energy is, in fact, distributed over multiple harmonics of the fundamental frequency, we apply a modified MLE in which peak frequency is determined based on cumulative power from the first two harmonic bands of the range of pulse rate:

$$\hat{f} = \arg\max_{f} \left\| \sum_{i=0}^{N} r(i)e^{2\pi f T_s i} \right\|^2 + \left\| \sum_{i=0}^{N} r(i)e^{2\pi 2 f T_s i} \right\|^2. \tag{10.9}$$

10.4 Implementation

In this section, we discuss our implementation of a vital monitoring system based on narrowband received power measurements. We first introduce the RF-sensing hardware and the devices providing the ground truth for pulse and respiration rates. We then present the software developed for data collection, followed by multiple experimental setups.

10.4.1 Hardware

We present Cerberus, an RF-sensing hardware composed of a pair of inexpensive wireless nodes configured as a transmitter and a receiver. Each node includes an RF sub-system based on Texas Instruments (TI) CC1200 sub-GHz narrowband transceivers. The two TI CC1200 transceivers on each node are configured to operate in the 915-MHz ISM band. They are connected via an SPI interface to a low-power Nordic nRF52840 BLE SoC microcontroller. Fig. 10.3 shows a Cerberus node used in the experiment. Received power measurements from Cerberus are transferred to computer for storage and post-processing.

FIGURE 10.3 The RF sensor node used in the experiments in this chapter adds multiple-antenna capability to the system presented in [45].

We make use of commercial sensors to collect the ground truth. For pulse rate, we use a Polar H10, which consists of a heart-rate sensor and a chest strap. In the experiment, the chest strap should be first connected to a smart-phone app via Bluetooth and then wrapped around the chest to capture cardiac activity. The device that provides the ground truth for respiration rate is the Vernier Go

Direct Respiration Belt, which is connected via Bluetooth to the computer for data recording. These two ground-truth sensors are shown in Figs. 10.4(a) and 10.4(b).

(a)   **(b)**

FIGURE 10.4 Devices used to collect ground truth for (a) pulse rate (Polar H10), (b) respiration rate (Vernier Go Direct Respiration Belt).

10.4.2   Software

We develop software to measure received power from a pair of stationary Cerberus nodes. Since each Cerberus node contains two TI CC1200 transceivers; the two on the transmitter side send packets in a time-division multiple-access (TDMA) fashion to avoid collisions. The two CC1200 on the receiver node, on the other hand, keep listening and capture data simultaneously. In addition, wireless channels sometimes suffer from an inability to measure respiration due to multipath fading. Therefore, we adopt channel hopping as a frequency diversity method to ensure that at least one channel observes the breathing and pulse-induced changes. In detail, four channels across the 915-MHz ISM band with 6-MHz channel spacing are used, and thus the center frequencies correspond to 905, 911, 917, and 923 MHz. The transmission interval is set to be five ms for each frequency channel. The protocol of the firmware is that the two TI CC1200 would first transmit in sequence and then follow the same fashion after hopping into another channel. At the receiver side, whenever either TI CC1200 transmits the packet, the receiver μC is capable of capturing the first 32 complex-baseband (IQ) samples from the signal and using them to calculate the received power.

The heart-rate sensor, Polar H10, obtains the pulse rate by calculating the inverse of the time interval between two heart beats, while the Vernier Go Direct Respiration Belt utilizes a force sensor to extract the respiration feature. Both the heart-rate sensor and the respiration belt have a fixed sampling rate of 1 Hz, where the sample represents the time interval and the force measurement, respectively.

10.4.3   Experimental setup

For respiration-rate monitoring, we ran experiments with two human subject volunteers. The subject lies on a cot consistent with the setup details of Fig. 10.5.

For evaluating the respiration rate monitoring system, we present three types of experimental designs:

- Controlled breathing: The subject is required to breathe based on a fixed frequency ranging from 10 to 30 bpm. In this case, the subject listens to a metronome app that provides a beat at the requested breathing rate. Supposing the frequency is set as f_m Hz, then the ground truth for respiration rate should be $60 f_m$ breaths per minute (bpm). The estimates of respiration rate can be compared to the given frequencies.
- Short-term uncontrolled breathing: The user is free from the breathing restrictions and thus can breathe as they normally do while resting. Here, the experimental results are evaluated using the ground truth given by the Vernier Respiration Belt.
- Long-term uncontrolled breathing: More experiments are conducted lasting for 30 minutes. The purpose of this design is to prove the high probability of long-term reliable respiration-rate estimation of our received-power-based monitoring system.

For pulse-rate monitoring, two subjects in three different environments are tested to evaluate our system. The three settings are: (1) a research laboratory, (2) a conference room, and (3) a bedroom at home. All three are furnished with typical furnishings, including chairs, desks, and other equipment. The first two rooms have an approximate area of 56 m², while the bedroom has a comparatively smaller area of 20 m². In each experiment, only a single user is present in the room and required to lie down on a cot. The cot elevates the person 15 cm above the floor, and the two directional antennas are set at 30 cm from the ground, separated one m from each other and directed at the chest of the subject as shown in Fig. 10.5. In the third environment, we follow the first type of setup and separate the two nodes by 80 cm due to the limitation of the whole area. The receiver node of our system is connected to a laptop for received power-data collection. Each experiment lasts for two to five minutes.

10.5   Experimental results

We present in this section quantitative experimental results for vital signs monitoring using received power measurements. We collect power measurements with a single pair of transceivers. For breathing-rate estimation, we evaluate the accuracy and reliability of the approach with three different experimental designs: (1) controlled breathing; (2) short-term uncontrolled breathing; and (3) long-term uncontrolled breathing. For pulse-rate detection, the performance of the proposed system is explored in three different scenarios: (1) a research laboratory, (2) a conference room, and (3) a bedroom. For each experimental setting, two subjects are involved in the experiment individually. The overall performance of our system is given by the RMSE. In this remainder of this section, we compare the estimates to the ground truth, which is fol-

1 m

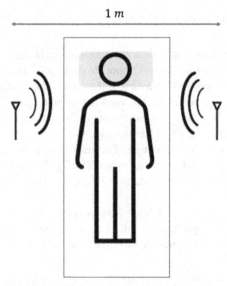

FIGURE 10.5 Experimental setup: (1) the subject lies on the cot which elevates the person 15 cm above the floor; (2) A pair of transceivers are set on each side at 30 cm from the ground.

lowed by the evaluation indicator RMSE with regard to various settings and users.

10.5.1 Breathing-rate accuracy

10.5.1.1 Respiration-rate estimation with controlled breathing

In the breathing monitoring experiment, users lies on a cot and adjust their breathing to match the frequency set via the metronome app. The user is the only one in the bedroom during the test. This restriction protects the received power measurements from being interfered with by other people's motions. Each of the two TI CC1200s at the receiver captures the received power coming from the two TI CC1200s at the transmitter via four different frequency channels. We call these the $4 \times 2 \times 2 = 16$ *channels*.

We first perform DC removal on each channel (subtracting the average value) for a 150-sec period during which the subject is breathing at 10 bpm, and we plot each channel's power signal in Fig. 10.6. It can be seen that each channel is able to capture the breathing-induced changes, and each has unique characteristics. In addition, the received data is corrupted by noise, both at frequencies higher than the breathing signal and at a slowly varying mean. We address this issue by applying a Hampel filter and a bandpass filter. The idea of the Hampel filter is to utilize a generalized median filter with flexible parameter tuning [56] for outlier removal. In addition, a fourth-order Butterworth bandpass filter (BPF) is used. As a normal breathing rate is in the 0.1 to 0.4 Hz range, we set

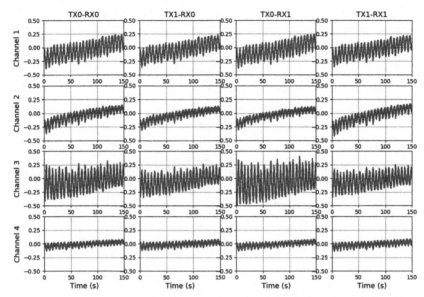

FIGURE 10.6 DC-removed received power measurements collected in 16 channels before filtering.

the low and high cut-off frequencies of the BPF to 0.05 Hz and 0.5 Hz, respectively, to account for monitoring during more extreme cases. In combination, the filtering is intended to cancel out the slowly varying mean, as well as the effects of thermal noise, interfering RF signals, and other power changes due to higher-frequency movements. The filtered received-power measurements are presented in Fig. 10.7. We note that 25 cycles are seen in 150 seconds, which corresponds to approximately 10 bpm. Thus we can conclude that the breathing pattern is captured by the received power measurements.

We compute the PSD of each channel, calculated from 30-sec windows of data and averages of the PSDs. For the data from a 30-sec window of the 16 different channels in Fig. 10.7, the average PSD is shown in Fig. 10.8. We can see that the peak lies around $f_{peak} = 0.168$ Hz, while the true respiration rate $f_{true} = 0.167$ Hz, represented by the blue (dark gray in print version) avertical line, is close to f_{peak}. Hence, the breathing pattern is observable and easily extracted. We ultimately estimate respiration rate from the peak frequency of the average PSD of the 16 channels.

Fig. 10.9 displays the estimated respiration rate vs. the ground truth for the controlled breathing experiments, showing results from both users over four different breathing rates. The estimated respiration rate stays close to the ground-truth value as the beeping frequency increases from 10 bpm to 20 bpm. We plot the RMSE as a function of the breathing rate in Fig. 10.10. The RMSE is less than 0.45 bpm across both users and all breathing rates.

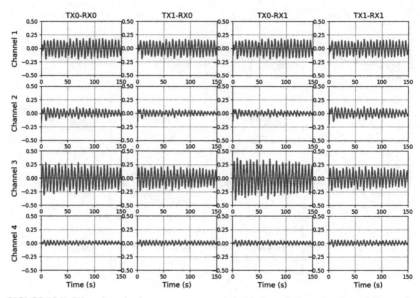

FIGURE 10.7 Filtered received power measurements in 16 channels through a Hampel filter and a fourth-order Butterworth bandpass filter.

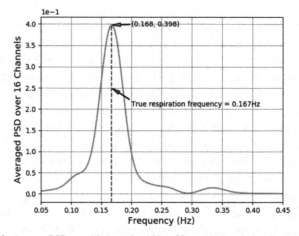

FIGURE 10.8 Average PSD over 16 channels within a 30-sec window during the controlled breathing experiment for User 1 with the ground truth = 10 bpm.

10.5.1.2   Respiration-rate estimation with short-term uncontrolled breathing

In this section, we present results from the short-term uncontrolled breathing experiments. The experimental setting is unchanged, but no metronome is used. Instead, we use the Vernier Go Direct Respiration Belt to measure the ground-truth breathing rate. An example received-power signal is plotted in Fig. 10.11.

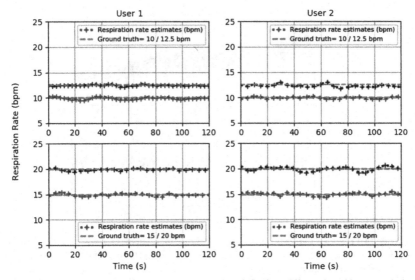

FIGURE 10.9 Estimated respiration rate vs. ground truth for four different breathing rates during controlled breathing experiments, for (Left) User 1 and (Right) User 2.

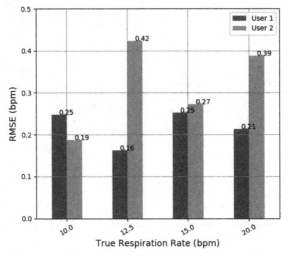

FIGURE 10.10 Breathing-rate RMSE over two subjects (Left: User 1; Right: User 2) in the controlled breathing experiments.

It shows us that the peak-to-peak amplitude of the received filtered signal is about 0.1 dB. The filtered received-power signal in Fig. 10.11(b) makes the breathing cycles more clear. Using the same estimator as presented before, respiration-rate estimates for the two subjects are plotted in Fig. 10.12, which shows that the respiration estimates follow the breathing rates as they rise and

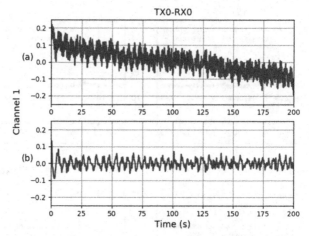

FIGURE 10.11 (a) DC-removed received-power measurements collected by TX0-RX0 pair on channel 1; (b) Filtered received-power measurements through a Hampel filter and a fourth-order Butterworth bandpass filter.

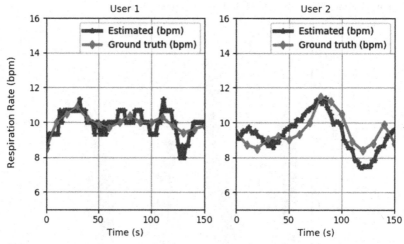

FIGURE 10.12 Estimated respiration rate vs. ground truth in the short-term uncontrolled breathing experiments.

fall over the course of the 150-sec experiments. The computed RMSE values of the two subjects are 0.462 and 0.973 bpm, respectively.

10.5.1.3 Respiration-rate estimation with long-term uncontrolled breathing

Next, we collect measurements for a longer period (30 min) with the same experimental uncontrolled breathing setup.

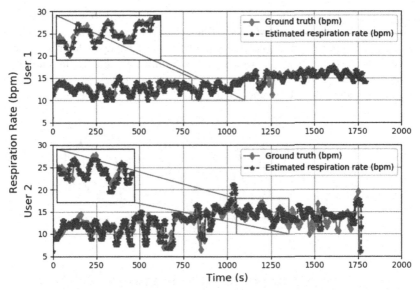

FIGURE 10.13 Estimated respiration rate vs. ground truth in long-term uncontrolled breathing experiment.

The estimated respiration rate vs. ground truth over 30 min is presented in Fig. 10.13. As shown, the proposed system generally tracks the rising and falling respiration rate over a long period of time. There are no sudden failures in tracking. Notably, the received power-based approach is able to track respiration rate changes up to eight bpm within a few tens of seconds, for example, at around 500 sec for User 2. Thus, it is possible for our system to detect breathing anomalies. The RMSE results show a slightly higher RMSE error of 1.87 bpm regarding the second user, whereas the RMSE error of the data collected with the first user remains as 0.87 bpm, lower than 1.0 bpm. We note that the system does not respond quickly to changes in the breathing rate of User 2 between 1,300–1,700 s, which accounts for the higher RMSE for this user.

10.5.2   Pulse-rate accuracy

In this subsection, we evaluate the performance of pulse-rate estimation in three different environments including a conference room, a research laboratory, and a bedroom. The differences between them are the typical furniture included and the area covered. The experiments are conducted with the subject lying on a cot with two directional antennas separated by 1 m in the first two rooms and 80 cm in the final bedroom setting. The average-removed and filtered received-power measurements by one antenna pair in one channel are shown in Fig. 10.14. We can observe from Fig. 10.14 that the periodic breathing cycles are now invisible due to the filtering. Similar to the denoising scheme in respiration-rate estimation, a Hampel filter and bandpass filter are applied to remove outliers, channel noise, and other interference. Rather than the Butterworth filter used in respiration-rate monitoring, a eighth-order Chebyshev bandpass filter is applied. This is because the pulse-induced signal is significantly smaller in amplitude than the breathing-induced signal, and the Chebyshev filter's steeper roll-off is needed to reduce the interference from the breathing-induced signal.

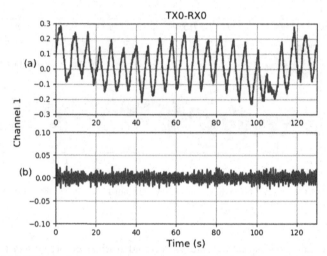

FIGURE 10.14 Received-power measurements collected in the bedroom setting (shown by TX0-RX0 pair on channel 1): (a) Mean removed received power measurements; (b) Filtered received power measurements by a Hampel filter and a eighth-order Chebyshev bandpass filter.

In Fig. 10.15, we present the average PSD over all 16 channels, as well as the signal when we superimpose the first two PSD harmonics as given in (10.9). The 30-sec received power measurements used for PSD calculation are collected in the bedroom setting. The first plot in Fig. 10.15 indicates that the signal below 0.84 Hz, the typical low frequency of pulse rates, is canceled out, and thus becomes zero. The peak at around 1 Hz can be attributed to the breathing harmonic. The peaks shown at about 2.5 Hz and 3.6 Hz, however, stand for the pulse harmonics. To accommodate the fact that the pulse under the skin repeats like a impulse instead of a sinusoid, two pulse harmonics are combined to the original pulse-rate range for better monitoring of pulse signal. The superimposed PSD over two harmonics is presented in the second plot of Fig. 10.15, where the amplitude of the peak is enhanced by the two harmonics. Notably, the peak frequency $f = 1.221$ Hz of the superimposed PSD is much closer to the true pulse rate than the one $f = 1.001$ in the average PSD. Therefore, superimposition of PSD over two harmonics is capable of eliminating the impact due to the breathing signal interference.

Fig. 10.16 quantifies the ability of the system to track the changes in pulse rate over time. First, the results have shown that our algorithm is able to estimate the pulse rate for users who have different pulse-rate ranges. The pulse-rate

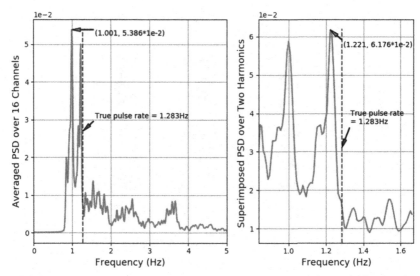

FIGURE 10.15   (Left) Averaged PSD over all channels within a 30-sec window; (Right) Superimposition of PSD over two harmonics for the same data.

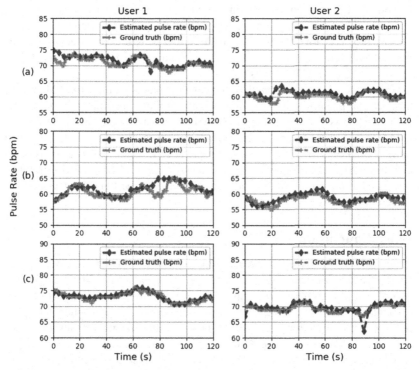

FIGURE 10.16   Estimated respiration rate vs ground truth in three settings: (a) a conference room; (b) a research laboratory; (c) a bedroom.

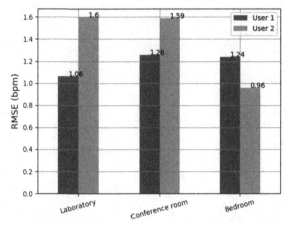

FIGURE 10.17 RMSE results of the two subjects (Left: User 1; Right: User 2) in three different settings: (a) a conference room; (b) a research laboratory; (c) a bedroom.

ranges described in the estimates include 65–75 bpm, 55–65 bpm, 65–75 bpm, and 70–80 bpm. Also, the pulse-rate estimates in Fig. 10.16 support the claim that our system is capable of tracking the increases and decreases of a pulse rate. However, there are short periods during which the pulse-rate estimate becomes very high or very low. Further investigation will be required to track other motions of a person to determine if the jumps are caused by larger motions of their body, or if there is another cause.

Given the 34-cm wavelength at the center frequency in use (920 MHz) and the fact that pulse rate-induced skin vibrations may move the skin on the order of mm, it is quite surprising to be able to observe a person's pulse rate in the measured received power. To summarize, the results indicate that received power-based pulse-rate monitoring is capable of tracking pulse-rate variations over time. Also, the proposed system can stay within 1.6 bpm of RMSE, when compared to the pulse-rate ground truth provided by the Polar H10, as demonstrated in Fig. 10.17.

10.6 Conclusion

In this chapter, we presented a received power-based vital signs monitoring system that is able of detecting the respiration and pulse rates using a single pair of low-cost radio transceivers employing a RF-bandwidth on the order of kHz. Channel diversity is utilized during data collection to better capture vital sign patterns. The estimation algorithm proposed in this chapter first leverages denoising technique to increase SNR of the breathing and pulse signal. Then the superimposed PSD from multiple channels is obtained for respiration-rate estimation. In addition, harmonics of pulse signals are superimposed over the computed PSD to enhance the probability of pulse-induced skin-variation detection. The monitor finally determines respiration or pulse rate via the peak frequency of the resulting PSD.

The developed system was tested in three different experimental designs for respiration rate monitoring: (1) fixed-frequency breathing; (2) free breathing; (3) long-term free breathing. The experimental results demonstrate that our system can achieve accurate respiration-rate estimation with the capability to track changes over long periods. For pulse-rate estimation, we empirically show through the experiments conducted in three different environments that our low-cost RF-based vital signs monitoring system performs as well as reported results from several state-of-the-art systems, while using three orders of magnitude less bandwidth. The experimental results provide an important proof-of-concept to show that low-cost and low-bandwidth sensing is possible and may be an enabler for ubiquitous RF sensing.

References

[1] Chronic diseases in America, Online, https://www.cdc.gov/chronicdisease/resources/ infographic/chronic-diseases.htm, 2019, (Accessed 10 June 2020).
[2] W. Wang, A.C. den Brinker, S. Stuijk, G. de Haan, Algorithmic principles of remote ppg, IEEE Transactions on Biomedical Engineering 64 (7) (2017) 1479–1491.
[3] W. Wang, S. Stuijk, G. de Haan, A novel algorithm for remote photoplethysmography: spatial subspace rotation, IEEE Transactions on Biomedical Engineering 63 (9) (2016) 1974–1984.
[4] M. Kumar, A. Veeraraghavan, A. Sabharwal, Distanceppg: robust non-contact vital signs monitoring using a camera, Biomedical Optics Express 6 (5) (2015) 1565–1588.
[5] E.M. Nowara, D. McDuff, A. Veeraraghavan, A meta-analysis of the impact of skin tone and gender on non-contact photoplethysmography measurements, in: The IEEE/CVF Conference on Computer Vision and Pattern Recognition (CVPR) Workshops, 2020, pp. 284–285.
[6] R. Sinhal, K. Singh, A. Shankar, Estimating vital signs through non-contact video-based approaches: a survey, in: 2017 International Conference on Recent Innovations in Signal Processing and Embedded Systems (RISE), 2017, pp. 139–141.
[7] A.D. Droitcour, O. Boric-Lubecke, G.T.A. Kovacs, Signal-to-noise ratio in Doppler radar system for heart and respiratory rate measurements, IEEE Transactions on Microwave Theory and Techniques 57 (10) (2009) 2498–2507.
[8] E.M. Staderini, Uwb radars in medicine, IEEE Aerospace and Electronic Systems Magazine 17 (1) (2002) 13–18.
[9] G. Vinci, S. Lindner, F. Barbon, S. Mann, M. Hofmann, A. Duda, R. Weigel, A. Koelpin, Six-port radar sensor for remote respiration rate and heartbeat vital-sign monitoring, IEEE Transactions on Microwave Theory and Techniques 61 (5) (2013) 2093–2100.
[10] L. Ren, Y.S. Koo, H. Wang, Y. Wang, Q. Liu, A.E. Fathy, Noncontact multiple heartbeats detection and subject localization using uwb impulse Doppler radar, IEEE Microwave and Wireless Components Letters 25 (10) (2015) 690–692.
[11] S.S. Ram, Y. Li, A. Lin, H. Ling, Doppler-based detection and tracking of humans in indoor environments, Journal of the Franklin Institute 345 (6) (2008) 679–699, advances in Indoor Radar Imaging.
[12] X. Liu, J. Cao, S. Tang, J. Wen, P. Guo, Contactless respiration monitoring via off-the-shelf wifi devices, IEEE Transactions on Mobile Computing 15 (10) (2016) 2466–2479.
[13] C. Li, J. Lin, Random body movement cancellation in Doppler radar vital sign detection, IEEE Transactions on Microwave Theory and Techniques 56 (12) (2008) 3143–3152.
[14] C. Gu, C. Li, J. Lin, J. Long, J. Huangfu, L. Ran, Instrument-based noncontact Doppler radar vital sign detection system using heterodyne digital quadrature demodulation architecture, IEEE Transactions on Instrumentation and Measurement 59 (6) (2010) 1580–1588.

[15] C. Li, J. Cummings, J. Lam, E. Graves, W. Wu, Radar remote monitoring of vital signs, IEEE Microwave Magazine 10 (1) (2009) 47–56.

[16] R. Fletcher, Jing Han, Low-cost differential front-end for Doppler radar vital sign monitoring, in: 2009 IEEE MTT-S International Microwave Symposium Digest, 2009, pp. 1325–1328.

[17] C. Gu, G. Wang, T. Inoue, C. Li, Doppler radar vital sign detection with random body movement cancellation based on adaptive phase compensation, in: 2013 IEEE MTT-S International Microwave Symposium Digest (MTT), 2013, pp. 1–3.

[18] W.Z. Li, Z.G. Li, H. Lv, G. Lu, Y. Zhang, X. Jing, S. Li, J. Wang, a new method for non-line-of-sight vital sign monitoring based on developed adaptive line enhancer using low centre frequency uwb radar, 2013.

[19] A. Lazaro, D. Girbau, R. Villarino, Analysis of vital signs monitoring using an ir-uwb radar, Progress in Electromagnetics Research 100 (2010) 265–284.

[20] J. Li, L. Liu, Z. Zeng, F. Liu, Advanced signal processing for vital sign extraction with applications in uwb radar detection of trapped victims in complex environments, IEEE Journal of Selected Topics in Applied Earth Observations and Remote Sensing 7 (3) (2014) 783–791.

[21] S.K. Leem, F. Khan, S.H. Cho, Vital sign monitoring and mobile phone usage detection using ir-uwb radar for intended use in car crash prevention, Sensors (Basel, Switzerland) 17 (2017).

[22] H. Yiğitler, O. Kaltiokallio, R. Hostettler, A.S. Abrar, R. Jäntti, N. Patwari, S. Särkkä, Rss models for respiration rate monitoring, IEEE Transactions on Mobile Computing 19 (3) (2020) 680–696.

[23] S. Wang, A. Pohl, T. Jaeschke, M. Czaplik, M. Köny, S. Leonhardt, N. Pohl, A novel ultra-wideband 80 ghz fmcw radar system for contactless monitoring of vital signs, in: 2015 37th Annual International Conference of the IEEE Engineering in Medicine and Biology Society (EMBC), 2015, pp. 4978–4981.

[24] M. He, Y. Nian, Y. Gong, Novel signal processing method for vital sign monitoring using fmcw radar, Biomedical Signal Processing and Control 33 (2017) 335–345.

[25] N. Hafner, I. Mostafanezhad, V.M. Lubecke, O. Boric-Lubecke, A. Host-Madsen, Non-contact cardiopulmonary sensing with a baby monitor, in: 2007 29th Annual International Conference of the IEEE Engineering in Medicine and Biology Society, 2007, pp. 2300–2302.

[26] C. Li, J. Lin, Y. Xiao, Robust overnight monitoring of human vital sign by a non-contact respiration and heartbeat detector, conference proceedings, Annual International Conference of the IEEE Engineering in Medicine and Biology Society. IEEE Engineering in Medicine and Biology Society. Conference 1 (2006) 2235–2238.

[27] A. Ahmad, J.C. Roh, D. Wang, A. Dubey, Vital signs monitoring of multiple people using a fmcw millimeter-wave sensor, in: 2018 IEEE Radar Conference (RadarConf18), 2018, pp. 1450–1455.

[28] D.T. Petkie, C. Benton, E. Bryan, Millimeter wave radar for remote measurement of vital signs, in: 2009 IEEE Radar Conference, 2009, pp. 1–3.

[29] M. Alizadeh, G. Shaker, J.C.M.D. Almeida, P.P. Morita, S. Safavi-Naeini, Remote monitoring of human vital signs using mm-wave fmcw radar, IEEE Access 7 (2019) 54958–54968.

[30] T. Rappaport, R. Heath, R. Daniels, J. Murdock, Millimeter Wave Wireless Communications, Prentice Hall, 2015, includes bibliographical references (pages 585-651) and index.

[31] T. Rappaport, Wireless Communications: Principles and Practice, 2nd edition, Prentice Hall PTR, Upper Saddle River, NJ, USA, 2001.

[32] O. Boric-Lubecke, V.M. Lubecke, A.D. Droitcour, B. Park, A. Singh, Doppler Radar Physiological Sensing, John Wiley & Sons, 2016.

[33] D. Halperin, W. Hu, A. Sheth, D. Wetherall, Tool release: gathering 802.11n traces with channel state information, Computer Communication Review 41 (1) (2011) 53.

[34] J. Lee, Y. Su, C. Shen, A comparative study of wireless protocols: bluetooth, uwb, zigbee, and Wi-Fi, in: IECON 2007–33rd Annual Conference of the IEEE, Industrial Electronics Society, 2007, pp. 46–51.

[35] A.S. Abrar, A. Luong, P. Hillyard, N. Patwari, Pulse rate monitoring using narrowband received signal strength measurements, in: 1st ACM Workshop on Device-Free Human Sensing (DFHS 2019), 2019.

[36] Y. Xie, Z. Li, M. Li, Precise power delay profiling with commodity wifi, in: Proceedings of the 21st Annual International Conference on Mobile Computing and Networking, MobiCom'15, Association for Computing Machinery, New York, NY, USA, 2015, pp. 53–64.

[37] K. Woyach, D. Puccinelli, M. Haenggi, Sensorless sensing in wireless networks: implementation and measurements, in: WiNMee 2006, 2006, pp. 1–8.

[38] M. Youssef, M. Mah, A. Agrawala, Challenges: device-free passive localization for wireless environments, in: MobiCom'07: ACM Int'l Conf. Mobile Computing and Networking, 2007, pp. 222–229.

[39] N. Patwari, P. Agrawal, Effects of correlated shadowing: connectivity, localization, and RF tomography, in: IEEE/ACM Int'l Conf. on Information Processing in Sensor Networks (IPSN'08), 2008, pp. 82–93.

[40] N. Patwari, J. Wilson, S. Ananthanarayanan P.R, S.K. Kasera, D. Westenskow, Monitoring breathing via signal strength in wireless networks, Tech. Rep., arXiv:1109.3898v1 [cs.NI], Sept 2011.

[41] N. Patwari, J. Wilson, S. Ananthanarayanan P.R., S.K. Kasera, D.R. Westenskow, Monitoring breathing via signal strength in wireless networks, IEEE Transactions on Mobile Computing 13 (8) (2014) 1774–1786.

[42] N. Patwari, L. Brewer, Q. Tate, O. Kaltiokallio, M. Bocca, Breathfinding: a wireless network that monitors and locates breathing in a home, Tech. Rep., arXiv:1302.3820 [cs.HC], Feb. 2013.

[43] H. Abdelnasser, K.A. Harras, M. Youssef, UbiBreathe: a ubiquitous non-invasive WiFi-based breathing estimator, CoRR, arXiv:1505.02388 [abs], 2015.

[44] Z. Yang, P.H. Pathak, Y. Zeng, X. Liran, P. Mohapatra, Monitoring vital signs using millimeter wave, in: Proceedings of the 17th ACM International Symposium on Mobile Ad Hoc Networking and Computing, MobiHoc'16, Association for Computing Machinery, New York, NY, USA, 2016, pp. 211–220.

[45] A. Luong, A.S. Abrar, T. Schmid, N. Patwari, Rss step size: 1 db is not enough!, in: Proceedings of the 3rd Workshop on Hot Topics in Wireless, 2016.

[46] N. Patwari, L. Brewer, Q. Tate, O. Kaltiokallio, M. Bocca, Breathfinding: a wireless network that monitors and locates breathing in a home, IEEE Journal of Selected Topics in Signal Processing 8 (1) (2014) 30–42.

[47] O. Kaltiokallio, H. Yiğitler, R. Jäntti, N. Patwari, Non-invasive respiration rate monitoring using a single cots tx-rx pair, in: IPSN-14 Proceedings of the 13th International Symposium on Information Processing in Sensor Networks, 2014, pp. 59–69.

[48] A.S. Abrar, N. Patwari, S.K. Kasera, Quantifying Interference-Assisted Signal Strength Surveillance of Sound Vibrations, Tech. Rep., 2020.

[49] J. Liu, Y. Wang, Y. Chen, J. Yang, X. Chen, J. Cheng, Tracking vital signs during sleep leveraging off-the-shelf wifi, in: ACM Intl. Symposium on Mobile Ad Hoc Networking and Computing, 2015, pp. 267–276.

[50] J. Liu, Y. Chen, Y. Wang, X. Chen, J. Cheng, J. Yang, Monitoring vital signs and postures during sleep using wifi signals, IEEE Internet of Things Journal 5 (3) (2018) 2071–2084.

[51] X. Wang, C. Yang, S. Mao, Resbeat: resilient breathing beats monitoring with realtime bimodal csi data, in: GLOBECOM 2017–2017 IEEE Global Communications Conference, 2017, pp. 1–6.

[52] X. Wang, C. Yang, S. Mao, Phasebeat: exploiting csi phase data for vital sign monitoring with commodity wifi devices, in: 2017 IEEE 37th International Conference on Distributed Computing Systems (ICDCS), 2017, pp. 1230–1239.

[53] P. Hillyard, A. Luong, A.S. Abrar, N. Patwari, K. Sundar, R. Farney, J. Burch, C.A. Porucznik, S.H. Pollard, Comparing respiratory monitoring performance of commercial wireless devices, 2018.

[54] M. Abramowitz, I.A. Stegun, Handbook of Mathematical Functions with Formulas, Graphs, and Mathematical Tables, vol. 55, US Government Printing Office, 1948.

[55] H. Yiğitler, O. Kaltiokallio, R. Hostettler, A.S. Abrar, R. Jäntti, N. Patwari, S. Särkkä, Rss models for respiration rate monitoring, IEEE Transactions on Mobile Computing 19 (3) (2019) 680–696.

[56] R.K. Pearson, Y. Neuvo, J. Astola, M. Gabbouj, Generalized hampel filters, EURASIP Journal on Advances in Signal Processing 2016 (1) (2016) 1–18.

Chapter 11

WiFi CSI-based vital signs monitoring

Daqing Zhang[a], Youwei Zeng[a], Fusang Zhang[a], and Jie Xiong[b]
[a]*School of Electronics Engineering and Computer Science, Peking University, Beijing, China,* [b]*College of Information and Computer Sciences, University of Massachusetts, Amherst, Amherst, MA, United States*

Contents

11.1 Introduction

Respiration monitoring plays a critical role in daily health care. Long-term respiration monitoring is widely used to track the progression of illness and also to predict emergencies that require immediate clinical attention, such as cardiac arrest [1]. Among many technologies employed for respiration monitoring, WiFi-based solutions have emerged in recent years due to the pervasive deployment of WiFi infrastructure at home and in public areas. The intrinsic nature of sensor-free and contact-free makes WiFi-based solutions particularly appealing in the current COVID-19 pandemic scenarios as compared to traditional sensor-based solutions.

The rational behind WiFi-based respiration monitoring is that the WiFi receiver (e.g., mobile phone, laptop, etc) captures the WiFi signal reflected off the human body, which contains the subtle displacement information caused by human respiration. The signal waveform varies with the exhalation and inhalation process periodically, and the respiration information can thus be extracted by analyzing the WiFi signal variation. Based on this basic idea, Liu et al. [2] explore the feasibility of employing WiFi signals to detect human respiration. Soon after that, in 2016, Wang et al. [3] found that the performance of WiFi-based respiration sensing is not always stable and is highly dependent on target location and orientation. This work triggered the following research to not just focus on sensing accuracy but also on robustness and practicality. Subsequent work [4,5] investigates the sensing stability and sensing range to further enhance the performance.

In this chapter, we first review the history of WiFi-based respiration monitoring and then introduce the theoretical models to help people to deeply understand the underlying mechanisms of WiFi sensing. The state-of-the-art solutions proposed to address the instability and limited range issues are introduced by taking human respiration monitoring as an example application. Finally, we present the approach to enable multi-person vital signs monitoring, which we believe is an important step towards real-life adoption of WiFi-based respiration monitoring.

11.2 An historic review of WiFi-based human respiration monitoring

11.2.1 RSS-based respiration monitoring

The Received Signal Strength (RSS) characterizes the attenuation of RF signals during the propagation process, and RSS readings are widely available on mainstream commodity hardware. The pioneer work [6] studying the feasibility of employing RSS signals for respiration monitoring was proposed in 2013. It makes use of links formed by a mesh network of multiple IEEE 802.15.4 nodes to estimate the respiration rate of a single person in the environment [6]. Later, in 2014, a more cost-effective solution was developed by O.J. Kaltiokallio et al. [7] in which only one pair of Zigbee nodes was required to monitor the respiration rate. In 2015, Abdelnasser et al. successfully realized respiration monitoring using commodity WiFi hardware [8]. Though promising, several issues were also pointed out including the coarse resolution and severe hardware noise. Therefore, the performance of the RSS-based solutions is limited in terms of accuracy and stability.

11.2.2 CSI-based respiration monitoring

One exciting opportunity emerged when the 802.11n WiFi standard was officially released in 2009 [9]. To support the multiple-input and multiple-output (MIMO) feature of 802.11n, channel state information (CSI) became available on commodity WiFi hardware. Soon thereafter, in 2011, the CSI tool [10] was developed to retrieve CSI readings from an Intel 5300 network interface card (NIC). Compared to RSS, CSI readings have the unique advantage of finer resolution and lower noise, and therefore, the sensing granularity and performance stability are significantly improved. Furthermore, RSS contains just the signal amplitude, but CSI contains amplitude and phase information, which can both be utilized for respiration monitoring.

Wi-Sleep [11] was among the first systems that can extract rhythmic respiration patterns from WiFi-CSI signals. In 2015, Liu et al. [12] claimed to monitor not just the respiration rate but also the heartbeat rate during sleep by analyzing the power spectral density (PSD) of CSI amplitude. Wu et al. [13] further extended respiration monitoring from sleeping to standing postures. While interesting progress had been achieved, researchers also found one critical issue with respiration monitoring: The performance is unstable and at certain locations (i.e., the "blind spots"), e.g., the performance can severely degrade even when the target is close to the WiFi transceivers. During the time period 2016–2018, the Fresnel zone models were introduced by the Peking University team in [3,14–16] to explain the underlying mechanism of WiFi-based respiration sensing. The Fresnel zone models mathematically characterize the relationship between the human chest displacements and the corresponding signal variations. People can therefore rely on these models to fully understand why the "blind spots" occur and propose solutions [4,5,17] to address this issue.

Leveraging the overlapped Fresnel zone model, Wang et al. [18] employed multiple transceiver pairs to overcome the blind-spot issue. In 2018, Zeng et al. discovered that the phase and amplitude are complementary to each other for respiration monitoring [4]. That is to say, when the sensing performance is bad using phase, it would be good with amplitude and vice versa. Therefore, amplitude and phase can be employed together to remove sensing "blind spots". Another interesting solution was proposed by Niu et al. [17] in 2018, who proposed to add a man-made "virtual" multipath to boost the respiration sensing performance at the "blind spots" to address the issue.

In addition to sensing respiration for one person, an exciting research direction in this area is to enable multi-target respiration monitoring. Different from traditional sensor-based solutions in which multi-target sensing is not an issue, it is very challenging to achieve multi-target respiration monitoring with WiFi because the reflection signals from multiple targets can be mixed at the receiver, and it is difficult to separate them to obtain the respiratory information of each individual. Liu et al. [12] presented the first attempt in this direction and demonstrated the feasibility of multi-person respiration monitoring by applying Fast Fourier Transform (FFT) on the CSI amplitude readings. Wang et al. [19] achieved multi-person respiration monitoring by applying the root-MUSIC algorithm [20] on the CSI phase readings. A similar method was also adopted in TR-BREATH [21] to achieve multi-target respiration monitoring. Wang et al. [22] further leveraged the tensor decomposition technique to extract the respiration patterns of multiple targets from the CSI phase readings. Though promising, one basic assumption for the aforementioned systems to work is that the respiration rates of different persons are distinct, and the performance degrades when different people have similar respiration rates. Furthermore, these systems can only obtain the average respiration rate over a period of time but are not able to extract the detailed respiration pattern over time. Inspired by the Fresnel zone theories [3,14–16], Yang et al. [23] successfully managed to monitor multi-person respiration by optimizing the deployment of WiFi transceivers so that each transceiver pair's transmission is only affected by one target. However, this method requires knowing the accurate location of each target in advance, and if the target changes the location, the deployment stops working and needs to be updated. In 2020, by leveraging the widely available antenna array with commodity WiFi hardware, Zeng et al. [24] proposed to model the multi-person respiration sensing as a BSS (Blind Source Separation) problem and solved it using the independent component analysis (ICA) method to separate and extract the respiration information of each individual person; even multiple targets have similar respiration rates and are physically closely located.

Besides respiration sensing for multiple subjects, another important research direction in this area is to increase the sensing range. Due to the nature of employing weak reflection signals for respiration sensing, the WiFi sensing range is very limited, being much smaller than the communication range. In 2019, Zeng et al. [5] pushed the respiration sensing range from room level (two–four m) to house level (eight–nine m) by introducing a new metric called the CSI ratio, which is defined as the ratio of CSI readings collected from two co-located antennas hosted on the same Wi-i hardware. With an increased sensing range, through-wall respiration sensing was also demonstrated with WiFi signals [5].

Other attempts in this area include [25,26] and [27]. By synthesizing a wider-bandwidth WiFi radio, Shi et al. [25] demonstrated they can detect a person's respiration rate in dynamic ambient environment. With the synthesized WiFi signal, the authors can identify the path reflected by the breathing person and then analyze the periodicity of the signal-power measurements only from this path to infer the respiration rate. Zhang et al. [26] calibrated the time-varying phase offset by using cables and splitters, and then it exploited the variation of accurate CSI phase to track human breath. One interesting observation worth mentioning is that, by comparing the respiration monitoring performance using different RF signals including WiFi RSS, Zigbee RSS, WiFi CSI, and UWB-IR [27], Hillyard et al. found that WiFi-CSI readings presented the most robust respiration monitoring performance.

With significant progress in this area, accurate and robust WiFi respiration monitoring has become a reality. However, challenges still exist. For example, while monitoring the respiration rate of a sleeping person is not an issue anymore, monitoring the respiration of a person running on a treadmill remains challenging due to the large self-interference movements from other body parts.

Severe interference from surrounding people is another critical issue that needs to be tackled before WiFi-based respiration monitoring can be widely adopted in real life.

11.3 The principle of WiFi CSI-based respiration monitoring

11.3.1 The basics of WiFi CSI

Due to reflection, diffraction, and scattering from surrounding objects, wireless signals (e.g., WiFi) arrive at the receiver from the transmitter through multiple paths. Channel State Information (CSI) is used to quantify the wireless propagation channel between the transmit–receive (Tx-Rx) pair. In essence, CSI characterizes the *Channel Frequency Response* (CFR) of each subcarrier between each antenna pair. For carrier frequency f, $X(f,t)$ and $Y(f,t)$ are the frequency domain representations of the transmitted and received signals, respectively. Letting $H(f,t)$ be the complex-valued CFR for carrier frequency f measured at time t, we have the relationship $Y(f,t) = H(f,t) \times X(f,t)$ [28]. As CSI on WiFi hardware is measured on each OFDM subcarrier, one CSI measurement is a CFR value between an antenna pair on a certain OFDM subcarrier frequency at a particular time stamp.

If an RF signal arrives at the receiver through L different paths, then the total CSI is the linear superposition of the CSI of all paths, which can be represented as:

$$H(f,t) = \delta(t)e^{-j\phi(t)} \sum_{i=1}^{L} A_i e^{-j \cdot 2\pi d_i(t)/\lambda}, \qquad (11.1)$$

where $\delta(t)$ is the amplitude impulse noise, $\phi(t)$ is time-varying phase shift, L is the number of paths, A_i is the complex attenuation, and $d_i(t)$ is the signal propagation length of the ith path.

11.3.2 Modeling human respiration

The respiration process of a person consists of periodic inflation and deflation of the lungs as shown in Fig. 11.1. The chest displacement during respiration is around 4.2–5.4 mm in the front dimension, 2.5–2.6 mm in the back dimension, and 0.6–1.1 mm in the mediolateral dimension. For a deep inspiratory breath hold (DIBH), this displacement can be increased up to 12.6 mm in anteroposterior dimension [29]. We model the human chest as a varying size flat-cylinder, as shown in Fig. 11.2, where the outer and inner cylinder surfaces correspond to the chest positions for exhalation and inhalation, respectively.

11.3.3 Fresnel diffraction and reflection sensing models

In this section, we quantify the relationship between the human chest motion and the variation of WiFi CSI by employing the Fresnel diffraction and reflection

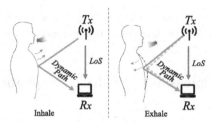

FIGURE 11.1 An example of how RF signal paths change with body chest motions during the respiration process.

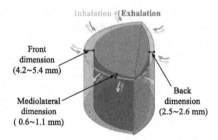

FIGURE 11.2 Human chest modeling.

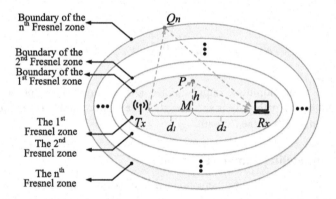

FIGURE 11.3 Geometry of the Fresnel zones.

sensing models. In free space, Fresnel zones are concentric ellipses with the transmitter T_x and receiver R_x as two focal points. The innermost ellipse is the called the First Fresnel Zone (FFZ), and more than 70% of the signal energy is transferred to the receiver via this zone. The nth Fresnel zone corresponds to the area between the $(n-1)^{th}$ and n^{th} ellipses. For a given RF wavelength λ, the Fresnel zones can be represented as follows:

$$|T_x Q_n| + |Q_n R_x| - |T_x R_x| = n\lambda/2, \tag{11.2}$$

where Q_n is a point on the nth ellipse. When a point P is inside the FFZ, the distance from P to the line-of-sight (LoS) path formed by the transceiver pair is h, as shown in Fig. 11.3. M is the intersection point of the vertical line and the LoS path. The path $T_x R_x$ is then divided into two segments: $d_1 = |T_x M|$ and $d_2 = |M R_x|$. The path difference Δd between the path $T_x P R_x$ and the LoS path $T_x R_x$ is calculated by:

$$\begin{aligned}
\Delta d &= |T_x P| + |P R_x| - |T_x R_x| \\
&= \sqrt{(d_1)^2 + (h^2)} + \sqrt{(d_2)^2 + (h^2)} - (d_1 + d_2) \\
&= d_1 \sqrt{1 + (h/d_1)^2} + d_2 \sqrt{1 + (h/d_2)^2} - (d_1 + d_2).
\end{aligned} \qquad (11.3)$$

This path difference can be used to derive the phase difference between the target reflection path and the direct LoS path as the following:

$$\varphi = \frac{2\pi \Delta d}{\lambda}. \qquad (11.4)$$

When the human target is located outside of the FFZ, reflection dominates. On the other hand, when the target is inside the FFZ, diffraction dominates. We discuss the effects of diffraction and reflection, respectively, next.

Diffraction effect: When the target P is located inside the FFZ, we have $h \ll d_1$ and $h \ll d_2$. Therefore, $(h/d_1)^2 \ll 1$ and $(h/d_2)^2 \ll 1$. By applying the approximation $\sqrt{1+x} \approx 1 + \frac{x}{2}$ when $x \ll 1$, Eq. (11.3) can be simplified as:

$$\Delta d \approx \frac{h^2}{2} \frac{(d_1 + d_2)}{d_1 d_2}. \qquad (11.5)$$

The corresponding phase difference induced by this path-length difference Δd can then be expressed as:

$$\varphi = \frac{2\pi \Delta d}{\lambda} = \pi h^2 \frac{(d_1 + d_2)}{\lambda d_1 d_2}. \qquad (11.6)$$

The Fresnel diffraction parameter v is defined as follows:

$$v = h \sqrt{\frac{2(d_1 + d_2)}{\lambda d_1 d_2}} = \sqrt{\frac{2}{\pi} \varphi}. \qquad (11.7)$$

With this parameter, the signal amplitude at the receiver due to diffraction can be expressed as [30]:

$$F(v) = \frac{1+j}{2} \cdot \int_v^\infty exp(\frac{-j\pi z^2}{2}) dz, \qquad (11.8)$$

where $F(v)$ is known as the Fresnel integral. The diffraction gain induced by the target P can thus be expressed as:

$$Gain_{Dif} = 20log|F(v)|. \qquad (11.9)$$

Reflection effect: When the target P is located outside of the FFZ, reflection dominates. The reflection signal at the receiver is a superposition of both static path and dynamic path signals. In the respiration monitoring scenario, the signal reflected from a surrounding wall is a static path signal, while the signal reflected from the human chest is a dynamic path signal. The received signal can then be denoted as:

$$H(t) = H_s + H_d(t) = H_s + A(t)\exp\left(-j\frac{2\pi\,\Delta d}{\lambda}\right), \qquad (11.10)$$

where the static component H_s is the sum of static path signals and the dynamic component $H_d(t)$ is induced by the moving human chest. $A(t)$ is the complex-valued representation of the signal amplitude and phase of the dynamic path, and $e^{-j2\pi\frac{\Delta d}{\lambda}}$ is the phase change caused by the length change of the dynamic path. Note that, when the length of the reflected signal path changes by λ, its phase rotates 2π. Hence the receiving signal $H(t)$ has a time-varying signal amplitude whose power can be represented as:

$$|H(\theta)|^2 = |H_s|^2 + |H_d|^2 + 2|H_s||H_d|\cos\theta, \qquad (11.11)$$

where θ is the phase difference between the static signal H_s and dynamic vector H_d. Based on this formula, we know that when an object crosses the Fresnel zone boundaries, the amplitude of the receiving signal varies like a sinusoidal wave.

To demonstrate this effect, we let a cylinder move across multiple Fresnel zone boundaries as shown in Fig. 11.4. According to the diffraction and reflection models [3,16], we plot the amplitude of the received signal. When the cylinder is located outside of the FFZ, the signal amplitude varies like a sinusoidal wave. The signal amplitude varies in the shape of a "W" during the process of crossing the FFZ.

Now, we explain why a subtle human chest displacement during the respiration process can cause such a large signal variation. If the chest displacement is 0.5 cm, for the wavelength (λ) of 5.7 cm, the corresponding phase change of the signal path during the process is around 63° ($2\pi \times \frac{2\Delta d}{\lambda} = 2\pi \times \frac{2\times 0.5}{5.7} = 63°$). Note that the resultant signal received at the receiver is a superposition of both static and dynamic signals. For a same amount of phase variation (63°) of the dynamic signal, the amount of signal variation of the resultant signal depends on the location of the target. We choose four different locations and let a cylinder vibrate with a small displacement at these locations to mimic the respiration of a person. As shown in Fig. 11.5, at Location 1, the cylinder vibration leads to a small signal-amplitude variation, and each respiration cycle

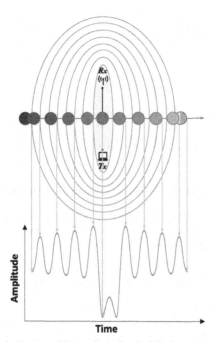

FIGURE 11.4 Amplitude changes of the resultant signal while the target moves across the Fresnel zones.

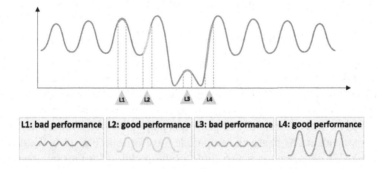

FIGURE 11.5 Amplitude of the resultant receiving signal for various target locations.

corresponds to two peaks. At Location 2, the same cylinder vibration leads to a much larger signal-amplitude variation, and each respiration corresponds to just one peak. For respiration monitoring, a better performance can be achieved at Location 2 because a clearer respiration pattern can be obtained. Similarly, bad and good performance can be achieved at Location 3 and Location 4 respectively.

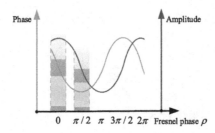

FIGURE 11.6 CSI phase and amplitude waveforms with respect to various target locations.

11.4 Robust single-person respiration monitoring

In this section, we focus on single-person respiration monitoring and present the state-of-the-art solutions to address the two issues discussed in Sect. 11.3: (i) the "blind spot" issue and (ii) short sensing range.

11.4.1 Removing "blind spots" for respiration monitoring

We present two methods to address the "blind spot" issue: (i) exploiting the complementarity of CSI amplitude and phase to ensure respiration monitoring performance; (ii) adding a "virtual" multipath to ensure large signal variation.

11.4.1.1 Exploiting complementarity of CSI amplitude and phase

Based on Eq. (11.10), the phase of CSI reading can be denoted as [4]:

$$\angle H_s - \arcsin \frac{|H_d| \sin \rho}{\sqrt{|H_s|^2 + |H_d|^2 + 2|H_s||H_d| \cos \rho}}, \tag{11.12}$$

where ρ is the Fresnel phase, which is defined as the phase difference between the dynamic signal vector H_d and static signal vector H_s. Mathematically, the CSI phase in Eq. (11.12) changes in the form of a sinusoidal-like wave.

Based on the definition of CSI phase and amplitude, we conduct simulations and plot the ideal CSI phase and amplitude waveforms in Fig. 11.6. We observe that, the CSI phase changes very little in the middle of a Fresnel zone (e.g., $\rho = \pi/2$ and $3\pi/2$), but more significantly at the boundary of a Fresnel zone (e.g., $\rho = 0, \pi$ and 2π). This suggests that, when CSI phase is employed for respiration sensing, the best subject location would be right at the boundary of a Fresnel zone and the worst location would be in the middle of a Fresnel zone. Interestingly, we observe that a bad location for sensing with amplitude turns out to be a good position for sensing with phase, and vice versa as shown in Fig. 11.6.

However, the CSI measurement in commodity WiFi suffers from the time-varying phase offset, which makes phase readings totally random across consecutive CSI samples. Fortunately, the time-varying random phase offsets are

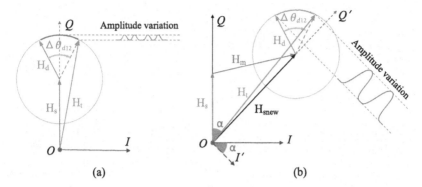

FIGURE 11.7 The effect of adding a multipath: (a) without a multipath; (b) with a multipath.

the same at different antennas on one WiFi card [31] since they share the same RF oscillator. Thus, we can apply conjugate multiplication of CSI between two antennas to remove the random phase offset. After the phase offset is removed, clean phase readings are obtained. Then, both phase and amplitude readings can be employed for respiration sensing. The complementary property between phase and amplitude in terms of sensing performance helps to ensure reliable human respiration monitoring without "blind spots".

11.4.1.2 Adding 'a 'virtual" multipath

Instead of eliminating the "blind spots" by combining both amplitude and phase for sensing, a novel method was proposed to significantly boost the respiration sensing performance at the "blind spots" by adding a man-made "virtual" multipath [17]. From the previous subsection, we know the Fresnel phase ρ (i.e., the phase difference between the dynamic and static component) is the key factor affecting the sensing performance. Therefore, we propose to a leverage multipath, which is commonly considered harmful in the existing sensing literature, to tune the phase of the static component for better sensing performance.

As shown in Fig. 11.7, when a new static multipath is introduced, the original static component H_s and the new multipath component H_m together form a new static component H_{snew}. The dynamic vector H_d is not changed but now rotates around the new static vector H_{snew}. From the perspective of vector transformation, adding such a multipath actually transforms the original IQ vector space to a new I'Q' vector space as shown in Fig. 11.7(b). After introducing a static virtual multipath, the original static vector H_s is converted to the new static vector H_{snew}, making the variation of the dynamic vector more distinguishable. With this multipath added, we successfully turn a "bad" position into a "good" position for sensing. That is to say, we can create a static multipath to improve the sensing performance at a particular location. In practice, we can either employ a physical object to create a multipath or add a "virtual" multipath value purely in software.

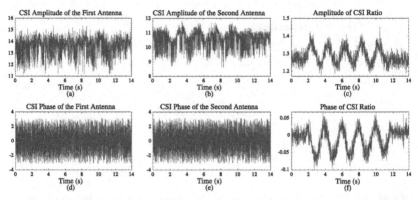

FIGURE 11.8 Comparison of raw CSI and CSI ratios when they are employed to sense the plate movements.

11.4.2 Pushing the sensing range of respiration monitoring

In this subsection, we aim to increase the respiration sensing range by constructing a new base signal called the CSI ratio.

11.4.2.1 The CSI-ratio model

The CSI ratio is defined as the quotient of CSI readings between two antennas of a receiver as:

$$
\begin{aligned}
H_{ratio}(f, t) &= \frac{H_1(f, t)}{H_2(f, t)} \\
&= \frac{\delta(t)e^{-j\phi(t)}(H_{s,1} + A_1 e^{-j2\pi d_1(t)/\lambda})}{\delta(t)e^{-j\phi(t)}(H_{s,2} + A_2 e^{-j2\pi d_2(t)/\lambda})} \\
&= \frac{H_{s,1} + A_1 e^{-j2\pi d_1(t)/\lambda}}{H_{s,2} + A_2 e^{-j2\pi d_2(t)/\lambda}},
\end{aligned}
\tag{11.13}
$$

where $H_1(f, t)$ is the complex-valued CSI of the first antenna and $H_2(f, t)$ is that of the second antenna. With the division operation, it can be seen that the noise in the original CSI amplitude and the time-varying phase offset are successfully canceled out, thus providing a high-SNR new base signal for sensing. Fig. 11.8 presents an example of sensing the plate movement to visualize the effect of noise cancellation with CSI ratio. Apparently, the CSI ratio of two antennas eliminates the noise but still retains the signal variation pattern corresponding to the plate movements.

Now let us take a deeper look at how to utilize the CSI ratio for human sensing. Prior work presents an interesting observation: When the target moves a short distance, although the target-reflection path length at each antenna changes, the target-reflection path-length difference between two close-by antennas $d_2(t) - d_1(t)$ can be seen as a constant [4]. With this observation,

Eq. (11.13) can be rewritten as:

$$H_{ratio}(f,t) = \frac{A_1 e^{-j2\pi d_1(t)/\lambda} + H_{s,1}}{(A_2 e^{-j2\pi(d_2(t)-d_1(t))/\lambda})e^{-j2\pi d_1(t)/\lambda} + H_{s,2}}$$

$$= \frac{\mathcal{A}\mathcal{Z} + \mathcal{B}}{\mathcal{C}\mathcal{Z} + \mathcal{D}},$$

$$(11.14)$$

where the coefficient $\mathcal{A}, \mathcal{B}, \mathcal{C}, \mathcal{D}$ are complex constants, and $\mathcal{Z} = e^{-j2\pi d_1(t)/\lambda}$ is a unit complex variable, whose phase is closely related to the reflection path length. Mathematically, Eq. (11.14) is exactly in the form of a Möbius Transformation (also known as fractional transformation) [32]. To study the properties of CSI ratio, we rewrite Eq. (11.14) as:

$$H_{ratio}(f,t) = \left(\frac{\mathcal{B}\mathcal{C} - \mathcal{A}\mathcal{D}}{\mathcal{C}^2}\right)\frac{1}{\mathcal{Z} + \mathcal{D}/\mathcal{C}} + \frac{\mathcal{A}}{\mathcal{C}}$$

$$= \frac{\beta e^{j\theta}}{\mathcal{Z} + \alpha} + \gamma,$$

$$(11.15)$$

where α and γ are complex numbers; β and θ are real numbers. We can see from Eq. (11.15) that the Möbius transformation can be decomposed into four basic operations, namely, translation, rotation, scaling, and inversion. The translation, rotation, and scaling preserve the geometry shape and rotation direction (clockwise or counterclockwise) of a circle. The inversion operation preserves the shape of a circle, however, it may change the rotational direction [5]. When the magnitude of the static component is larger than that of the dynamic component (i.e., $|\mathcal{D}| > |\mathcal{C}|$), which is usually the case in human sensing [5], the inversion operation still preserves the rotational direction.

Through rigorous mathematical derivation and real-life experiments in [5], we can obtain the following properties for human sensing from the CSI-ratio model in Eq. (11.14):

1. The CSI ratio preserves the relationship between target movements and the corresponding CSI changes. Specifically, if the reflection path length is changed by one wavelength, due to target movement, the CSI ratio would rotate for a full circle in the complex plane. If the path length is changed by less than one wavelength, the CSI ratio just rotates a small part (an arc) along the circle.
2. If the reflection path length increases, CSI ratio rotates clockwise when the magnitude of the static component is larger than that of the dynamic component. If the path length decreases, CSI ratio rotates counterclockwise.
3. The amplitude and phase of the CSI ratio is complementary to each other for respiration sensing.

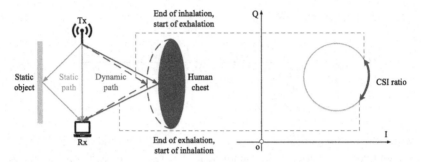

FIGURE 11.9 Conceptual illustration of applying CSI ratio to sense single-person respiration.

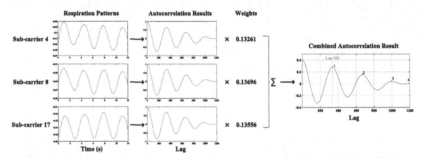

FIGURE 11.10 Example of multiple sub-carriers combining. In this case, we use the weighted sum of three selected sub-carriers' autocorrelation results to obtain a final one. The estimated respiration rate is $\frac{60}{335/100} = 17.9$ bpm.

11.4.2.2 Applying CSI ratio to single-person respiration sensing

Since the chest displacement caused by respiration is between 0.6 mm to 12.6 mm, the reflection path length changes less than one wavelength ($\lambda = 5.7$ cm for 5.24 GHz signal), then the trajectory of the CSI ratio during respiration is just an arc (part of a full circle). As shown in Fig. 11.9, when a subject inhales and exhales during respiration, the length of the dynamic reflection path increases and decreases accordingly. The CSI ratio rotates along the arc clockwise or counterclockwise, corresponding to inhalation and exhalation, respectively. Note that the commodity WiFi card provides CSI on multiple sub-carriers, e.g., 30 sub-carriers for Intel 5300 card. We then apply the Savitzky–Golay filter [33] to smooth the CSI ratio data of each sub-carrier and use them for further processing.

As shown in Eq. (11.13), the amplitude noise and time-varying phase offset are effectively canceled out by the division operation. Apparently, compared to the raw CSI reading from a single antenna, the CSI ratio between two adjacent antennas contains less noise and is thus more sensitive in detecting weak reflection signals, leading to a longer sensing range. To extract respiration pattern from a time series of complex-valued CSI ratio data **x**, we linearly combine its

real part (I component) and imaginary part (Q component) by assigning weights
$\cos \alpha$ and $\sin \alpha$ as:

$$\mathbf{y} = \begin{bmatrix} \cos \alpha & \sin \alpha \end{bmatrix} \begin{bmatrix} \Re(\mathbf{x}) & \Im(\mathbf{x}) \end{bmatrix}^{T}, \tag{11.16}$$

where $\Re(\mathbf{x})$ is the real part of $\mathbf{x}$ and $\Im(\mathbf{x})$ is the imaginary part of $\mathbf{x}$. By varying α from 0 to 2π at a fixed step size, we can generate various combination
of candidates. The one that has the maximal periodicity is selected from these
candidates following the method presented in [5]. Note that, as amplitude and
phase are complementary to each other and when both of them are utilized, the
number of "blind spots" is significantly reduced.

We further estimate respiration rate by combining results from multiple sub-
carriers. First, for each sub-carrier, we calculate the autocorrelation of its respiration pattern, where the autocorrelation function describes the similarity of a
signal to a shifted version of itself [34]. Second, we combine the sub-carriers by
employing a weighted sum of each sub-carrier's autocorrelation result, where
the weight of each sub-carrier is the respiration pattern's BNR value [5]. The
first peak of the combined autocorrelation result is the component describing the
periodicity of respiration. Also, the shift of the first peak divided by the sampling
rate is the estimated period for one respiration cycle. Fig. 11.10 shows an example of respiration-rate estimation by combining multiple sub-carriers. Three
sub-carriers (i.e., 4, 8, and 17) are selected to participate in the weighted sum operation. As shown in the final combined autocorrelation result, the lag (shift) of
the first peak labeled with 1 is 335. For a sampling rate of 100 Hz, the estimated
respiration rate is calculated as $\frac{60}{335/100} = 17.9$ bpm.

11.4.2.3   Evaluation

We employ a pair of GIGABYTE mini-PCs, each equipped with one Intel 5300
WiFi card as a transceiver. One antenna is employed at the transmitter (Tx),
and two antennas are used at the receiver (Rx). The carrier frequency of the
WiFi channel is set as 5.24 GHz, and the transmitter is set to broadcast 100
packets per second. We collect CSI data at the receiver using the CSI tool [10]
and process it with MATLAB® at a DELL Precision 5520 laptop. We collect
the ground-truth respiration rates of a human subject with a commercial device
(Neulog Respiration Monitor Belt logger sensor NUL-236).

In the experiments, the human subject is instructed to sit or lie down facing
the WiFi transceiver pair and breathe naturally. To demonstrate the effectiveness and robustness of the presented methods, we conduct experiments in three
different scenarios:

1. Scenario 1: As shown in Fig. 11.11 (a), the human subject is located at various distances ranging –nine 9 m with respect to the WiFi transceivers placed
   at a height of one m in a large office room with a size of 7.5 m × 9 m;
2. Scenario 2: As shown in Fig. 11.11 (b), both WiFi transceivers are mounted
   on the ceiling, far away from the human subject;

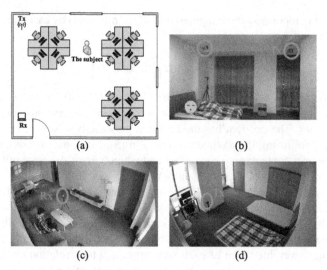

FIGURE 11.11 Experimental setup in three different environments: (a) sitting far away from the transceivers; (b) sleeping scenario; (c) and (d) NLoS scenario.

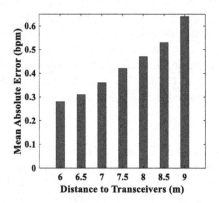

FIGURE 11.12 The mean absolute error of respiration rate vs. distance to transceivers.

3. Scenario 3: As shown in Fig. 11.11 (c) and (d), the WiFi transmitter and receiver are placed in two different rooms separated by a wall.

   The experiment results are presented here:

1. We plot the mean absolute error of the respiration rate when the distance between the human subject and WiFi transceivers is increased from 4 m to 9 m in Fig. 11.12. As expected, the error increases with the distance. This is because the reflection signal is weaker when the distance is increased. Even when the distance is eight m, the mean absolute error is still less than 0.5 bpm. This result shows that the proposed method is able to achieve a sensing range of eight–nine m, large enough for house-level sens-

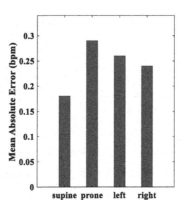

FIGURE 11.13 The mean absolute error of respiration rate vs. the sleeping postures.

ing. In contrast, the traditional approaches have a much shorter sensing range (two–four m).
2. Fig. 11.13 shows the mean absolute error of the respiration rate for various sleeping postures. For all four postures (supine, prone, left, and right), the mean absolute error is always less than 0.3 bpm, which demonstrates the effectiveness and robustness of the proposed method. It can also be observed that the respiration rate is more accurate in a supine posture than in other postures. We believe this is because in a supine posture, the subject's chest faces the transceivers and therefore induces larger signal fluctuations compared to other postures.
3. For the challenging NLoS scenario with WiFi transmitter and receiver placed in two different rooms, the achieved mean absolute error is still as small as 0.34 bpm. The error at the same distance of five m for the NLoS scenario is slightly larger than that for the LoS scenario. This is because the signal reflected off the human subject becomes even weaker after penetrating the wall.

11.5 Robust multi-person respiration monitoring

11.5.1 Modeling of CSI-based multi-person respiration sensing

In this subsection, we show how to model multi-person respiration sensing with WiFi CSI.

11.5.1.1 Effects of multi-person respiration on WiFi CSI

Fig. 11.14 presents a typical WiFi-based multi-person respiration sensing scenario, where the Wi–Fi transceivers are placed at fixed locations and multiple persons are present breathing normally. In this scenario, the WiFi signals are affected by multiple targets simultaneously. Let's first study the effect of multi-

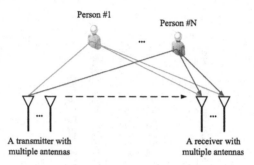

FIGURE 11.14 A typical WiFi-based multi-person respiration sensing scenario.

person respiration on WiFi CSI with just one transmitting antenna and one receiving antenna. Based on Eq. (11.1), we can represent the CSI in multi-person scenario as follows:

$$H(f,t) = \delta(t)e^{-j\phi(t)}\left(\sum_{i=1}^{N} A_i e^{-j\cdot 2\pi d_i(t)/\lambda} + b\right), \qquad (11.17)$$

where $\delta(t)$ is the amplitude impulse noise, $\phi(t)$ is the time-varying phase offset, N is the number of persons to monitor, $d_i(t)$ is the path length of the signal bounced off the chest of the ith person, and b is the combined static signal. When the ith person breathes, the reflection path length $d_i(t)$ varies corresponding to his/her chest movement, and $H(f,t)$ changes accordingly. The key to achieve multi-person respiration sensing is to extract the respiration signal $e^{-j\cdot 2\pi d_i(t)/\lambda}$ corresponding to the respiration of each individual person from Eq. (11.17).

We now consider the scenario when multiple antennas are available at the WiFi transceivers. Two key observations identified in [5] help us derive the formula of CSI from various antenna pairs. First, for each person, with a tiny movement (e.g., a few millimeters), although the reflection path length obtained from each antenna pair (one at the transmitter and one at the receiver) changes, the path-length difference between two antenna pairs can be considered as a constant. Second, for commodity WiFi devices, the time-varying phase offsets in Eq. (11.17) are the same across different antennas at the same device since they share a same RF oscillator. With these two observations, we can denote the CSI readings from two different antenna pairs as follows:

$$\begin{cases} H_1(f,t) = \delta(t)e^{-j\phi(t)}(\sum_{i=1}^{N} A_{1,i}e^{-j2\pi\frac{d_{1,i}(t)}{\lambda}} + b_1) \\ H_2(f,t) = \delta(t)e^{-j\phi(t)}(\sum_{i=1}^{N}(A_{2,i}e^{-j2\pi\frac{d_{2,i}(t)-d_{1,i}(t)}{\lambda}})e^{-j2\pi\frac{d_{1,i}(t)}{\lambda}} + b_2), \end{cases}$$
$$(11.18)$$

where $H_1(f,t)$ is the CSI from one antenna pair, $H_2(f,t)$ is the CSI from another antenna pair, and $d_{2,i}(t) - d_{1,i}(t)$ is the path-length difference between two antenna pairs for the ith person, which is a constant.

11.5.1.2 Modeling multi-person respiration sensing as a blind-source separation problem

The goal of multi-person respiration sensing is to extract the source respiration signal $s_i(f, t) = e^{-j2\pi d_{1,i}(t)/\lambda}$ ($i = 1, 2, ..., N$) corresponding to the respiration of each individual from Eq. (11.18). This is quite similar to the well-known blind-source separation (BSS) problem that aims to recover each individual source signal from a given mixed signal without knowing how they are mixed together [35]. In essence, the BSS problem can be efficiently solved by a technique called Independent Component Analysis (ICA), if three assumptions are satisfied [36]: (1) The signal sources are non-Gaussian; (2) the signal sources are mutually statistically independent; and (3) the mixture is of linear nature. All the three assumptions are satisfied in multi-person respiration sensing, and therefore the problem can be modeled as the BSS problem, and the source signals can be separated by the ICA method.

The basic ideal sounds promising. However, as shown in Eq. (11.18), it is noticed that the CSI readings obtained from commodity WiFi hardware suffer from amplitude noise $\delta(t)$ and time-varying phase offset $\phi(t)$. To apply the ICA method for WiFi-based multi-person respiration sensing, we need to eliminate the noise and random phase offset. To achieve this goal, we propose a novel signal-transformation technique, which contains two steps: (1) We construct a so-called "reference CSI" that has the same form as $H_0(f, t) = \delta(t)e^{-j\phi(t)}b_0$; and (2) for CSI $H_k(f, t)$ from the kth antenna pairs, we normalize it by the "reference CSI" as follows:

$$\frac{H_k(f, t)}{H_0(f, t)} = \frac{\delta(t)e^{-j\phi(t)}(\sum_{i=1}^{N} a_{k,i}s_i(f, t) + b_k)}{\delta(t)e^{-j\phi(t)}b_0} = \sum_{i=1}^{N} \frac{a_{k,i}}{b_0}s_i(f, t) + \frac{b_k}{b_0},$$

(11.19)

where $a_{k,i}$ is the coefficient of the respiration signal $s_i(f, t) = e^{-j2\pi d_{1,i}(t)/\lambda}$ in CSI collected from the kth antenna pair. We can clearly observe that the noise in original CSI is successfully canceled out by the signal transformation operation, while the linear mixture nature of WiFi CSIs are still kept.

To construct such a special "reference CSI", we employ a weighted sum of all CSI readings from M different antenna pairs as follows:

$$H_{sum}(f, t) = \sum_{k=1}^{M} g_k H_k(f, t)$$

$$= \delta(t)e^{-j\phi(t)}(\sum_{i=1}^{N}\sum_{k=1}^{M} a_{k,i}g_k s_i(f, t) + \sum_{k=1}^{M} g_k b_k),$$

(11.20)

where M is the number of antenna pairs, N is the number of persons, and g_k ($k = 1, 2, ..., M$) is the weight. Mathematically, the "reference CSI" can be obtained from Eq. (11.20):

$$H_{ref}(f, t) = \delta(t)e^{-j\phi(t)} \sum_{k=1}^{M} g_k b_k, \qquad (11.21)$$

if the following condition is satisfied:

$$\begin{cases} \sum_{k=1}^{M} a_{k,1} g_k = 0 \\ \sum_{k=1}^{M} a_{k,2} g_k = 0 \\ \quad \vdots \\ \sum_{k=1}^{M} a_{k,N} g_k = 0 \end{cases} \qquad (11.22)$$

Obviously, Eq. (11.22) is an homogeneous system of linear equations that has N equations and M variables ($g_1, g_2, ..., g_M$). Given that $M > N$, there must be some nontrivial solutions to satisfy Eq. (11.22).

In practice, we apply the genetic algorithm to search for such a solution by minimizing the RER (Respiration Energy Ratio) of $H_{sum}(f, t)$, where the RER is defined as the ratio of respiration-related energy to the total signal energy [24]. We subtract the background CSI measured when there is no target in the environment. In this way, the CSI after signal transformation is exactly the linear mixture of the respiration signal of each person, and the ICA method can then be applied to separate the source signal $s_i(f, t) = e^{-j2\pi d_{1,i}(t)/\lambda}$. After that, we extract the respiration pattern of each individual target from the complex-valued signal by applying the Principle Component Analysis (PCA) method presented in [37]. The respiration rate can be estimated by counting the peaks in a specific window of respiration signals using the method proposed in [4].

11.5.2 The advantages of our approach

In this subsection, we will explain why our approach outperforms the existing approaches that rely on the spectral analysis of the CSI amplitude.

11.5.2.1 Reducing the noise in CSI amplitude

As shown in Eq. (11.22), most of the noise in CSI amplitude is canceled out by the division operation. Compared to the original CSI amplitude, the proposed new signal ratio has a higher SNR (Signal-to-Noise Ratio), which is thus more sensitive in detecting subtle respiration-induced chest displacement.

11.5.2.2 Eliminating the "blind spots"

Now, we explain why the "blind spots" do not exist in our systems. In multi-person scenarios, the signals reflected from multiple persons are mixed together in the CSI. When the complex-valued CSI is projected to its real-valued amplitude, for each respiration signal, different projection axes have different capabil-

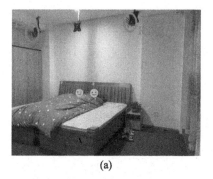

(a)   (b)

FIGURE 11.15 The experimental setup in two scenarios: (a) All subjects are sleeping on a bed; (b) each subject is sitting on a couch or chair.

ities in terms of sensing respiration [24]. Traditional approaches simply project the complex-valued CSI into its amplitude, thus all respiration signals share the same project axis. However, a good projection axis for the respiration signal of one person may be a bad one for that of another person, thus resulting in the "blind spots" issue. In contrast to prior work that projects the mixed respiration signals onto one axis, our approach separates the respiration signals of multiple persons and projects each respiration signal on its unique best axis to extract the respiration pattern. Therefore, the "blind spots" issue is well-addressed.

11.5.2.3   Resolving similar respiration rates

Based on the literature on blind-source separation, our approach can efficiently extract the source respiration signals as long as they are mutually independent of each other. Such condition holds true even when two persons have similar respiration rates. This is because breathing is a natural physical activity and two persons' respiration processes have an extremely small chance to be fully synchronized over time. In other words, two people might have similar respiration rates but generally with different phases.

11.5.3   Evaluation

11.5.3.1   Experimental setup

We use two GigaByte mini-PCs equipped with Intel 5300 WiFi cards as transceivers. Both the transmitter and receiver are equipped with three omni-directional antennas. The frequency of WiFi channel is set to 5.24 GHz with a bandwidth of 20 MHz. The transmitter sends 200 packets per sec at a transmission power of 15 dBm. We collect CSI data at the receiver using the CSI tool [10] and process it with MATLAB on a DELL Precision 5520 laptop in real time. To obtain the ground-truth respiration of multiple persons, we ask each of them to wear a commercial Neulog respiration belt.

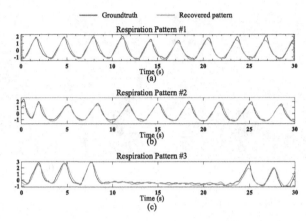

FIGURE 11.16 An example of detected respiration patterns and the ground-truth patterns in a three-person scenario.

In our experiments, each of the human subjects is instructed to sit or lie down facing the WiFi transceiver pair and breathe naturally. We recruited 21 participants, i.e., 12 males and 9 females, aged from 14 to 57. Their natural respiration rates range from 14 bpm to 22 bpm. To demonstrate the effectiveness and robustness of the proposed system, we conduct experiments in two different environments:

1. Environment 1: As shown in Fig. 11.15 (a), multiple subjects are sleeping on a bed with a size of 1.9 m × 2.4 m in a bedroom;
2. Environment 2: As shown in Fig. 11.15 (b), multiple subjects are sitting in a couch or chair in a living room with a size of 4 m × 7.5 m.

Both environments are equipped with furniture and electrical appliances, which create rich multipaths. For the sleeping scenario in Fig. 11.15 (a), we mount the transceivers on the ceiling with a spacing of 2.7 m between them. For the sitting scenario in Fig. 11.15 (b), we place the transceivers on two tripods.

11.5.3.2 Experiment results

The experiment results are presented using multi-person scenario as examples, detailed respiration patterns, and the mean absolute respiration-rate errors will be reported:

1. Fig. 11.16 shows three detected respiration patterns and the corresponding ground-truth patterns when three persons sit on the couch. From Fig. 11.16(a) and Fig. 11.16(b), we can see that the proposed method is able to separate two respiration patterns even when they are very similar to each other. Fig. 11.16(c) shows that the proposed method can correctly detect irregular respiration patterns caused by health issues such as apnea that occur in the time interval from the 9th to 22nd second.

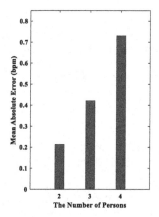

FIGURE 11.17 The mean absolute respiration rate error vs. the number of persons.

2. We plot the mean absolute errors of respiration rate for two-person, three-person, four-person, and five-person scenarios in Fig. 11.17. The results show that the mean absolute error is 0.21 bpm for the two-person scenario and it increases to 0.42 bpm for the three-person scenario. Please note that a larger than 1-bpm error is usually unacceptable for respiration monitoring. For the four-person scenario, the proposed method can still achieve an error of 0.73 bpm. However, in the five-person scenario, an error larger than 1 bpm is observed. As expected, the error increases with the number of persons monitored. The experiment results demonstrate the effectiveness and robustness of the proposed method for respiration monitoring in two-person, three-person, and four-person scenarios.

11.6 Summary

In this chapter, we first provide a historical review of WiFi-based contact-free human respiration monitoring solutions, and then we introduce the Fresnel reflection and diffraction model to show how human respiration monitoring can be achieved by using WiFi signals. To address the problem of "blind spots" and limited sensing range, we propose a series of solutions including exploiting the complementary property of amplitude and phase, adding "virtual" multipath, and creating a new base signal called CSI ratio. We further leverage multiple antennas available in commodity WiFi cards to achieve accurate multi-person respiration sensing for the first time.

While great progress has been made in WiFi-based respiration sensing, several challenges still need to be addressed in future work, such as how to robustly monitor human respiration when the targets are moving, and how to ensure the robustness of human respiration sensing performance when the targets are in different locations and orientations.

Acknowledgments

This research is supported by the EU CHIST-ERA RadioSense Project, the EU Horizon 2020 research and innovation programme IDEA-FAST (No. 853981), the Project 2019BD005 supported by PKU-Baidu Fund, the National Natural Science Foundation of China (No. 61802373), the Youth Innovation Promotion Association, Chinese Academy of Sciences (No. 2020109).

References

[1] A.N. Pedersen, S. Korreman, H. Nyström, L. Specht, Breathing adapted radiotherapy of breast cancer: reduction of cardiac and pulmonary doses using voluntary inspiration breath-hold, Radiotherapy and Oncology 72 (1) (2004) 53–60.
[2] X. Liu, J. Cao, S. Tang, J. Wen, P. Guo, Contactless respiration monitoring via off-the-shelf wifi devices, IEEE Transactions on Mobile Computing 15 (10) (2015) 2466–2479.
[3] H. Wang, D. Zhang, J. Ma, Y. Wang, Y. Wang, D. Wu, T. Gu, B. Xie, Human respiration detection with commodity wifi devices: do user location and body orientation matter?, in: Proceedings of the 2016 ACM International Joint Conference on Pervasive and Ubiquitous Computing, 2016, pp. 25–36.
[4] Y. Zeng, D. Wu, R. Gao, T. Gu, D. Zhang, Fullbreathe: full human respiration detection exploiting complementarity of csi phase and amplitude of wifi signals, Proceedings of the ACM on Interactive, Mobile, Wearable and Ubiquitous Technologies 2 (3) (2018) 1–19.
[5] Y. Zeng, D. Wu, J. Xiong, E. Yi, R. Gao, D. Zhang, Farsense: pushing the range limit of wifi-based respiration sensing with csi ratio of two antennas, Proceedings of the ACM on Interactive, Mobile, Wearable and Ubiquitous Technologies 3 (3) (2019) 1–26.
[6] N. Patwari, J. Wilson, S. Ananthanarayanan, S.K. Kasera, D.R. Westenskow, Monitoring breathing via signal strength in wireless networks, IEEE Transactions on Mobile Computing 13 (8) (2013) 1774–1786.
[7] O. Kaltiokallio, H. Yiğitler, R. Jäntti, N. Patwari, Non-invasive respiration rate monitoring using a single cots tx-rx pair, in: IPSN-14 Proceedings of the 13th International Symposium on Information Processing in Sensor Networks, IEEE, 2014, pp. 59–69.
[8] H. Abdelnasser, K.A. Harras, M. Youssef, Ubibreathe: a ubiquitous non-invasive wifi-based breathing estimator, in: Proceedings of the 16th ACM International Symposium on Mobile Ad Hoc Networking and Computing, 2015, pp. 277–286.
[9] E. Perahia, R. Stacey, Next Generation Wireless LANs: 802.11 n and 802.11 ac, Cambridge University Press, 2013.
[10] D. Halperin, W. Hu, A. Sheth, D. Wetherall, Tool release: gathering 802.11 n traces with channel state information, ACM SIGCOMM Computer Communication Review 41 (1) (2011) 53.
[11] X. Liu, J. Cao, S. Tang, J. Wen, Wi-sleep: contactless sleep monitoring via wifi signals, in: 2014 IEEE Real-Time Systems Symposium, IEEE, 2014, pp. 346–355.
[12] J. Liu, Y. Wang, Y. Chen, J. Yang, X. Chen, J. Cheng, Tracking vital signs during sleep leveraging off-the-shelf wifi, in: Proceedings of the 16th ACM International Symposium on Mobile Ad Hoc Networking and Computing, 2015, pp. 267–276.
[13] C. Wu, Z. Yang, Z. Zhou, X. Liu, Y. Liu, J. Cao, Non-invasive detection of moving and stationary human with wifi, IEEE Journal on Selected Areas in Communications 33 (11) (2015) 2329–2342.
[14] D. Wu, D. Zhang, C. Xu, H. Wang, X. Li, Device-free wifi human sensing: from pattern-based to model-based approaches, IEEE Communications Magazine 55 (10) (2017) 91–97.
[15] D. Zhang, H. Wang, D. Wu, Toward centimeter-scale human activity sensing with wi-fi signals, Computer 50 (1) (2017) 48–57.
[16] F. Zhang, D. Zhang, J. Xiong, H. Wang, K. Niu, B. Jin, Y. Wang, From Fresnel diffraction model to fine-grained human respiration sensing with commodity wi-fi devices, Proc. ACM Interact. Mob. Wearable Ubiquitous Technol. 2 (1) (2018) 53:1–53:23.

[17] K. Niu, F. Zhang, J. Xiong, X. Li, E. Yi, D. Zhang, Boosting fine-grained activity sensing by embracing wireless multipath effects, in: Proceedings of the 14th International Conference on Emerging Networking EXperiments and Technologies, 2018, pp. 139–151.

[18] P. Wang, B. Guo, T. Xin, Z. Wang, Z. Yu Tinysense, Multi-user respiration detection using wi-fi csi signals, in: 2017 IEEE 19th International Conference on e-Health Networking, Applications and Services (Healthcom), IEEE, 2017, pp. 1–6.

[19] X. Wang, C. Yang, S. Mao, Phasebeat: exploiting csi phase data for vital sign monitoring with commodity wifi devices, in: 2017 IEEE 37th International Conference on Distributed Computing Systems (ICDCS), IEEE, 2017, pp. 1230–1239.

[20] B.D. Rao, K.S. Hari, Performance analysis of root-music, IEEE Transactions on Acoustics, Speech, and Signal Processing 37 (12) (1989) 1939–1949.

[21] C. Chen, Y. Han, Y. Chen, H.-Q. Lai, F. Zhang, B. Wang, K.R. Liu, Tr-breath: time-reversal breathing rate estimation and detection, IEEE Transactions on Biomedical Engineering 65 (3) (2017) 489–501.

[22] X. Wang, C. Yang, S. Mao Tensorbeat, Tensor decomposition for monitoring multiperson breathing beats with commodity wifi, ACM Transactions on Intelligent Systems and Technology (TIST) 9 (1) (2017) 1–27.

[23] Y. Yang, J. Cao, X. Liu, K. Xing, Multi-person sleeping respiration monitoring with cots wifi devices, in: 2018 IEEE 15th International Conference on Mobile Ad Hoc and Sensor Systems (MASS), IEEE, 2018, pp. 37–45.

[24] Y. Zeng, D. Wu, J. Xiong, J. Liu, Z. Liu, D. Zhang, Multisense: enabling multi-person respiration sensing with commodity wifi, Proceedings of the ACM on Interactive, Mobile, Wearable and Ubiquitous Technologies 4 (3) (2020) 1–29.

[25] S. Shi, Y. Xie, M. Li, A.X. Liu, J. Zhao, Synthesizing wider wifi bandwidth for respiration rate monitoring in dynamic environments, in: IEEE INFOCOM 2019-IEEE Conference on Computer Communications, IEEE, 2019, pp. 181–189.

[26] D. Zhang, Y. Hu, Y. Chen, B. Zeng, Breathtrack: tracking indoor human breath status via commodity wifi, IEEE Internet of Things Journal 6 (2) (2019) 3899–3911.

[27] P. Hillyard, A. Luong, A.S. Abrar, N. Patwari, K. Sundar, R. Farney, J. Burch, C. Porucznik, S.H. Pollard, Experience: cross-technology radio respiratory monitoring performance study, in: Proceedings of the 24th Annual International Conference on Mobile Computing and Networking, 2018, pp. 487–496.

[28] W. Wang, A.X. Liu, M. Shahzad, K. Ling, S. Lu, Understanding and modeling of wifi signal based human activity recognition, in: Proceedings of the 21st Annual International Conference on Mobile Computing and Networking, MobiCom'15, Association for Computing Machinery, New York, NY, USA, 2015, pp. 65–76.

[29] C. Lowanichkiattikul, M. Dhanachai, C. Sitathanee, S. Khachonkham, P. Khaothong, Impact of chest wall motion caused by respiration in adjuvant radiotherapy for postoperative breast cancer patients, SpringerPlus 5 (1) (2016) 1–8.

[30] A.F. Molisch, Wireless communications, John Wiley and Sons, Chichester, UK.

[31] M. Kotaru, K. Joshi, D. Bharadia, S. Katti, Spotfi: decimeter level localization using wifi, in: Proceedings of the 2015 ACM Conference on Special Interest Group on Data Communication, 2015, pp. 269–282.

[32] H. Schwerdtfeger, Geometry of Complex Numbers: Circle Geometry, Moebius Transformation, Non-euclidean Geometry, Courier Corporation, 1979.

[33] K. Baba, L. Bahi, L. Ouadif, Enhancing geophysical signals through the use of Savitzky-Golay filtering method, Geofísica Internacional 53 (4) (2014) 399–409.

[34] G.E. Box, G.M. Jenkins, G.C. Reinsel, G.M. Ljung, Time Series Analysis: Forecasting and Control, John Wiley & Sons, 2015.

[35] F. Abrard, Y. Deville, A time–frequency blind signal separation method applicable to underdetermined mixtures of dependent sources, Signal Processing 85 (7) (2005) 1389–1403.

[36] V.D. Sánchez A, Frontiers of research in bss/ica, Neurocomputing 49 (1–4) (2002) 7–23.

[37] A.D. Droitcour, et al., Non-contact measurement of heart and respiration rates with a single-chip microwave Doppler radar, Ph.D. thesis, Citeseer, 2006.

Chapter 12

RFID-based vital signs monitoring

Yuanqing Zheng and Yanwen Wang

The Hong Kong Polytechnic University, Hong Kong, China

Contents

12.1 Introduction

In recent years, we have witnessed the surge of RFID systems and their applications, such as user interaction [20,38,40], localization [45], tracking [33,34,46], supply-chain systems [30–32], and vital signs monitoring [19]. As one of the key vital signs, respiration has been very important and informative for inferring the health conditions of users. For example, respiration monitoring can help assess the general personal health, give clues to chronic diseases, and track progress towards recovery [5,19]. Scientific studies show that a deep breath reduces blood pressure and stress, while shallow breath and unconscious hold of breath indicate chronic stress [6]. People may have irregular breathing patterns alternating between fast and slow with occasional pauses [7]. In such cases, a convenient non-intrusive respiration monitoring system would be helpful to enable innovative healthcare applications.

Current respiration-monitoring technologies however are inconvenient and intrusive. For instance, breath monitoring devices typically require the person to attach a nasal probe or a wear chest band [9]. Such monitoring devices restrict body movements and cause inconvenience. As such, people may feel uncomfortable when wearing nasal probes or forget to wear chest bands. Parents are concerned about the safety of breath-monitoring devices for their newborns. Sensor-embedded wearable devices (e.g., smart watches) can monitor some vital signs (e.g., heart rate), but they cannot accurately monitor breathing.

Recent wireless sensing technologies explore the possibility of capturing vital signs (e.g., breathing patterns) using wireless signals [9,13,17]. Such technologies transmit wireless signals to a subject and capture periodic changes by reflected radio waves caused by chest motion during inhalation and exhalation. These systems however require customized Doppler radios, which are not readily available on the market. More importantly, when multiple users gathered, the reflected radio waves from the users may cause interference and dramatically decrease monitoring accuracy. Even when monitoring a single subject, because various parts of the body can simultaneously reflect radio waves, it remains challenging to accurately monitor a user's breathing patterns.

We ask the following question: *Can we use COTS systems to simultaneously monitor the breathing of multiple users?* We leverage COTS RFID systems to implement a breath-monitoring system named *TagBreathe* to accurately monitor breathing, even in the presence of multiple users. The *TagBreathe* system consists of several passive commodity RFID tags that are attached to users' clothes and a commodity RFID reader that captures reflected radio waves from the tags. Researchers have embedded RFID chips into yarns that can be woven and knitted to make fabrics for RFID clothing [8,28]. Unlike traditional battery-powered sensor motes, once attached to or embedded into cloths, the light thin RFID tags are hardly intrusive for users.

Due to chest motion during breathing, the distance between the tags to the reader varies periodically. When a user inhales, the distance decreases, while when the user exhales the distance increases. Such miniature variations in distance translate into small changes in data streams of low-level information (e.g., phase values) reported by the reader. *TagBreathe* aggregates the low-level information and extracts periodic patterns from the data stream. Compatible with the standard EPC protocol, *TagBreathe* benefits from the collision arbitration and naturally avoids the interferences of multiple backscatter tags [4].

Implementing RFID-based respiration monitoring systems however entails substantial challenges. Based on careful characterization of low-level data reported by commodity readers, we first preprocess the streams of collected phase values. Due to frequency hopping, the phase values dramatically change when the reader hops to the next frequency channels. Instead of directly tracking body movements with the raw phase values, we calculate the displacement during two consecutive phase readings measured in the same frequency channel and continuously track the body movements with the displacement values. We then analyze the frequency domain and extract rhythmic breathing signals from the displacement values. To better extract weak breathing signals, we further form an array of multiple tags and fuse the displacement values reported by multiple tags to improve the monitoring performance.

We consolidate the above techniques and implement *TagBreathe* with COTS RFID systems. We evaluate *TagBreathe* in various scenarios under various experiment parameter settings. The experiment results show that *TagBreathe* is able to simultaneously monitor the breathing of multiple users with high accuracies. In particular, *TagBreathe* can achieve a high breathing-rate detection accuracy of less than one breath per minute error on average for various breathing rates.

The contribution of our work [19] can be summarized as follows. First, *TagBreathe* is the first system that can monitor breathing using COTS RFID systems. Second, we design and implement signal-extraction algorithms to measure the breathing signals based on careful characterization of the low-level data reported by COTS RFID reader. Third, we extensively evaluate the *TagBreathe* system under various practical scenarios and demonstrate the feasibility of monitoring breathing wirelessly using COTS RFID systems.

The rest of this chapter is organized as follows. We briefly describe the background of vital signs monitoring using RFID systems in Sect. 12.2. We next focus on our work on respiration monitoring using RFID systems. We give detailed description of key technical components in Sect. 12.3. We present the implementation and evaluation results in Sect. 12.4. Section 12.5 summarizes and concludes this chapter.

12.2 Background

12.2.1 Literature review

12.2.1.1 RFID sensing

Researchers have proposed to develop various RFID sensing applications using COTS RFID systems in the literature [11,12,15,16,21,22,33,37,38,45,52,53]. Tagoram [45] leverages COTS RFID systems to accurately pinpoint RFID-labeled items that support automatic luggage handling and enables innovative human–computer interactions. Tagball [21] labels physical objects with passive RFID tags and tracks the orientation and attitude of the objects to enable 3D human–computer interaction. Tagyro [40] overcomes practical challenges, such as imperfect radiation pattern of practical tags, and enables orientation tracking in 3D space. R# [14] estimates the number of humans in various locations by examining backscattered radio waves. Tadar [46] tracks users' movement by building antenna arrays with RFID tags and analyzes their reflected radio signals. OTrack [33] tracks the ordering of luggage with received signal strength of RFID tags. STPP [34] analyzes phase profiles with a mobile reader and calculates spatial ordering to determine the relative locations of tags. MobiTagBot [35] carefully handles the multipath reflections and leverages frequency hopping to achieve accurate sorting of tags with a mobile reader. TACT [39] leverages predeployed RFID tags to monitor and recognize user's activities in the environment. NodTrack [41] detects the nodding movements of drivers, which is a good indicator of fatigue suggesting drowsy driving. ER-Rhythm [47] captures the exercise locomotion rhythm of users by simultaneously monitoring body movement and respiration rhythm using RFID systems. These works build on COTS RFID systems and deliver the benefit of wireless sensing to users in a cost-effective manner.

12.2.1.2 Wireless-based vital signs monitoring

Recent studies monitor breathing rates by measuring wireless signals reflected from the human body using Doppler radars or ultra-wideband radars [9,13,17, 50]. These devices typically transmit wireless signals to a user and capture miniature changes in reflected radio waves due to chest motion during breathing. To detect the small movements, such systems need high sampling rates which incur high power consumption. Vital-Radio [9] builds active radios to transmit frequency modulated carrier waves and detect small shifts in reflected frequency due to body movements. Those systems however require customized high-end active radios, which are not readily available on the market and incur prohibitive cost. These systems cannot effectively differentiate whether the reflected signals are indeed from the subject under monitoring or due to the movements of other objects in the environment. As a result, those systems suffer low detection accuracy since movements in the environment would reflect radio waves and affect detection results. Recent works [10,23,29,36] leverage WiFi signals (e.g., received signal strength, channel state information) to estimate breathing rates. The WiFi-based monitoring approaches suffer low accuracy in the presence of multiple users.

12.2.1.3 RFID-based vital signs monitoring

Recent works leverage RFID technologies to develop vital signs monitoring systems. CRH [51] develops a vital signs monitoring system using COTS RFID tags that can simultaneously monitor respiration and heartbeat of users. CRH collects the temporal phase information from the tag array near or on the body to extract respiration and heartbeat signals. CRH develops a sequence of signal processing techniques to process the collected RFID phase values to monitor respiration and heartbeats. However, it is extremely challenging to accurately monitor heartbeats using COTS RFID systems due to the very small displacement caused by heartbeats. We are not aware of any previous research that can monitor blood pressure or body temperature using RFID systems probably because there are mature alternative solutions to monitor the two vital signs (e.g., heart rate sensor on smart watch, temperature sensor). AutoTag [42] presents an unsupervised detection system for apnea using commodity RFID tags with a recurrent variational autoencoder. AutoTag [42] also studies how to address the practical challenge of frequency hopping in RFID systems. A respiration monitoring system designed for driving environment is presented in [44] that overcomes the influence of frequency hopping, random sampling, and vehicle vibration. A drowsy-driving detection system has been developed with RFID systems for the driving environment [43] by analyzing the phase difference between RFID tags attached to the back of a hat worn by a driver. RF-RMM [49] presents an RFID-based approach to accurately and continuously monitor the respiration of mobile users by addressing the technical challenges involved in tracking tiny respiration movements, while eliminating the effect of large body movement. Previous work [48] improves the robustness of respiration monitoring systems in dynamic environments with rich multi-path components. A recent paper [24] compares previous works and describes research opportunities and challenges in developing RFID-based vital signs monitoring systems.

12.2.1.4 Respiration monitoring system

Respiration monitoring systems typically require users to attach nasal probes or wear chest bands [9]. For instance, capnometers [25] require the attachment of nasal cannula to patients to monitor their breathing [10]. Wearable devices (e.g., smart watches) are equipped with sensors but cannot accurately monitor breathing patterns [9]. Recent wireless sensing technologies have explored the feasibility of extracting breathing patterns by measuring the reflected wireless signals from a human body [9,13,17]. Such technologies transmit wireless signals to a human subject and capture changes in the reflected radio waves due to breathing. By increasing sampling rates, recent work improves the detection resolution. To detect the small chest movements, the FMCW based systems need a GHz sampling rate which incurs a high manufacturing cost. Instead, Doppler radios send radio signals and measure the frequency shift to estimate the speed [9]. Those systems however require customized Doppler radios which are not readily available. More importantly, such monitoring systems may fail in the presence of multiple users since reflected radio signals from multiple users mix in the air and interfere with each other. As a result, previous wireless sensing systems can only monitor one user and perform poorly with multiple human subjects.

12.2.2 RFID physical-layer measurement

Recent years have witnessed the development of accurate localization and tracking with COTS RFID systems [45]. Unlike the traditional signal strength-based localization schemes [26], recent schemes track the location of RFID tags based on the low-level phase values of backscattered radio waves. COTS RFID readers measure low-level information at the RFID physical layer and output phase values for each identified tag. Although the phase measurements are subject to noise, the high resolution and high sampling rate of COTS RFID systems provide opportunities to achieve accurate localization. By combining the multiple readings of phase measurements, recent works cancel out the noise and achieve cm-level accuracy [45]. Instead of aiming at higher localization accuracy, *Tag-Breathe* draws strength from these works and extends to breath monitoring.

Passive RFID tags feature the factors of low cost, lightweight, and small dimensions, which are ideal for non-intrusive breath monitoring. Once seamlessly attached to a user's cloth, a passive lightweight tag is hardly intrusive to the user. Fashion and textile industries have started to embed RFID chips into fabrics for RFID clothing, which provides opportunities for non-intrusive cost-effective healthcare monitoring [8,28].

Passive tags harvest energy from a commodity reader and backscatters incoming radio waves to send messages. The commodity reader measures the phase information of the backscattered signal at the physical layer and reports the phase value for each successfully identified tag. Suppose that the distance between the reader antenna and the tag is d. As radio waves traverse back and forth between the reader antenna and the tag in backscatter communication, the radio waves traverse a total distance of $2d$. Then the reader outputs a phase value of the backscatter radio wave as follows

$$\theta = \left(\frac{2\pi}{\lambda} \times 2d + c \right) \mod 2\pi, \tag{12.1}$$

where λ is the wavelength and c is a constant phase offset that captures the influence of reader and tag circuits independent of the distance between the reader antenna and the tag. The phase value repeats with a period of 2π radians every $\lambda/2$ in the distance of backscatter communication. COTS RFID readers (e.g., Impinj R420 [1]) can be programmed to output the low-level data for application development. In the presence of multiple tags, readers avoid collisions with the standard RFID protocol and report phase values without interference. *TagBreathe* aims to leverage the COTS RFID systems to measure the phase values from the RFID tags attached to users' clothes and simultaneously monitor breathing of multiple users.

12.3 Respiration monitoring using RFID systems

12.3.1 System overview

The inspiration of *TagBreathe* is that radio waves reflect off commodity tags attached to the human body and thus the phases of the radio waves will be modulated by the body movements of breathing. By carefully analyzing the data streams of phase values, we aim to extract the breathing signals with the following three key techniques, which we will elaborate in the following sections.

1) Phase Measurement and Preprocessing. COTS RFID readers measure the phase values of backscattered signals from the tags attached to users' clothes. COTS RFID readers report the phase value of each tag identification, as well as the time stamp of the phase measurement. *TagBreathe* continuously reads the phase values and groups the phase values according to the various users in the presence of multiple users. The changes in phase values indicate the body movements.

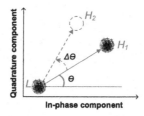

FIGURE 12.1 Illustration of low-level data measurements.

2) Breath Signal Extraction. *TagBreathe* extracts the breathing signals based on the fact that the phase values are influenced by the periodic changes introduced by the chest during inhalation and exhalation. We leverage our prior knowledge of human breathing rates and adopt a low-pass filter to extract the breathing signals.

3) Enhance Monitoring with Multiple Tags. To mitigate the impact of packet loss and blockage of line-of-sight path in practice, *TagBreathe* attaches multiple tags to each user. We use multiple antennas to ensure a full coverage of the monitoring area. The antennas are scheduled by commodity readers and work in a round robin manner without antenna-to-antenna interference. *TagBreathe* combines the multiple data streams from the array of tags to enhance monitoring robustness in practical scenarios.

12.3.2   Low-level data characterization

COTS RFID readers report low-level information (e.g., received signal strength, phase value, Doppler shift, etc.) for each successfully identified tag to support various applications. RFID tags modulate incoming radio signals by either reflecting or absorbing the radio signals, which results in two possible states (i.e., *High* (H) and *Low* (L)) [18,27]. The physical-layer symbols (denoted by crosses in Fig. 12.1) exhibit two clusters (i.e., H_1 and L) in the constellation map as illustrated in the figure. The magnitude of vector $\overrightarrow{LH_1}$ measures the received signal strength, while θ measures the phase value of the backscatter signals. Due to the Doppler frequency shift, one symbol cluster may rotate (e.g., from H_1 to H_2) in the constellation map during one packet transmission. As illustrated in the figure, $\Delta\theta$ measures the phase rotation which indicates the Doppler frequency shift in the constellation map [3].

In the initial experiment, a user equipped with a passive tag on his or her clothes breathes naturally, while sitting two-m away from a reader's antenna. We collect the low-level readings using the Impinj R420 reader for 25 seconds. The data sampling rate was around 64 Hz, i.e., 64 readings per second. The sampling rate is sufficiently high to capture the relatively low breathing rates of human subjects (e.g., around 10–20 breaths per minute). The low-level data reports the received signal strength, raw phase value, raw Doppler shift, time stamp, and the tag ID. We note that the experimental setting is static and without

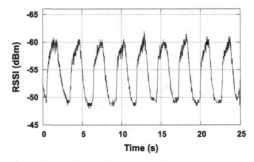

FIGURE 12.2 Raw RSSI readings during the measurements.

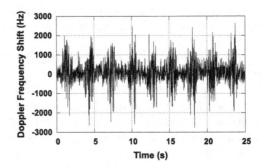

FIGURE 12.3 Raw Doppler frequency shift during the measurements.

much interference or background noise. In addition, the communication range is rather short to ensure strong backscatter signals from attached tags. Generally, it is challenging to monitor vital sign signals with RFID RSSI and Doppler frequency in practical scenarios.

1) Received signal strength. Fig. 12.2 plots the collected received signal-strength indicator (RSSI) data which shows the clear trends of periodic changes in the RSSI readings. That is because, when the user inhales, the body gets closer to the reader's antenna and thus the strength of backscatter signals increases, while, when the user exhales, the RSSI readings decrease. The initial experiment shows that it is possible to track the breathing patterns with RSSI readings in the ideal scenario. The limitations of RSSI measurement are its high sensitivity and low resolution. For instance, the resolution of the COTS reader is only 0.5 dBm. Due to the low resolution and sensitivity of RSSI readings, it is hard to precisely extract the subtle body movements in more challenging working scenarios.

2) Doppler frequency shift. Fig. 12.3 plots the raw Doppler frequency shifts whose envelope roughly tracks periodic changes. COTS readers calculate the Doppler frequency shift with the phase rotation during one backscatter packet transmission as follows [1]

$$f = \frac{\Delta\theta}{4\pi\,\Delta T},\tag{12.2}$$

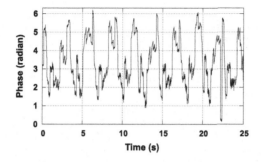

FIGURE 12.4 Raw phase values during the measurements.

where f denotes the Doppler frequency shift, $\Delta\theta$ denotes the phase rotation during one backscatter packet transmission, and ΔT denotes the duration of the packet transmission. In the figure, we see that, although the raw Doppler frequency shifts are noisy, we can still observe some periodic changes in the measurement data. The limit of Doppler frequency-shift measurement is that, because the duration of a packet transmission ΔT is relatively small, the measured phase rotation $\Delta\theta$ during the one packet transmission is not reliable and may be subject to noise in practice. As such, the Doppler frequency shift requires relatively high moving speeds of RFID-labeled objects to generate notable $\Delta\theta$ during the short duration of packet transmissions.

3) Phase values. Fig. 12.4 plots the phase values during the measurement. Due to the channel frequency hopping, the phase values discontinuously changes when the reader hops to the next channel, even when the tag is static. As specified in the standard EPC protocol [4], COTS readers hop among frequency channels to mitigate frequency selective fading and co-channel interference. Fig. 12.5 plots the channel indexes that show that the reader hops among ten frequency channels and resides in each channel for around 0.2 s. When the reader hops to a neighboring channel, the wavelength λ and the phase offset c in Eq. (12.1) also change, leading to discontinuity of phase values every 0.2 s [40]. A fixed-frequency channel may not be supported by commodity readers in some regions (e.g., the US, Singapore, Hong Kong) [4], since a continuous radio wave at a certain frequency may cause co-channel interference and violate radio-frequency regulations.

To continuously track body movements without interruption by channel hopping, we first group the phase values according to channel indexes. Then, we calculate the displacement during two consecutive phase readings in each channel, according to Eq. (12.1). Because the body movement speed is relatively low and the sampling rate is high (e.g., > 60 Hz), the tag displacement during two consecutive phase readings is within a half of a radio wavelength. Thus, we calculate the displacement during two consecutive phase readings as follows

$$\Delta d_{i+1} = d_{i+1} - d_i = \frac{\lambda}{4\pi}(\theta_{i+1} - \theta_i), \qquad (12.3)$$

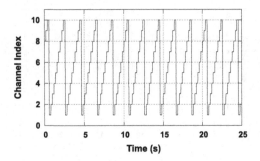

FIGURE 12.5 Channel hopping during the measurements.

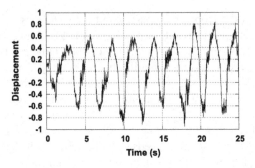

FIGURE 12.6 Displacement values during the measurements.

where Δd_{i+1} denotes the displacement at time $i+1$ and θ_{i+1} and θ_i denote the two consecutive phase readings measured in the same frequency channel. Then, we measure the total displacement during N readings as follows

$$D_j = \sum_{i=1}^{N} \Delta d_{i+j}. \tag{12.4}$$

We normalize the displacement values and plot the results in Fig. 12.6. We see that the displacement values are not influenced by the frequency hopping and track the periodic body movement due mainly to breathing.

12.3.3 Breath signal extraction

We analyze the displacement values collected during the measurements with the Fourier transform (FFT). Fig. 12.7 shows the FFT results. In the figure, the peak of the FFT output corresponds to the breathing rate. One of the pitfalls of the Fourier transform for a window size of w seconds is that it has a resolution of $1/w$. In other words, a wider window provides better frequency resolution but poor time resolution. In our initial experiment, since the window size is 25 sec, the frequency resolution is 0.04 Hz which corresponds to 2.4 breaths per minute.

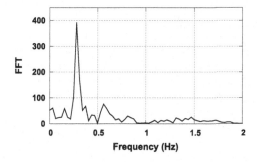

FIGURE 12.7 FFT for displacement values collected during the measurements.

Instead of measuring the breathing rate by finding the peak of FFT outputs, we apply an FFT-based low pass filter to filter out high-frequency noises and then extract the breathing signals. We note that the typical breathing rate for a healthy person at rest is around 12–20 breaths per minute and generally lower than 40 breaths per minute. Thus, we first apply the FFT to convert the time-domain displacement values to the frequency domain and set the cutoff frequency of the low pass filter at 0.67 Hz. After that, we use an inverse FFT (IFFT) to convert back to the time-domain displacement values. Fig. 12.8 plots the extracted breathing signals after applying the low pass filter. In the figure, we see that noise is successfully filtered out. The extracted signal exhibits clear trends, and we can apply time domain analysis to study the breathing signal. A finite impulse response (FIR) low-pass filter can also be adopted to extract breathing signals.

To monitor breathing rates, we detect the zero crossings as plotted in Fig. 12.8. We record the time stamps of the zero crossing events as t_i and calculate the instant breathing rate as follows

$$\overline{f_{BR}}(t_i) = \frac{M - 1}{2(t_i - t_{i-M})},\tag{12.5}$$

where M denotes the number of buffered zero crossings. To enhance the robustness, we buffer seven zero crossings which correspond to three breaths to calculate the breathing rates for real-time visualization.

12.3.4 Enhance monitoring with sensor fusion of multiple tags

In the following, we describe how to enhance breath monitoring with sensor fusion of data streams from multiple RFID tags. Intuitively, instead of only using one tag for each user, we form an array of tags by attaching multiple RFID tags to each user to improve signal strengths and ensure that some tags can be read to mitigate the impact of the blockage of line-of-sight paths. To this end, we aggregate the data streams from the tags and fuse them so that the raw

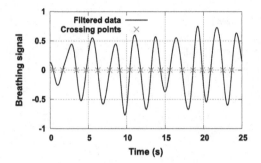

FIGURE 12.8 Extracted breathing signal from the measurements.

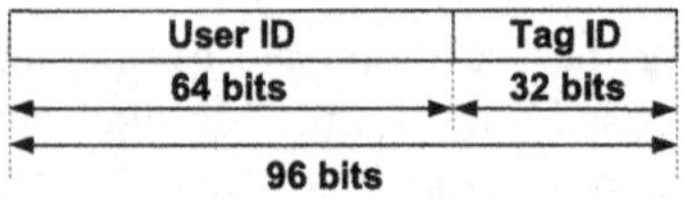

User ID	Tag ID
64 bits	32 bits

96 bits

FIGURE 12.9 *TagBreathe* overwrites 96-bit tag ID with 64-bit user ID and 32-bit short tag ID.

data streams reenforce each other and enhance the periodic signals due to body movements. By doing so, we improve the monitoring performance, especially in the extraction of weak breathing signals.

We attach an array of tags consisting of n RFID tags, $\{T_1, T_2, \ldots, T_n\}$ to a user. We overwrite the 96-bit tag ID with a 64-bit user ID followed by a 32-bit short tag ID as shown in Fig. 12.9. The 64-bit user ID enables us to differentiate users and group the n tags for each user. The 32-bit tag ID enables us to differentiate the tags and calculate the displacement values of each tag. By doing so, once a tag is identified, the low-level data can be classified according to the user ID and the tag ID. We group the sensor data for the same user according to the 64-bit user ID and fuse n data streams from $\{T_1, T_2, \ldots, T_n\}$ to enhance the monitoring performance. In the presence of multiple users, the low-level data can be fused for the multiple users by examining user IDs. Thus, we focus on the sensor fusion of multiple tags for the same user. Note that overwriting tag IDs is a standard RFID operation supported by commodity RFID systems (e.g., Impinj R420). If the overwriting operation is not supported, the reader can build a mapping table to map and lookup 96-bit tag IDs to user IDs and short tag IDs.

In the sensor fusion, one may extract breathing signals from each data stream and fuse the results in the final fusion phase. Instead, we carry out low-level data fusion by fusing the raw data before extracting breath signals. That is because we can effectively improve signal strength by fusing raw data, which substantially enhances signal extraction especially when the signals are weak. Furthermore, because breathing-signal extraction involves relatively high computational overhead, the low-level sensor fusion cuts the computational overhead otherwise involved in the multiple extraction operations.

In the process of data fusion, we first calculate the displacement values for each tag according to Eq. (12.3). The displacement value for tag T_j collected during the time period $[t, t + \Delta t]$ is denoted as $\Delta d_{i+1}^{T_j}(t + \delta t), 1 \leq j \leq n, 0 \leq \delta t \leq \Delta t$. We aggregate the displacement values from n tags for the time interval $[t, t + \Delta t]$ and fuse them together as follows

$$\Delta \mathbf{d}(t) = \sum_{j=1}^{n} \sum_{\delta t=0}^{\Delta t} \Delta d_{i+1}^{T_j}(t + \delta t). \tag{12.6}$$

Then, we measure the total displacement during N time intervals (i.e., $[t, t + N\Delta t]$) as follows

$$\Delta \mathbf{D}(t) = \sum_{i=0}^{N} \Delta \mathbf{d}(t + i\Delta t). \tag{12.7}$$

We process the fused data stream $\Delta \mathbf{D}(t)$ sampled at every Δt and extract breathing signals using the breathing signal extraction algorithms (Sect. 12.3.3).

Fig. 12.10 illustrates the overall workflow. *TagBreathe* interrogates multiple RFID tags attached to users and collects low-level data from those tags (Sect. 12.3.2). *TagBreathe* groups the readings according to user IDs and carries out raw data fusion by synthesizing multiple data streams (Sect. 12.3.4). *TagBreathe* analyzes the synthesized data stream and extracts breathing signals for each user (Sect. 12.3.3).

12.3.5 Discussion

1) Tag placement. After many trials with various users, we find that some users breath with their chests, while others breath with their abdomens. To better capture the breathing patterns, we place three tags on the upper body of each user: one on the chest, one on the lower abdomen, and one in between. Note that, when a user inhales or exhales, the three tags' relative displacements to the reader's antenna simultaneously decrease and increase, which allows us to constructively fuse the sensor data and enhance the breathing signals.

2) Enhancement with RSSI and Doppler frequency shift. TagBreathe mainly uses the displacement values inferred from the phase values to monitor breathing. In the low-level data characterization (Sect. 12.3.2), we notice that, although RSSI and Doppler frequency shifts have their limits, they are also informative in the experiments. One possible enhancement is to fuse the RSSI and Doppler frequency shift with the phase values to improve the monitoring accuracy.

3) Multiple antennas. To increase communication coverage, a commodity reader can be connected to multiple antennas (e.g., four antenna ports for one Impinj R420 [1] reader). The reader coordinates the multiple antennas with the round-robin scheduling and avoids the inter-antenna interference. In other

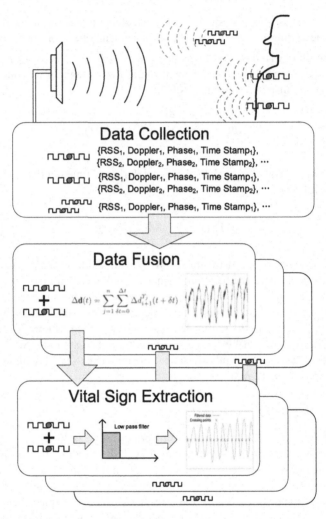

FIGURE 12.10 Illustration of *TagBreathe* workflow.

words, only one antenna will be powered up at a time, and the power consumption of the RFID system will not increase with the number of antennas. Note that *TagBreathe* does not strictly require multiple antennas to monitor breathing. *TagBreathe* works well as long as the tags can be read successfully. The low-level data is reported with the antenna port information. Thus, *TagBreathe* fuses the data according to the antenna port. Since the antennas are distributed geographically, the data qualities of antennas vary across various users in different locations. *TagBreathe* evaluates the data quality in terms of received signal strength and data sampling rate and extracts breathing signals with the data reported by the optimal antenna for each user.

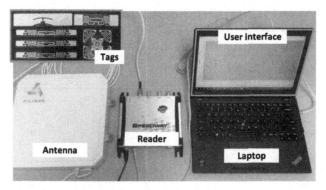

FIGURE 12.11 *TagBreathe* prototype system.

4) Respiration rates. In the evaluation, we mainly evaluate the performance with relatively low respiration rates. One may tune the parameters to increase the monitoring range of respiration rates, e.g., higher than 20 breaths per minute.

12.4 Implementation and evaluation

12.4.1 Implementation

We implement *TagBreathe* using COTS RFID systems.

1) Hardware. We use the Impinj Speedway R420 RFID reader [1] to interrogate commodity passive tags [2]. Fig. 12.11 depicts the prototype system. We evaluate various types of commodity passive tags (e.g., Alien 9640, Alien 9652, Impinj H47 tags). Since the performance with different tags was comparable, we report the experiment results with the Alien 9640 passive tags. The RFID system operates at the Ultra-High Frequency (UHF) band between 902 MHz and 928 MHz [1]. Both the reader and the tags follow the standard EPC protocol, which arbitrates tag-to-tag collisions at the MAC layer. We configure the transmission power to 30 dBm. The reader supports upto four directional antennas. In our prototype implementation, we adopt the Alien ALR-8696-C circular polarized antenna with the antenna gain of 8.5 dBic. As the communication range is around 10 m, the reader can cover a relatively large area with multiple antennas. The reader sends the low-level data with time stamps to the laptop (Leveno X240) via an Ethernet cable. *TagBreathe* processes the low-level data and extracts breathing signals in realtime.

2) Software. We implement *TagBreathe* based on the LLRP Toolkit (LTK) [3,45] to configure the commodity reader and read the low-level data. The LTK communicates with the reader following the LLRP protocol [2]. We program the reader to continuously identify tags in the communication range and report the low-level data (e.g., received signal strength, phase value, Doppler shift, etc). The low-level data are processed with the *TagBreathe* algorithms implemented in Java. We implement all the components (e.g., preprocessing, sensor fusion,

TABLE 12.1 System parameters and default experiment settings.

Parameter	Range	Default
Channel	channel 1–10	Hopping
Tx power	15–30 dBm	30 dBm
Distance	1–6 m	4 m
Orientation	0° (front)–180° (back)	front
Number of users	1–4 users	1 user
Tags per user	1–3 tags	3 tags
Breathing rate	5–20 bpm	10 bpm
Posture	Sitting, Standing, Lying	Sitting
Propagation path	with/without LOS path	with LOS path

breath signal extraction, etc.) which execute in a pipelined manner. The results are computed and visualized as shown in Fig. 12.11, which shows the extracted breathing signals in real time.

12.4.2 Experiment setting

To evaluate the performance of *TagBreathe* , we recruit four volunteers. The volunteers wear their everyday clothes (e.g., T-shirts, jackets, baggy cloths) made of various fabrics. Various numbers of COTS tags are attached to different parts of their cloths during the experiments. To evaluate the accuracy of *TagBreathe* , we use a breathing metronome application [10] to instruct the participants to regulate their breaths to evaluate the accuracy of breathing rate estimate of *TagBreathe* . We carry out the experiments in a standard office building. The office environment contains furniture, including desks and chairs, and electric appliances, including laptops and fans. We experiment with varied communication distances and various different postures. The users breathe naturally following the instructions of the metronome application. Table 12.1 summarizes key system parameters and default experiment settings.

12.4.3 Experiment results

The measurement accuracy is one of the most important metrics for the breathing rate monitoring. We calculate the accuracy as follows

$$Accuracy = 1 - \frac{|\hat{R} - R|}{R}, \tag{12.8}$$

where $\hat{R}$ is our measurement result and R is the actual breathing rate, respectively.

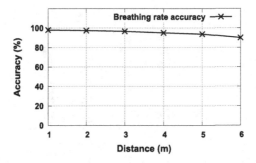

FIGURE 12.12 Breathing rate accuracy at various distances.

1) Impact of distance. We first evaluate the accuracy of breath monitoring at various distances from the reader's antenna to the tags attached to users. In the experiment, we fix the location of the antenna one m above the ground and ask users to sit at different locations at distances ranging 1–6 m away from the antenna. In each experiment, the user breathes according to a metronome mobile application running on a smartphone. The breathing rates range over 5–20 breaths per minute (bpm). Each experiment lasts for two minutes. We continuously measure the breathing signals and compute the average breathing rates using *TagBreathe* . We repeat the experiments for 100 times. During the experiments, we collect around 500,000 low-level readings in total.

We compare the experiment results of *TagBreathe* with the ground truth and plot the accuracies at various distances in Fig. 12.12. According to the experimental results, the accuracy of breathing rate measurement is 98.0% at one m. Although the accuracy decreases slightly as the distance increases, the experimental results show that the accuracy remains higher than 90.0% throughout the experiments. As the communication range increases, the reader observes weaker backscatter signal strengths and lower reading rates of low-level data, which negatively affect the breathing signal extraction.

2) Impact of the number of users. We evaluate the performance of *TagBreathe* with multiple users. The users sit side-by-side four m from the antenna. Each user wears three commodity passive tags. A commodity reader reads the low-level data from all the tags and groups the data according to the user IDs extracted from the tag IDs. We evaluate the monitoring accuracies with various different number of users in Fig. 12.13. According to the experimental results, the breathing rate accuracies with various different numbers of users remain around 95.0%. Thanks to the RFID collision avoidance protocol, the backscattered signals from multiple users do not interfere with each other. Moreover, the experimental results indicate that the reading rate of the commodity reader is sufficiently high to simultaneously monitor breathing even with four users wearing 12 tags.

3) Impact of contending tags. To evaluate the impact of lower reading rates due to an increasing number of tags in presence, we label daily items with RFID

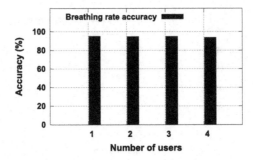

FIGURE 12.13 Breathing rate accuracy with various number of users.

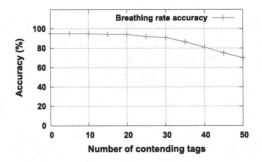

FIGURE 12.14 Breathing rate accuracy with various numbers of contending tags.

tags and place the RFID-labeled items in the communication range of the commodity reader. In the same way as the breath monitoring tags are attached to users, the item-labeling tags in the communication range contend for wireless channels following the standard EPC protocol. As a result, the reading rates decrease as the number of contending tags increases in the communication range. In the experiments, a user wears three tags and sits in front of the antenna. We vary the number of RFID tags and repeat the experiments with various numbers of tags. The commodity reader continuously monitors all the tags and reads low-level data from them, including the ones used for labeling the everyday items. Before the experiments, we overwrite the tag IDs of the three breath monitoring tags so that the reader can identify them and monitor breathing rates. Fig. 12.14 plots the measurement accuracy with multiple numbers of contending RFID tags present. According to the experimental results, we find that *TagBreathe* is able to achieve an accuracy of 91.0% even with 30 contending tags in the communication range. The main reason is because the total reading rate is sufficiently high, and *TagBreathe* can collect sufficient readings from the breath monitoring tags. The accuracy decreases when more contending tags are present, which leads to lower reading rates of three breath monitoring tags.

4) Impact of tag orientation and line-of-sight path. The performance of RFID systems is influenced by the antenna orientations as well as blockages

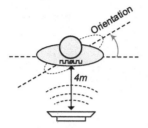

FIGURE 12.15 User orientation relative to antenna.

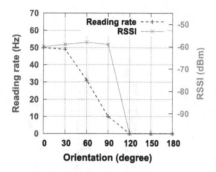

FIGURE 12.16 Reading rate and RSSI at various orientations.

of line-of-sight paths by the human body. We evaluate the monitoring accuracy at various antenna orientations. For evaluation purposes, we only connect one directional antenna to the reader to intentionally limit the coverage of the RFID system. In practical usage scenarios, we note that multiple antennas can be connected to fully cover the monitoring area. As illustrated in Fig. 12.15, a user is four-m away from the antenna and repeats the experiments at with various orientations. The user first faces the antenna (i.e., 0°) and rotates counter-clockwise until the user faces in the opposite direction (i.e., 180°). In Fig. 12.16, we measure the reading rate of low-level data, as well as the RSSI of backscatter signals at multiple orientations. We observe that, as long as there are line-of-sight paths between the tags and the antenna (i.e., [0°,90°]), the RSSI of the backscatter signal does not change much. On the other hand, the reading rate decreases from 50 Hz when the user faces to the antenna to 10 Hz when the user rotates to 90°. When the user further rotates (e.g., [120°, 180°]), as the line-of-sight path is blocked by the user's body, the reader cannot identify the tag or read low-level data any more. In such cases (> 90°), *TagBreathe* does not report breath monitoring results.

In practical usage scenarios, to increase the reader coverage and fully enable breath monitoring in the environment, a commodity reader can connect multiple antennas to ensure line-of-sight paths to the tags in practice. Note that the reader can coordinate the multiple antennas with the round-robin scheduling and avoid

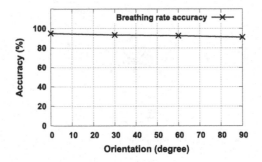

FIGURE 12.17 Accuracy at various orientations.

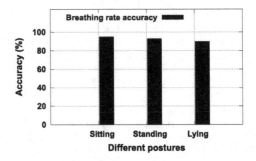

FIGURE 12.18 Accuracy at different postures.

the inter-antenna interference. Moreover, only one antenna will be powered up at a time, and the overall power consumption of the RFID system will not increase with the number of antennas.

Fig. 12.17 plots the measurement accuracy with multiple tag orientations with line-of-sight paths (i.e., $< 90°$). According to the experimental results, when the user faces to the antenna, the measurement accuracy is above 90%. The accuracy decreases from 90% to 85% as the user rotates to 90°.

4) Impact of different postures. We evaluate the monitoring accuracy with various postures, i.e., sitting, standing, and lying. The location of the antenna is fixed around one-m above the ground and the communication range between the tags and the reader remains unchanged throughout the experiments. As such, the results are mainly influenced by the orientation of the tags to the antenna, as well as the breathing behaviors with various different postures. According to the experiment results, the monitoring accuracy remains above 90.0% across different postures as shown in Fig. 12.18.

12.5 Conclusion

In this chapter, we describe vital signs monitoring with RFID systems. We focus on our work of respiration monitoring as an example to describe how COTS

RFID systems can be used to monitor vital signs. We introduce the key characteristics of RFID systems and describe the key lessons learned in the design and implementation of *TagBreathe* , which wirelessly senses breaths of multiple users with COTS RFID systems. Experiment results show that it is indeed possible to monitor respiration using RFID systems and achieve very high monitoring accuracy. To further improve the accuracy, one may deploy multiple readers to increase the coverage and mitigate the impact of blockage of line-of-sight paths. We believe RFID systems can potentially be used to monitor other vital signals, such as body temperature, blood pressure, and the pulse rate, which are worthy of further development and improvement.

References

[1] Impinj, http://www.impinj.com.

[2] Alien Technology, http://www.alientechnology.com.

[3] LLRP Toolkit, http://www.llrp.org.

[4] EPCgloable C1G2 Standard, http://www.gs1.org.

[5] Cleveland Clinic, http://my.clevelandclinic.org.

[6] Breathe Deep to Lower Blood Pressure, http://www.nbcnews.com/id/14122841.

[7] Your Newborn Baby's Breathing Noises, http://www.webmd.com/parenting/baby/your-newborn-babys-breathing-noises.

[8] Researchers Unveil Breakthrough in Weaving NFC Chips into Cloths, http://www.nfcworld.com.

[9] F. Adib, H. Mao, Z. Kabelac, D. Katabi, R. Miller, Smart homes that monitoring breathing and heart rate, in: ACM CHI, 2015.

[10] H. Abdelnasser, K. Harras, M. Youssef, UbiBreathe: A ubiquitous non-invasive WiFi-based breathing estimator, in: ACM MobiHoc, 2015.

[11] K. Bu, B. Xiao, Q. Xiao, S. Chen, Efficient misplaced-tag pinpointing in large RFID systems, IEEE Transactions on Parallel and Distributed Systems 23 (11) (2012) 2094–2106.

[12] S. Chen, M. Zhang, B. Xiao, Efficient information collection protocols for sensor-augmented RFID networks, in: IEEE INFOCOM, 2011.

[13] A. Droitcour, O. Boric-Lubecke, G. Kovacs, Signal-to-noise ratio in Doppler radar system for heart and respiratory rate measurements, IEEE Transactions on Microwave Theory and Techniques 57 (10) (2009) 2498–2507.

[14] H. Ding, J. Han, A.X. Liu, J. Zhao, P. Yang, W. Xi, Z. Jiang, Human object estimation via backscattered radio frequency signal, in: IEEE INFOCOM, 2015.

[15] H. Ding, L. Shangguan, Z. Yang, J. Han, Z. Zhou, P. Yang, W. Xi, J. Zhao FEMO, A platform for free-weight exercise monitoring with RFIDs, in: ACM SenSys, 2015.

[16] C. Duan, L. Yang, Y. Liu, Accurate spatial calibration of RFID antennas via spinning tags, in: IEEE ICDCS, 2015.

[17] R. Fletcher, J. Han, Low-cost differential front-end for Doppler radar vital sign monitoring, in: IEEE MTT-S, 2009.

[18] Y. Hou, J. Ou, Y. Zheng, M. Li, PLACE: Physical layer cardinality estimation for large-scale RFID systems, in: IEEE INFOCOM, 2015.

[19] Y. Hou, Y. Wang, Y. Zheng, Monitor breathing with commodity RFID systems, in: IEEE ICDCS, 2017.

[20] H. Jin, J. Wang, Z. Yang, S. Kumar, J. Hong, WiSh: towards a wireless shape-aware world using passive RFIDs, in: ACM MobiSys, 2018.

[21] Q. Lin, L. Yang, Y. Sun, T. Liu, X.-Y. Li, Y. Liu, Beyond one-dollar mouse: a battery-free device for 3D human-computer interaction via RFID tags, in: IEEE INFOCOM, 2015.

[22] X. Liu, K. Li, H. Qi, B. Xiao, X. Xie, Fast counting the key tags in anonymous RFID systems, in: IEEE ICNP, 2014.

[23] X. Liu, J. Cao, S. Tang, J. Wen, P. Guo, Contactless respiration monitoring via off-the-shelf WiFi devices, IEEE Transactions on Mobile Computing 15 (10) (2016) 2466–2479.

[24] X. Liu, J. Yin, Y. Liu, S. Zhang, S. Guo, K. Wang, Vital signs monitoring with RFID: opportunities and challenges, IEEE Network 33 (4) (2019) 126–132.

[25] M.L.R. Mogue, B. Rantala, Capnometers, Journal of Clinical Monitoring (1988).

[26] L. Ni, Y. Liu, Y. Lau, A. Patil. Landmarc, Indoor location sensing using active RFID, Wireless Networks 10 (6) (2004) 701–710.

[27] J. Ou, M. Li, Y. Zheng, Come and be served: parallel decoding for COTS RFID tags, in: ACM MobiCom, 2015.

[28] R. Nayak, A. Singh, R. Padhye, L. Wang, RFID in textile and clothing manufacturing: technology and challenges, Fashion and Textiles 2 (1) (2015) 9.

[29] R. Ravichandran, E. Saba, K. Chen, M. Goel, S. Gupta, S. Patel, WiBreathe: estimating respiration rate using wireless signals in natural settings in the home, in: IEEE PerCom, 2015.

[30] S. Qi, Y. Zheng, M. Li, L. Lu, Y. Liu, COLLECTOR: A secure RFID-enabled batch recall protocol, in: IEEE INFOCOM'14, 2014.

[31] S. Qi, Y. Zheng, M. Li, L. Lu, Y. Liu, Secure and private RFID-enabled third-party supply chain systems, IEEE Transactions on Computers 65 (11) (March 2016) 3413–3426.

[32] S. QI, Y. Zheng, M. Li, Y. Liu, J. Qiu, Scalable industry data access control in RFID-enabled supply chain, IEEE/ACM Transactions on Networking 24 (6) (March 2016) 3551–3564.

[33] L. Shangguan, Z. Li, Z. Yang, M. Li, Y. Liu, OTrack: Order tracking for luggage in mobile RFID systems, in: IEEE INFOCOM, 2013.

[34] L. Shangguan, Z. Yang, A.X. Liu, Z. Zhou, Y. Liu, Relative localization of RFID tags using spatial-temporal phase profiling, in: USENIX NSDI, 2015.

[35] L. Shangguan, K. Jamieson, The design and implementation of a mobile RFID tag sorting robot, in: ACM MobiSys, 2016.

[36] H. Wang, D. Zhang, J. Ma, Y. Wang, Y. Wang, D. Wu, T. Gu, B. Xie, Human respiration detection with commodity WiFi devices: do user location and body orientation matter?, in: ACM Ubicomp, 2016.

[37] J. Wang, F. Adib, R. Knepper, D. Katabi, D. Rus, RF-compass: Robot object manipulation using RFIDs, in: ACM MobiCom, 2013.

[38] J. Wang, D. Vasisht, D. Katabi, RF-IDraw: Virtual touch screen in the air using RF sensing, in: ACM SIGCOMM, 2014.

[39] Y. Wang, Y. Zheng, Modeling RFID signal reflection for contact-free activity recognition, in: ACM UbiComp, 2019.

[40] T. Wei, X. Zhang, Gyro in the air: tracking 3D orientation of batteryless Internet-of-things, in: ACM MobiCom, 2016.

[41] C. Yang, X. Wang, S. Mao, RFID-based driving fatigue detection, in: Globecom, 2019.

[42] C. Yang, X. Wang, S. Mao, Unsupervised detection of apnea using commodity RFID tags with a recurrent variational autoencoder, IEEE Access Journal, Special Section on Advanced Information Sensing and Learning Technologies for Data-centric Smart Health Applications 7 (1) (June 2019) 67526–67538.

[43] C. Yang, X. Wang, S. Mao, Unsupervised drowsy driving detection with RFID, IEEE Transactions on Vehicular Technology 69 (8) (August 2020) 8151–8163.

[44] C. Yang, X. Wang, S. Mao, Respiration monitoring with RFID in driving environments, IEEE Journal on Selected Areas in Communications: Special Issue on Internet of Things for In-Home Health Monitoring 39 (2) (Feb. 2021).

[45] L. Yang, Y. Chen, X.-Y. Li, C. Xiao, M. Li, Y. Liu, Tagoram: real-time tracking of mobile RFID tags to high precision using COTS devices, in: ACM MobiCom, 2014.

[46] L. Yang, Q. Lin, X. Li, T. Liu, Y. Liu, See through walls with COTS RFID system!, in: ACM MobiCom, 2015.

[47] Y. Yang, J. Cao, X. Liu, ER-rhythm: coupling exercise and respiration rhythm using lightweight COTS RFID, in: Proceedings of the ACM on Interactive, Mobile, Wearable and Ubiquitous Technologies (IMWUT)/Ubicomp, Dec 2019.

[48] Y. Yang, J. Cao, Robust RFID-based respiration monitoring in dynamic environments, in: IEEE SECON, 2020.

[49] S. Zhang, X. Liu, Y. Liu, B. Ding, S. Guo, Accurate respiration monitoring for mobile users with commercial RFID devices, IEEE Journal on Selected Areas in Communications (Sep. 2020).

[50] M. Zhao, F. Adib, D. Katabi, Emotion recognition using wireless signals, in: ACM Mobicom, 2016.

[51] R. Zhao, D. Wang, Q. Zhang, H. Chen, A. Huang, CRH: A contactless respiration and heart-beat monitoring system with COTS RFID tags, in: IEEE SECON, 2018.

[52] Y. Zheng, M. Li, Towards more efficient cardinality estimation for large-scale RFID systems, IEEE/ACM Transactions on Networking 22 (6) (December 2014) 1886–1896.

[53] Y. Zheng, M. Li, Read bulk data from computational RFIDs, IEEE/ACM Transactions on Networking 24 (5) (October 2016) 3098–3108.

Chapter 13

Acoustic-based vital signs monitoring

Xuyu Wang[a] and Shiwen Mao[b]

[a]*Department of Computer Science, California State University, Sacramento, CA, United States,*
[b]*Department of Electrical and Computer Engineering, Auburn University, Auburn, AL, United States*

Contents

13.1 Introduction

With the rapid development of mobile techniques and improvements in living standards, healthcare problems become more important [29,30]. It is reported that three-fourths of the total US healthcare cost are used to address chronic health conditions such as heart diseases, lung disorders, and diabetes [5]. Breathing signals are useful for physical health monitoring because such vital signs can offer important information for about problems, such as sudden infant death syndrome [9]. Traditional systems require a person to wear special devices, such as a pulse oximeter [25] or a capnometer [14], which are inconvenient and uncomfortable, especially for elders and infants. There is a compelling need for technologies that can enable contact-free, easily deployable, and long-term vital signs monitoring for healthcare.

Existing vital signal monitoring systems mainly focus on radio-frequency (RF) based techniques, which use radio-frequency (RF) signals to capture breathing and heart movements. Existing RF based schemes include radar-based and WiFi-based vital signs monitoring. In the radar-based category, the Doppler radar [7,20] and the ultra-wideband radar [24] are used to estimate breathing beats, where customized hardware is used operating on a high-frequency band. Moreover, Vital-Radio employs a frequency-odulated continuous-wave (FMCW) radar to estimate breathing and heart rates [2] with specially designed hardware operating on a large bandwidth from 5.46 GHz to 7.25 GHz. In the WiFi-based category, UbiBreathe can only monitor breathing signals by exploiting the received signal strength (RSS) of WiFi signals, which represents coarse channel information [1]. Another technique, mmVital [38], uses 60-GHz mm-wave (mmWave) signals for breathing and heart-rate estimations, while using a larger bandwidth of about 7 GHz and high-gain directional antennas at the transmitter and receiver. Recently, the authors in [11] exploited the amplitudes of WiFi CSI data to track the vital signs of a sleeping person. Our recently developed devices PhaseBeat [29,33] and TensorBeat [30] use the CSI phase-difference data instead for monitoring breathing and heartbeat for a single person and breathing for multiple persons. In addition, in some locations, using the CSI amplitude and phase difference together could improve the robustness of breathing monitoring [34]. Although RF-based techniques can effectively monitor vital signs over a long distance for healthcare, RF-based signals can be easily influenced by environmental factors, such as movements of other persons in the vicinity, and are thus not suitable for certain deployment scenarios.

To this end, the smartphones can be exploited for vial sign measurement with the built-in accelerometer, gyroscope [3] and microphone [22]. Usually, the smartphone needs to be placed near the body, or the person needs to wear special types of sensors that connect to the smartphone. Note that, for device-free and contact-free monitoring of vital signs, attached sensors should not be used. In a recent work [17], the authors propose to use the active sonar in the smartphone by leveraging the FMCW technique for breathing monitoring. However, the FMCW-based technique requires an accurate estimation of the distance between the smartphone and the chest of the person, before it can monitor the respiration of the person. When the body suddenly moves (e.g., rolling over in bed), the system needs to detect the new smartphone–chest distance, thus leading to additional time complexity. In addition, the LLAP system employs a continuous-wave (CW) radar to measure distance and achieves device-free hand tracking using the sonar phase information [28].

Motivated by these interesting studies, we employ the sonar phase data with a smartphone to monitor the periodic signal caused by the rises and falls of the chest (i.e., inhaling and exhaling). We find that the sonar phase information can track the periodic signal of breathing beats with high accuracy. Compared with other existing systems, such as Doppler shift and FMCW [17], the sonar phase-based scheme has lower latency and complexity. In addition, the sonar phase data is highly robust to different orientations, different distances, and different breathing rates of different persons.

We present a rigorous sonar phase analysis and prove that the sonar phase information can capture the breathing beats with the same frequency. Built upon the sonar phase analysis, we designedSonarBeat, a robust breathing monitoring system by using active sonar phase information with smartphones [31,32]. The SonarBeat system consists of four modules implemented in the smartphone, including signal generation, data extraction, received signal preprocessing, and breathing-rate estimation. First, it transmits an inaudible sound signal in the frequency range of 18–22 kHz by utilizing the smartphone speaker as a CW radar. Then, the signal is reflected by the chest of the person and is then received by the microphone of the same smartphone. The received signal is then processed to recover the breathing signal. We implement SonarBeat with two different smartphones and validate its performance with extensive experiments that involve five persons over a period of three months in three different environments, including an office scenario, a bedroom scenario, and a movie-theater scenario. The experimental results show that SonarBeat can achieve a low mean estimation error for breathing-rate estimation, with a medium error of 0.2 bpm. We also find that SonarBeat is highly robust to various experimental parameters and settings.

In the remainder of this chapter, we discuss related work in Sect. 13.2. Then, we present the sonar phase analysis in Sect. 13.3 and present the SonarBeat design in Sect. 13.4. We validate its performance in Sect. 13.5 and summarize this paper in Sect. 13.6.

13.2 Related work

The material in this chapter has two common aspects: sensing systems and mobile health systems based on acoustic signals with smartphones.

Sensing systems based on acoustic signal with smartphones: Acoustic sensing systems have attracted great attention within industrial fields and other researchers [10,26]. These systems are increasingly convenient for peoples' lives and healthcare, where acoustic sensing systems only need to be equipped with newly developed software for their smartphones to implement the new sensing functions. Traditionally, acoustic sensing systems can be classified into two categories: passive and active sensing systems. First, for passive acoustic sensing systems, the mainly focus is how to use the microphone to sense and recognize the surrounding audible signal [8]. Recently, the AAmouse system leverages inaudible sound pulses at different frequencies to transform a mobile device into a mouse based on the Doppler shifts, speed, and distance estimation [39]. Moreover, CAT system implements the distributed FMCW for device tracking used in VR/AR headsets, and this work mainly focuses on synchronizing two smartphones, while using the microphone as a mouse, which can interact with VR/AR headsets for higher accuracy localization [13]. In addition, an acoustic sensing technique is also used for wireless virtual keyboards with smartphones. Keystroke snooping [12] and Ubik systems [27] can obtain the sound signal with a smartphone's single or dual microphones, and leverage time-different-of-arrival (TDOA) measurements to monitor finger strokes on the table. Then, the strokes are transformed into related alphabets in the same position and computer keyboard. SilentWhistle is a light-weight indoor localization system that using acoustic sensing to obtain the users' locations [21]. Moreover, the Dhwani system builds an acoustics-based NFC system with smartphones with the technology of JamSecure, and it can provide a secure communication channel between devices based on an orthogonal frequency-division multiplexing (OFDM) technique for audible signals [18].

On the other hand, for active acoustic sensing systems [16], it can transfer a smartphone to an active sonar with the speaker and microphone using ultrasonic sound waves at 18 kHz to 22 kHz, which is close to the proposed SonarBeat system. OFDM–based sensing systems such as FingerIO can track the finger movement in a 2-D domain through tracking echoes of the fingers received by microphone to measure their positions [19]. The BatMapper system employs an acoustic sensing-ased system for fast, accurate floor plan construction using commodity smartphones [40]. Moreover, LLAP leverages the principal of CW radar to measure distance and implements device-free hand tracking [28]. This work is related to the SonarBeat system because both systems use the phased-based CW signal to sense movements. The difference between the two systems is that the SonarBeat system is more robust for different environments by the using adaptive median filter technique. On the other hand, the AudioGest system employs a pair of built-in speakers and a microphone to send inaudible sound and leverage the echos to sense the hand movement [23]. Thus, our SonarBeat system is motivated by the active inaudible sensing systems by transforming a smartphone into an active sonar with ultrasonic sound wave.

Mobile health systems: Mobile health applications and research have received great attention by the public [6]. Also, smartphones and other wearable device can provide people a more convenient way to monitor their health problems without the need of professional equipments [4]. Recently, the Burnout system leverages accelerometers to sense skeletal muscle vibrations; it does not require people to wear a suit embedded with sensors [15]. Moreover, the authors propose to capture depth video of a human subject with Kinect 2.0 to monitor people's heart rate and rhythm [37]. Moreover, wearable devices for monitoring exercise and the body is a trend of modern society for healthcare. The FEMO system achieves an integrated free-weight exercise monitoring service with RFID tags on the dumbbells and leverages the Doppler shift for recognition and assessment duringfree-weight exercise [15]. For vital signal monitoring, fine-grained sleep monitoring using a microphone in an earphone can record human breathing sounds to read people's vital heath signals when they are in sleeping [22]. This system is a passive audible method. Also, BreathListener uses acoustic devices on smartphones to obtain the fine-grained breathing waveform for drivers in real driving environments using deep learning techniques with the Generative Adversarial Network (GAN) [36]. Moreover, Apnea also leverages an active sonar with smartphone to monitor the breathing signal [17]. The work employs FMCW technique for breathing monitoring, which requires that the system acquires the distance between the smartphone and the chest of the person before tracking breathing. When the body of the person suddenly moves such as in turning over, the system needs to measure the new distance, thus encountering lrge time complexity. However, we proposed the sonar phase-based breathing monitoring for adapting to the movement of the body. In the following section, we will discuss the Sonarbeat system in detail.

13.3 Sonar phase analysis

We propose to use smartphones to monitor vital signs by using an inaudible sound signal in which the speaker and microphone of the smartphone emulate an active sonar system. In particular, the speaker transmits an inaudible sound signal in the frequency range of 18–22 kHz, in the form of a CW signal as $C(t) = A\cos(2\pi f t)$, where A is the amplitude and f is the frequency of the sound. Then the signal is reflected by the chest of the person and is received by the microphone at a sampling rate of 48 kHz. Because the speaker and microphone use the same frequency, there is no carrier frequency-offset (CFO) errors between the sender and receiver. Thus, we can exploit the phase of the inaudible signal data at the receiver to estimate the vital sign. The recently developed devices [19,28] use the phase of OFDM and CW sonar for finger tracking, while this paper focuses on exploiting CW waves for respiration monitoring.

To obtain the phase of the CW wave, we need to design a coherent detector structure in the receiver to down-convert the received sound signal to a baseband signal. The SonarBeat design is the first split the received sound signal into two identical copies. Then, these two copies are multiplied with the transmitted signal $C(t) = A\cos(2\pi f t)$ and its phase shifted version $C'(t) = -A\sin(2\pi f t)$. Finally, the corresponding in-phase and quadrature signals are obtained by using a low-pass filter to remove the high-frequency components.

We first present a simple analysis for the ideal case where there is no multipath effect (or, for the high signal-to-noise (SNR) regime and where the line-of-sight (LOS) component is the dominant part of the received signal). Under this assumption, the inaudible signal travels through a single path (i.e., from the speaker to the chest, and then from the chest to the microphone), and the propagation delay can be modeled as $d(t) = (D_0 + D\cos(2\pi f_b t))/c$, where D_0 is the constant distance of the path, D and f_b are the amplitude and frequency of the chest movements caused by breathing, respectively, and c is the speed of the sound. The received inaudible signal from this path can be modeled as $R(t) = A_r \cos(2\pi f t - 2\pi f d(t) - \theta)$, where A_r is the amplitude of the received inaudible signal and θ is a constant phase offset due to the delay in audio recording and playing. To estimate the phase of the inaudible signal, we need to remove the high frequency components. Multiplying the received signal with

$C(t) = A \cos(2\pi f t)$, we have

$$A_r \cos(2\pi f t - 2\pi f d(t) - \theta) \times A \cos(2\pi f t)$$
$$= \frac{A_r A}{2} (\cos(4\pi f t - 2\pi f d(t) - \theta) + \cos(-2\pi f d(t) - \theta)). \qquad (13.1)$$

The first term in Eq. (13.1) has a high frequency of $2f$, which can be removed with a properly designed low-pass filter. Thus, the I-component of the baseband is extracted as $I = \frac{A_r A}{2} \cos(-2\pi f d(t) - \theta)$. With a similar approach (i.e., multiplying by $C'(t)$ and removing the high-frequency component), we can estimate the Q-component of the baseband signal, i.e., $Q = \frac{A_r A}{2} \sin(-2\pi f d(t) - \theta)$. We then demodulate the phase of the inaudible signal data as

$$\varphi(t) = \arctan(Q/I) = -2\pi f d(t) - \theta$$
$$= -2\pi f (D_0 + D \cos(2\pi f_b t))/c - \theta. \qquad (13.2)$$

The phase of the inaudible signal, $\varphi(t)$, is a periodic signal with the same frequency as the breathing signal, under this single person scenario with only one propagation path. So the breathing rate f_b can be estimated from Eq. (13.2).

In reality, the received inaudible signal is a complex signal, as a sum of multiple reflected signals (e.g., on the chest, as well as on other surfaces) and interference signals from other sources. In the signal space, we can lump all the other signals into a static vector component rooted at the origin, while the desired signal is a smaller dynamic vector adding to the tip of the static vector. If there are no movements in the neighborhood, the static vector will be relatively constant, while the dynamic vector oscillates on the tip of the static vector following the chest movements (see [29,33]). To accurately monitor the breathing beats under a multipath and noisy environment, we need to remove the static vector and demodulate the phase information from the dynamic vector. In the following section, we show how to achieve this goal and introduce the design of the SonarBeat system in detail.

13.4 The SonarBeat system

13.4.1 SonarBeat system architecture

We designed the SonarBeat system for tracking the breathing beats of one person using active sonar phase data. Specifically, the SonarBeat system employs sonar phase data to monitor the periodic signal caused by the rises and falls of the chest when inhaling and exhaling. According to the sonar phase analysis, SonarBeat can effectively exploit the sonar phase information to monitor breathing signals for three reasons. First, the phase information can track the periodic signal of breathing beats with a high accuracy. Moreover, the phase information is sensitive to the small chest movements caused by breathing. Second, compared to other traditional methods, such as Doppler shift and FMCW [2], the

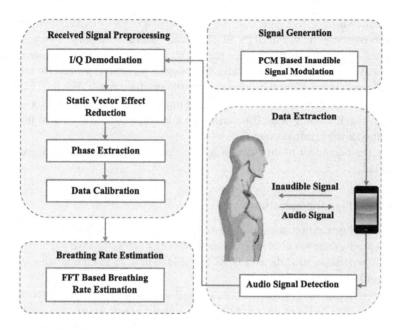

FIGURE 13.1 SonarBeat system architecture.

phase-based approach has a lower latency and complexity. Finally, the sonar phase data is robust to different orientations, different distances, different cloth thickness, and different breathing rates of different persons. It is also robust to large movements of the body, which only leads to a change of the stationary component of the phase data, and can be effectively removed by the proposed adaptive median filter method.

Fig. 13.1 depicts the SonarBeat system architecture. It includes four basic modules: (i) Signal Generation, (ii) Data Extraction, (iii) Received Signal Preprocessing, and (iv) Breathing-Rate Estimation. The Signal Generation module mainly implements a Pulse-code Modulation (PCM) of the inaudible signal, where a CW inaudible signal at 18 kHz to 22 kHz is generated and modulated with the PCM technique. The Data Extraction module detects the audio signal, employing a short-time Fourier transform (STFT) based method for audio signal detection. Then, a threshold-based method is proposed for detecting the beginning part of the received inaudible signal.

The Received Signal Preprocessing module consists of I/Q demodulation, static vector effect reduction, phase extraction, and data calibration. For I/Q demodulation, we first reduce the sampling rate from 48 kHz to 480 Hz, which can achieve a lower processing delay for real-time monitoring. Then, a coherent detector structure is used to down-convert the received signal to a baseband signal, while a low-pass filter is used to remove the high-frequency components and environmental noises to obtain the in-phase and quadrature components.

For static vector effect reduction, we propose an adaptive median filter method to remove the static vector of the in-phase and quadrature signals, which is suitable for real-time processing. For phase extraction, we derive the sonar phase information, which includes the breathing signal. Moreover, we need to unwrap the phase data to obtain the calibrated breathing phase result. For data calibration, we implement a median filter as a simple Low Pass FIR to remove the noise. The Breathing-Rate Estimation module employs an FFT-based method to estimate the breathing rate.

In the remainder of this section, we discuss the four modules of SonarBeat in detail.

13.4.2 Signal generation

The signal generation module uses one speaker of the smartphone as a transmitter to produce the inaudible signals for breathing monitoring. We implement the signal generation module as a PCM based inaudible signal modulation on the Android platform.

Specifically, the speaker generates an inaudible sound signal in the frequency range of 18–22 kHz in the form of a CW signal $C(t) = A\cos(2\pi f t)$. We produce the sampled analogy signal and then use the PCM technique to digitally represent the sampled CW signal. To generate a PCM stream, the amplitude of the analog CW signal is sampled at uniform intervals, where each sampled value is quantized. The PCM-based inaudible signal modulation is implemented with the AudioTrack class.

13.4.3 Data extraction

We use the microphone of the smartphone with a sampling rate of 48 kHz to receive the inaudible signal reflected from the chest of the person. In fact, the microphone is likely to record other sound signals with different frequencies from the surrounding environment. We implement an audio signal detection method to identify the beginning of the inaudible signal from the speaker as follows.

We propose an STFT-based method for audio signal detection. When the microphone receives the beginning of inaudible signal, there will be a large change in the power at the inaudible carrier frequency. A threshold-based method is proposed for detecting the beginning of the inaudible signal. Fig. 13.2 illustrates the STFT-based method for audio signal detection, where the carrier frequency is 20 kHz. We can see that, before 0.25 sec, the microphone only receives audio frequencies from the surrounding environment. After 0.25 sec, the microphone detects the inaudible signal since the magnitude of the 20 kHz spectrum is much stronger than that at other audio frequencies (see the bright yellow (white in print version), horizontal strip at 20 kHz). We adopt a window size of 512 points for estimating the spectrum in STFT. Moreover, we set a threshold of 200 for the power change to detect the beginning of the inaudible signal. In fact, if we de-

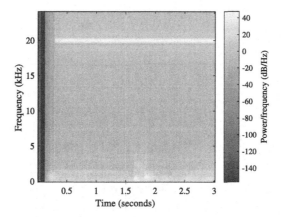

FIGURE 13.2 STFT based method for audio signal detection.

tect the power change with the threshold method, the beginning of the inaudible signal can be set as the ending location of the STFT chirp.

13.4.4 Received signal preprocessing

The Received Signal Preprocessing module consists of four components, including I/Q Demudulation, Static Vector Effect Reduction, Phase Extraction, and Data Calibration. We discuss the design of these components in the following.

13.4.4.1 I/Q demodulation

Before I/Q Demodulation, we need to down-sample the received signal $R(t) = A_r \cos(2\pi f t - 2\pi f d(t) - \theta)$ for reducing the computation complexity, which is necessary for real-time breathing monitoring. The original system sampling frequency of 48 kHz is reduced to 480 Hz with a down-sampling ratio of 100. Then, we implement the I/Q demodulation to get the I-component and Q-component of the baseband signal using the coherent detector structure. The design splits the received audio signal into two identical copies. Due to the down-sampling ratio of 100, these two copies should be multiplied with the signal $A \cos(2\pi \frac{f}{100} t)$ and its phase-shifted version $-A \sin(2\pi \frac{f}{100} t)$ to obtain the I-component and Q-component of the baseband signal. Because we use phase modulation for breathing monitoring, down-sampling only reduces the number of samples of the amplitude of breathing signal, rather than the phase information.

Finally, a low-pass filter is employed to obtain the corresponding in-phase and quadrature signals. The low-pass filter has a cutoff parameter of one Hz, a sampling rate of 480 Hz, and a resonance of two. This setting has been shown to be effective for removing the high-frequency components and audible environmental noises. From Figs. 13.3 and 13.4, we can see the raw I-component

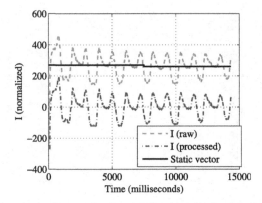

FIGURE 13.3 The adaptive filter median for removing the static vector in the baseband component I.

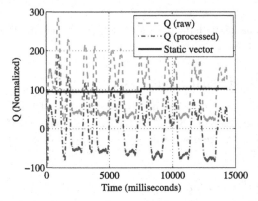

FIGURE 13.4 The adaptive filter median for removing the static vector for the baseband component Q.

and Q-component of the baseband signal after the low-pass filters (the dashed curves), which, however, still include their static vectors.

13.4.4.2   Static vector effect reduction

The performance of SonarBeat for breathing estimation largely depends on mitigating the effect of the static vector in multipath environments. This is because usually the static vector is much stronger than the dynamic vector that represents the small chest movements. It is difficult to detect the weak breathing signal if we directly use the received sound signal without removing the static vector. Recently, there are two methods being proposed for static vector effect reduction. The Dual-Differential Background Removal approach is used for hand tracking with 60-GHz millimeter wave (mmWave) signals [35]. The method is susceptible to environmental noise and long latency, which is not effective for breathing

---

Algorithm 1: The adaptive median filter method.

---

1 **Input:** One baseband signal component: $X(n) = I(n)$ or $Q(n)$ for $n = 0, 1, ..., N - 1$ and the window size w ;

2 **Output:** Real or Imaginary part after removing the static vector: $O(n)$ for $n = 0, 1, ..., N - 1$;

3 //Initialize ;

4 n_w: Number of windows;

5 r: Number of the remaining elements of $X(n)$, which cannot form a full W window;

6 $W[1, 2, ..., w]$: Sublists with window size w;

7 $R[1, 2, ..., r]$: Sublist with the remaining r elements;

8 //Find the median for each window;

9 **for** $i = 0 : n_w$ **do**

10   **if** $((i + 1) * w) <= N$ **then**

11          $W[1, 2, ..., w] \leftarrow X((w * i)$ to $((i + 1) * w - 1))$;

12          $M \leftarrow$ the median of $W[1, 2, ..., w]$,;

13   **for** $j = w * i : (i + 1) * w$ **do**

14              $O(j) = X(j) - M$;

15   **end**

16   **end**

17   **else if** $r! = 0$ **then**

18          $R[1, 2, ..., r] \leftarrow X((w * i)$ to $(w * i + r - 1))$;

19          $M \leftarrow$ the median of $R[1, 2, ..., r]$;

20   **for** $j = w * i : (i + 1) * w$ **do**

21              $O(j) = R(j) - M$;

22   **end**

23   **end**

24 **end**

25 Return $O(n)$ for $n = 0, 2, ..., N - 1$;

---

monitoring. The second scheme, Local Extreme Value Detection (LEVD) [28], is used to track hand movements. The method requires an empirical threshold for detecting the stationary vector, which is not robust for different environments and different persons.

In Algorithm 1, we present an adaptive median filter method for removing the static vector, which has a low latency and is robust for various different environments. The idea is to use a window to obtain the median for estimating the static vector. The only parameter is the window size w, which is robust for different environments. For one baseband signal component $I(n)$ or $Q(n)$, for $n = 0, 1, ..., N - 1$, we partition it into multiple non-overlapping sublists (each is denoted by $W[1, 2, ..., w]$) with a window size w and a single sublist $R[1, 2, ..., r]$ with a window size r, where r is the number of remaining elements of $I(n)$ or $Q(n)$ is not included in the previous W sublists. The sublists $W[1, 2, ..., w]$ and sublist $R[1, 2, ..., r]$ are used to estimate the medians for the first $n_w - 1$ windows and the last window, respectively, where $n_w = \lfloor N/w \rfloor$. Finally, the output $O(n)$, for $n = 0, 1, ..., N - 1$, can be obtained as in Steps 14 and 21 by removing the static vector. The proposed method is simple and robust for real-time processing of received data in different environments with a low delay.

Figs. 13.3–13.4 show the adaptive median filter method for removing the static vectors in the baseband signal components I and Q, respectively. We can see that the estimated static vector data can well represent the average amplitude information for the baseband signal components. After the adaptive median filter, the components I and Q are roughly centered at zero, making it easier to extract the breathing signal they carry.

13.4.4.3   Phase extraction

After removing the static vector based on the adaptive median filter method, we can extract the phase data in the I-Q plane, which only includes the dynamic breathing component. Let $O_I(t)$ and $O_Q(t)$ denote the output values of Algorithm 1 after removing static vectors for the corresponding I and Q data, respectively. The phase for the inaudible signal can be computed with (13.2), that is

$$\varphi(t) = \arctan\left(\frac{O_Q(t)}{O_I(t)}\right). \tag{13.3}$$

Using Eqs. (13.2) and (13.3), we find the phase values $\varphi(t)$ for the breathing signals reflected from the chest. Although the reflected breathing signal may still have multipath components, these multipath signals have the same breathing frequency but different phase shifts, each of which is a constant. Thus, the breathing rate is not affected by the dynamic multipath effect. This is different than hand tracking, which requires only one path from the smartphone and the hand. Thus, our SonarBeat can estimate the breathing rate using a single subcarrier rather than multiple subcarriers. Fig. 13.5 presents the respiration curve obtained from the phase data with its static vectors removed. It can be noticed that the magnitude of the breathing signal is large and is periodic if we can remove the sudden phase changes. Thus, we need to implement a data calibration scheme for the demodulated phase data to achieve better accuracy.

13.4.4.4   Data calibration

For data calibration, we implement a phase unwrapping scheme for recovering the right phase values, as well as a median filter for reducing the environmental noise. To estimate breathing rates, we need to obtain the right breathing curve for the phase data. Because the phase value will have a change of 2π for every wavelength distance, we implement the phase unwrapping to process the demodulated phase data to obtain the right breathing curve. Fig. 13.6 shows the respiration curve obtained from the phase data after phase unwrapping. It is a

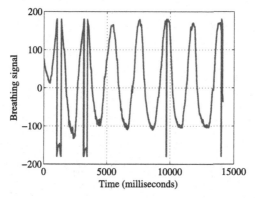

FIGURE 13.5 Respiration curve obtained from the phase data with the static vector effect removed.

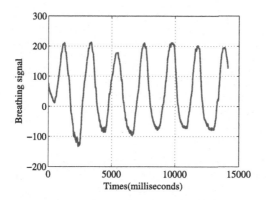

FIGURE 13.6 Respiration curve for phase data after unwrapping the phase data.

clean breathing signal with smaller environmental noises. On the other hand, we adopt the median filter method for removing environment noise, where the filter window size is set to 300. Fig. 13.7 presents the respiration curve for unwrapped phase data after the median filter, which is useful for accurate breathing rate estimation.

13.4.5 Breathing-rate estimation

SonarBeat has three stages for breathing monitoring by using the calibrated inaudible phase data. In the first stage of a 15-sec duration, we cannot effectively estimate the breathing rate because the phase data in a window can continually change when new data arrive. In the second stage, we exploit the 15-sec phase data to estimate the breathing rate in real time with the FFT-based method. In fact, the frequency resolution depends on the window size of FFT. If the window size becomes larger, the estimation accuracy will be higher, but the larger window size also leads to a lower time-domain resolution. Thus, for on-

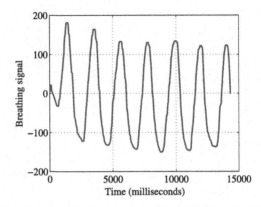

FIGURE 13.7 Respiration curve for unwrapped phase data after the median filter.

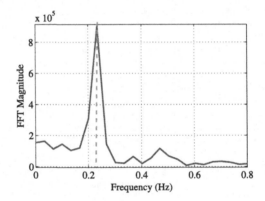

FIGURE 13.8 Respiration rate estimation based on FFT.

line breathing-rate estimation, we use the same window size, which is in fact equal to that of the STFT-based method. This balances the tradeoff between the frequency-domain resolution and the time-domain resolution. In the final stage, we use all the phase data from the three stages for breathing-beat estimation with the FFT-based method, which can achieve a higher estimation accuracy. Fig. 13.8 illustrates the FFT-based respiration-rate estimation. We can see that the estimated frequency for the breathing beats of a single person is 0.23 Hz, which is approximately the same as the true breathing rate measured by the NEULOG Respiration Monitor Belt Logger Sensor.

13.5 Experimental study

13.5.1 Implementation and test configuration

We prototype the SonarBeat system with two different smartphones, a Samsung Galaxy S6 and a Samsung Galaxy S7 Edge, both of which are based on the Android platform. Moreover, we implement all the signal processing algorithms in Java using the Android SDK. Both smartphones perform well for processing the audio data for real-time breathing-rate monitoring and display. The first edition of SonarBeat is implemented with the minimum version of Android 5.1.1 OS (API 21), so it works with all the more recent Android systems such as Android 6.0 and Android 7.0. For breathing monitoring, we only use one speaker and one microphone to transmit and receive the inaudible audio data, with the microphone and speaker fixed at the bottom on the smartphone. Furthermore, we use the AudioTrack class to play inaudible sound and the AudioRecord class to record sound. The buffer of the recording thread is set to 1,920 points with a sampling rate of 48 kHz. Therefore, we set the real-time signal processing unit to 1,920 points, which is about 40 ms.

We conducted extensive experiments with SonarBeat with five persons over three months. The test scenarios include an office, a bedroom, and a movie theater. The *office* is a 4.5×8.8 m^2 room. The room is crowded with tables and PCs, which form a complex propagation environment. In this office environment, we measure breathing monitoring data under various system parameters. The second environment is a *bedroom* of 3.9×6 m^2, where we test breathing monitoring for a single person. The third setup is a *movie theater* of a large 27×40 m^2 area, where many people are watching a movie, and there sources of audio interference from the movie and other people. For comparison, we use the NEULOG Respiration Monitor Belt Logger Sensor to record the ground truths of the breathing rates.

13.5.2   Performance of breathing-rate estimation

Fig. 13.9 presents the cumulative distribution functions (CDF) of estimation error in breathing-rate estimation with SonarBeat. For comparison, we also developed an LEVD based system [28], where the LEVD method is used for estimating the static vector and all other signal processing methods are the same as in SonarBeat. The LEVD-based system is used as a benchmark in the experiment. We find that SonarBeat and the LEVD-based method achieve a median error of 0.2 bpm and 0.3 bpm, respectively. This illustrates that both systems can effectively estimate breathing rates. However, it is worth noting that for SonarBeat, 95% of the test results have an estimated error under 0.5 bpm, while only 60% of the test results with the LEVD-based method have an estimated error under 0.5 bpm. Moreover, the maximum estimation error for SonarBeat and the LEVD-based method are 2.4 bpm and 5 bpm, respectively. This is because the LEVD-based method requires setting the empirical threshold based on the standard deviation of the baseband signal in a static environment. It is not robust in varying environments where the same threshold will not work. However, SonarBeat leverages the adaptive median filter method and is thus more robust to changes in the environment. Therefore, the SonarBeat system can achieve a higher and more stable breathing-rate estimation accuracy than the LEVDbased method.

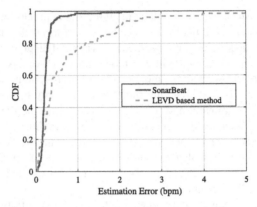

FIGURE 13.9 CDFs of estimation error in breathing rate estimation.

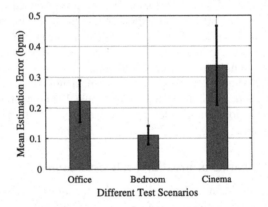

FIGURE 13.10 Mean estimation error for three different scenarios.

Fig. 13.10 presents the mean estimation errors for the three different scenarios, which are 0.22 bpm, 0.11 bpm, and 0.33 bpm for the office, bedroom, and cinema scenarios, respectively. We plot the 95% confidence intervals as error bars. It can be noticed that the mean estimation error for the bedroom case is the minimum because the bedroom has a better environment: with smaller noise and no sound interference from other persons. This shows that SonarBeat is suitable for breathing monitoring during sleeping and effectively detects apnea or abnormal breathing. For breathing monitoring in the office, the performance is worse than the bedroom. This is because there is a more complex propagation environment and interference from other people. Furthermore, loud noises from computers, air conditioners, and other equipment also influence the inaudible signal. For breathing monitoring in the cinema, we find that it has the largest mean estimation error and variance because of the more complex environment and louder noises. In fact, breathing monitoring in the cinema is still quite accu-

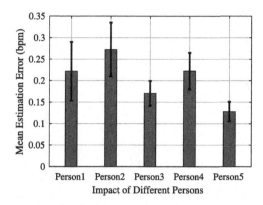

FIGURE 13.11 Impact of different test persons.

rate given the extremely adverse environment. These experiments validate that SonarBeat system is highly accurate and robust in several different scenarios.

13.5.3   Impact of various test factors

Figure 13.11 shows the impact of different groups of persons in the office scenario. In the experiment, we test five persons including three men and two women, respectively, where every volunteer wears the NEULOG Respiration unit for recording the benchmark result for breathing data. From Fig. 13.11, we can see that Persons 3, 4, and 5 have a lower mean estimation error. This is because they work out quite often and have a strong respiration, leading to stronger breathing signals. On the other hand, the other two persons have weaker breathing magnitudes, but their estimation errors are still under 0.5 bpm, which is acceptable. Thus, we can see that SonarBeat is adaptive for different groups of people.

Fig. 13.12 shows the impact of different distances between the chest and the smartphone. When the distance is increased, the accuracy for breathing estimation becomes lower. In particular, we can see that for a distance of 55 cm, the mean breathing-estimation error becomes 1 bpm with a larger variance. In this experiment, we find that ultrasound waves at 18 kHz to 22 kHz experience large attenuation, and the microphone will receive a lower power from the reflection of the chest movements if the distance is increased. Moreover, breathing-rate estimation with SonarBeat depends on I/Q demodulation. The magnitude of the I/Q components becomes weaker when the distance is increased, leading to higher errors. To improve the measurement distance, we leverage the parameter resonance of the low-pass filter to strengthen the amplitude of the inaudible signal near the cutoff frequency, thus improving the magnitudes of I/Q components. In the experiment, we set the cutoff frequency to 40 Hz for the sampling rate of 48 kHz. We can see that, under 50 cm, the proposed system can achieve very good accuracy.

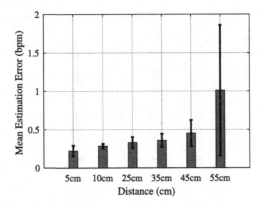

FIGURE 13.12 Impact of the distance between the chest and the smartphone.

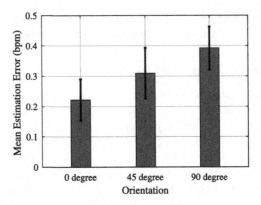

FIGURE 13.13 Impact of chest orientation relative to the smartphone.

Fig. 13.13 shows the impact of chest orientation relative to the smartphone in the office scenario, where we consider three cases including 0°, 45° and 90°. It can be noticed that at 0° direction with the front orientation relative to the smartphone, we can obtain the minimum mean estimation error at about 0.22 bpm. Moreover, at the 90° direction, the maximum mean estimation error of 0.39 bpm is achieved. The reason is that the received inaudible signal is the strongest when the person faces the smartphone speaker. It is most effective for monitoring the breathing signal for the chest movements in this case. Thus, we can obtain the higher estimation accuracy at the front direction.

Fig. 13.14 presents the estimation errors with a Samsung Galaxy S6 with Android 6.0 and a Samsung Galaxy S7 Edge working with the latest Version Android 7.0. The speaker and microphone are on the bottom of both smartphones. We can see that the Samsung Galaxy S7 Edge has a similar performance as the Samsung Galaxy S6, with a mean error of 0.21 bpm and 0.22 bpm, respectively.

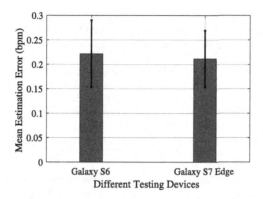

FIGURE 13.14 Estimation error results with two different smartphone platforms.

We find that the Samsung Galaxy S7 Edge has stronger processing power, and thus it can obtain better real-time performance than the Samsung Galaxy S6.

13.6   Conclusions

In this chapter, we discussed acoustic-based vital signs monitoring with smartphones and introduced several acoustic sensing applications. Then, we presented SonarBeat system, a smartphone based system that exploits phase-based sonar to monitor breathing beats. We first provided a sonar phase analysis and proved that the sonar phase-based method can extract breathing signals. We then described the SonarBeat design in detail, including its signal generation, data extraction, received signal preprocessing, and breathing-rate estimation modules. Finally, we implemented SonarBeat with two different smartphones and conducted an extensive experimental study with three test scenarios. The experimental results validated that SonarBeat can achieve superior performance for breathing beats estimation under various different environments.

Acknowledgment

This work is supported in part by the US National Science Foundation (NSF) under Grant ECCS-1923163, ECCS-1923717, and CNS-2105416, and through the Wireless Engineering Research and Education Center (WEREC) at Auburn University. Any opinions, findings, and conclusions or recommendations expressed in this material are those of the authors and do not necessarily reflect the views of the foundation.

References

[1]   H. Abdelnasser, K.A. Harras, M. Youssef, Ubibreathe: a ubiquitous non-invasive WiFi-based breathing estimator, in: Proc. IEEE MobiHoc'15, Hangzhou, China, 2015, pp. 277–286.
[2]   F. Adib, et al., Smart homes that monitor breathing and heart rate, in: Proc. ACM CHI'15, Seoul, Korea, 2015, pp. 837–846.

[3] H. Aly, M. Youssef, Zephyr: ubiquitous accurate multi-sensor fusion-based respiratory rate estimation using smartphones, in: Proc. IEEE INFOCOM'16, San Francisco, CA, 2016, pp. 1–9.

[4] Edgar A. Bernal, et al., Deep temporal multimodal fusion for medical procedure monitoring using wearable sensors, in: IEEE Transactions on Multimedia, 2017.

[5] O. Boric-Lubeke, V.M. Lubecke, Wireless house calls: using communications technology for health care and monitoring, IEEE Microwave Magazine 3 (3) (2002) 43–48.

[6] Charalampos A. Dimoulas, Audiovisual spatial-audio analysis by means of sound localization and imaging: a multimedia healthcare framework in abdominal sound mapping, IEEE Transactions on Multimedia 18 (10) (2016) 1969–1976.

[7] A. Droitcour, O. Boric-Lubecke, G. Kovacs, Signal-to-noise ratio in Doppler radar system for heart and respiratory rate measurements, IEEE Transactions on Microwave Theory and Techniques 57 (10) (2009) 2498–2507.

[8] Biying Fu, et al., Opportunities for activity recognition using ultrasound Doppler sensing on unmodified mobile phones, in: Proc. 2nd International Workshop on Sensor-Based Activity Recognition and Interaction, ACM, Rostock, Germany, 2015, p. 8.

[9] C. Hunt, F. Hauck, Sudden infant death syndrome, CMAJ. Canadian Medical Association Journal 174 (13) (2006) 1309–1310.

[10] Fan Li, et al., CondioSense: high-quality context-aware service for audio sensing system via active sonar, Personal and Ubiquitous Computing 22 (1) (2017) 17–29.

[11] J. Liu, et al., Tracking vital signs during sleep leveraging off-the-shelf WiFi, in: Proc. ACM Mobihoc'15, Hangzhou, China, 2015, p. 267276.

[12] Jian Liu, et al., Snooping keystrokes with mm-level audio ranging on a single phone, in: Proc. ACM Mobicom'15, ACM, Paris, France, 2015, pp. 142–154.

[13] Wenguang Mao, Jian He, Lili Qiu, CAT: high-precision acoustic motion tracking, in: Proc. ACM Mobicom'16, ACM, New York City, NY, 2016, pp. 491–492.

[14] M.L.R. Mogue, B. Rantala, Capnometers, Journal of Clinical Monitoring 4 (2) (1988) 115–121.

[15] Frank Mokaya, et al., Burnout: a wearable system for unobtrusive skeletal muscle fatigue estimation, in: Proc. IEEE/ACM ISPN'16, IEEE, Kobe, Japan, 2016, pp. 1–12.

[16] Rajalakshmi Nandakumar, Shyamnath Gollakota, Unleashing the power of active sonar, IEEE Pervasive Computing 16 (1) (2017) 11–15.

[17] Rajalakshmi Nandakumar, Shyamnath Gollakota, Nathaniel Watson, Contactless sleep apnea detection on smartphones, in: Proc. ACM MobiSys'15, ACM, Florence, Italy, 2015, pp. 45–57.

[18] Rajalakshmi Nandakumar, et al., Dhwani: secure peer-to-peer acoustic NFC, ACM SIG-COMM Computer Communication Review 43 (4) (2013) 63–74.

[19] Rajalakshmi Nandakumar, et al., Fingerio: using active sonar for finegrained finger tracking, in: Proc. ACM CHI'16, ACM, Santa Clara, CA, 2016, pp. 1515–1525.

[20] Phuc Nguyen, et al., Continuous and fine-grained breathing volume monitoring from afar using wireless signals, in: Proc. IEEE INFO-COM'16, San Francisco, CA, 2016, pp. 1–9.

[21] Chen Qiu, Matt W. Mutka, Silent whistle: effective indoor positioning with assistance from acoustic sensing on smartphones, in: A World of Wireless, Mobile and Multimedia Networks (WoWMoM), 2017 IEEE 18th International Symposium on, IEEE, 2017, pp. 1–6.

[22] Y. Ren, et al., Fine-grained sleep monitoring: hearing your breathing with smartphones, in: Proc. IEEE INFOCOM'15, Hong Kong, China, 2015, pp. 1194–1202.

[23] Wenjie Ruan, et al., AudioGest: enabling fine-grained hand gesture detection by decoding echo signal, in: Proc. ACM UBICOMP'16, ACM, Heidelberg, Germany, 2016, pp. 474–485.

[24] J. Salmi, A.F. Molisch, Propagation parameter estimation, modeling and measurements for ultrawideband mimo radar, IEEE Transactions on Microwave Theory and Techniques 59 (11) (2011) 4257–4267.

[25] N.H. Shariati, E. Zahedi, Comparison of selected parametric models for analysis of the photoplethysmographic signal, in: Proc. 1st IEEE Conf. Comput., Commun. Signal Process, Kuala Lumpur, Malaysia, 2005, pp. 169–172.

[26] Dan Stowell, et al., Detection and classification of acoustic scenes and events, IEEE Transactions on Multimedia 17 (10) (2015) 1733–1746.

[27] Junjue Wang, et al., Ubiquitous keyboard for small mobile devices: Harnessing multipath fading for fine-grained keystroke localization, in: Proc. ACM Mobisys' 14, ACM, Bretton Woods, NH, 2014, pp. 14–27.

[28] Wei Wang, Alex X. Liu, Ke Sun, Device-free gesture tracking using acoustic signals, in: Proc. ACM MobiCom' 16, ACM, New York City, NY, 2016, pp. 82–94.

[29] X. Wang, C. Yang, S. Mao, PhaseBeat: exploiting CSI phase data for vital sign monitoring with commodity WiFi devices, in: Proc. IEEE ICDCS 2017, Atlanta, GA, 2017, pp. 1–10.

[30] X. Wang, C. Yang, S. Mao, TensorBeat: tensor decomposition for monitoring multi-person breathing beats with commodity WiFi, in: ACM Transactions on Intelligent Systems and Technology, 2017.

[31] X. Wang, et al., Smartphone sonar based contact-free respiration rate monitoring, ACM Transactions on Computing for Healthcare 2 (2) (2021) 15, https://doi.org/10.1145/3436822.

[32] Xuyu Wang, Runze Huang, Shiwen Mao, SonarBeat: sonar phase for breathing beat monitoring with smartphones, in: 2017 26th International Conference on Computer Communication and Networks (ICCCN), IEEE, 2017, pp. 1–8.

[33] Xuyu Wang, Chao Yang, Shiwen Mao, On CSI-based vital sign monitoring using commodity WiFi, ACM Transactions on Computing for Healthcare 1 (3) (2020) 1–27.

[34] Xuyu Wang, Chao Yang, Shiwen Mao, Resilient respiration rate monitoring with realtime bimodal CSI data, IEEE Sensors Journal (2020).

[35] Teng Wei, Xinyu Zhang, mTrack: high-precision passive tracking using millimeter wave radios, in: Proc. ACM Mobicom' 15, ACM, Paris, France, 2015, pp. 117–129.

[36] Xiangyu Xu, et al., Breathlistener: fine-grained breathing monitoring in driving environments utilizing acoustic signals, in: Proceedings of the 17th Annual International Conference on Mobile Systems, Applications, and Services, 2019, pp. 54–66.

[37] Cheng Yang, Gene Cheung, Vladimir Stankovic, Estimating heart rate and rhythm via 3D motion tracking in depth video, in: IEEE Transactions on Multimedia, 2017.

[38] Z. Yang, et al., Monitoring vital signs using millimeter wave, in: Proc. IEEE MobiHoc' 16, Paderborn, Germany, 2016, pp. 211–220.

[39] Sangki Yun, Yi-Chao Chen, Lili Qiu, Turning a mobile device into a mouse in the air, in: Proc. ACM MobiSys' 15, ACM, 2015, pp. 15–29.

[40] Bing Zhou, et al., BatMapper: acoustic sensing based indoor floor plan construction using smartphones, in: Proceedings of the 15th Annual International Conference on Mobile Systems, Applications, and Services, ACM, 2017, pp. 42–55.

Chapter 14

RF and camera-based vital signs monitoring applications

Li Zhang[a], Changhong Fu[a], Changzhi Li[b], and Hong Hong[a]

[a]*Nanjing University of Science and Technology, Nanjing, China,* [b]*Texas Tech University, Lubbock, TX, United States*

Contents

14.1 The pros and cons of RF and camera sensors

14.1.1 RF sensor

Measurement of human physiological signs based radar has been explored with three types of electromagnetic radar systems; continuous-wave, frequency-modulated (FM), and ultra-wide band (UWB). Many studies have focused on using Doppler radar at various frequencies, output powers, and distances.

Several pros are associated with electromagnetic radar-based methods, which can be listed as follows:

Reliable results with stationary objects have been achieved. Even if there is some non-metallic material, such as wood, clothing, glass, or water, between the radar and the subjects being measured, the heart and lung signals can be extracted. There is a possibility to extract the cardiorespiratory signal from medium and long distances (up to 30 m) [3,4].

While RF-based sensors have several pros over current vital signs monitoring, there are limitations, including:

When the movement of the surface of interest is very small (small Doppler movements) or superimposed on other movements (such as the subject's head and arms), the radar-based method is prone to motion artifacts and noise. Radar-based methods restrict subjects' movement and need an exposed Region of Interest (ROI) to analyse, which is inappropriate for long-term monitoring [12]. At distances of greater than one m, radar-based methods are prone to degradation due to the motion artifacts and noise caused by the increased free-space loss. For example, SNR decreases to about 50% as the distance of the subject from the electromagnetic radar sensor increases from 0.25 m to 2.5 m [21]. Detecting the vital signs of multiple people at the same time is prone to interference with radar signals. It is more difficult to obtain vital sign data from people with high body fat. While studies of slim people have produced usable data, detection of data from overweight people is more challenging [14].

Based on the limitations shown, radar would likely need to be supplemented with other technologies for a complete remote non-contact vital signs monitoring system.

14.1.2 Video sensors

Besides the radar-based approaches already mentioned, contact-free monitoring of the pulse rate using videos of human faces, also named remote-photoplethysmography (rPPG) [5,17,23–25,29], is another user-friendly option.

Compared with the conventional cardiac monitoring by eletrocardiography (ECG) [6], photoplethysmography (PPG) [2,10], the common advantage of contact-free approaches (e.g., radar and video) is that they are desirable in cases where contact is not preferred, e.g. sensitive skin of neonates, skin-damage, discomfort, surveillance, and fitness activities.

In the initial stage, the rPPG methods [17,23] are also prone to the motion-induced noise, as the radar-based approaches. Later, this main concern was addressed from various perspectives [5,24,25,29]. Among them, [5,24] projected the observed RGB signal containing pulse signal to a plane that was orthogonal to the standardized skin-tone vector or intensity variation direction. As a result, the influence of motion can be minimized in the extraction of a pulse signal. For cases containing strong periodic motion, e.g., fitness videos, the motion-induced components sometimes dominate the spectrum of the observed signal and make it difficult to extract the pulse rate. The separate motion signal is extracted from the video and then eliminated from the observed RGB signals [25,29]. These approaches increase the robustness of the video-based rPPG methods in vital signs monitoring.

The cons of video-based rPPG approaches are mainly due to the illumination conditions. As demonstrated in [20], the vital signs detection by camera is not reliable enough when the illumination level of the environment is low (under 100 lx). Besides, privacy protection is another concern for the rPPG methods. People usually worry about the privacy leakage when they are monitored by camera. This prevents the rPPG methods from being practical for special applications, e.g., when the scenario is in a private space such as the bedroom and bathroom.

14.1.3 Complementarity of radar and camera

Based on the previous content, we can find that radars and cameras have been recently developed for non-contact vital sign detection, due to their numerous advantages. Both technologies are showing promising performance in non-contact remote vital signs detection. However, the various approaches have their own strengths and weaknesses in different environments. Video-based methods are very sensitive to light conditions, whereas RF-based methods are easily influenced by the subject's body movements. This makes it possible to exploit their complementarity to improve the measurement accuracy of the detection of vital signs signals.

Prof. Fathy provided a comparison between these two techniques, shedding some light on their capabilities and limitations [7,20].

14.1.3.1 Luminance robustness of radar

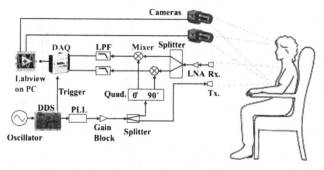

FIGURE 14.1 Block diagram of the experimental setup [20].

As shown in Fig. 14.1, a Step Frequency Continuous Wave (SFCW) radar and cameras were used to detect the vital sign signal from one subject simultaneously.

When a subject is 1 m away from the experimental setup at the position "front" and the illumination level is 500 lx, the demodulated signals spectrums from both radar and optical sensor are compared with the results of the corresponding reference sensor, as shown in Fig. 14.2. The respiratory and heart reference rates of a subject were 0.23 and 1.35 Hz, respectively, during the experiment. These differences are less than 1% using either technique in vital signs

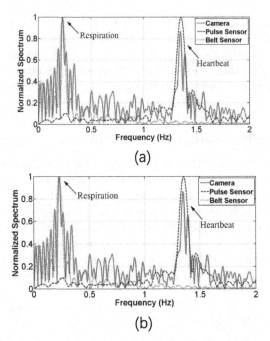

FIGURE 14.2 Spectra of demodulated signals from (a) SFCW radar and (b) camera with 520-nm filter when illumination level is 500 lx [20].

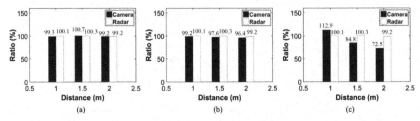

FIGURE 14.3 Comparison of ratio to reference using a SFCW radar and camera in RR detection when illumination level is (a) 500, (b) 200, and (c) 100 lx [20].

detection. Both radar and camera demonstrate high accuracies in respiratory rate (RR) and heartbeat rate (HR) monitoring under normal conditions. But the figure also shows that the spectrum of the heartbeat signal that gives the radar signal is very weak, and is often masked by the third harmonic of the breath signal.

However, vital signs detected by camera in low illumination level are not reliable. It is worthwhile mentioning that radar is not influenced by illumination conditions as demonstrated in Figs. 14.3 and 14.4. RR and HR are readily monitored with a camera when the lighting level is above 200 lx; when the lighting level drops to 100 lx, accuracies for RRs significantly degrade with the increase

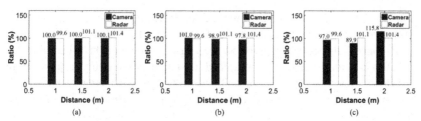

FIGURE 14.4 Comparison of ratio to reference using a SFCW radar and camera in HR detection when illumination level is (a) 500, (b) 200, and (c) 100 lx [20].

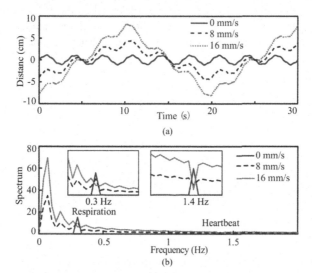

FIGURE 14.5 Impacts to the spectra of vital sign signals due to RBMs with various velocities. (a) Time-domain signals. (b) Spectra [16].

of subject distance. As shown in Fig. 14.4(c), the ratio of RR measurement to reference degrades from 97% at 1 m, to 89.93% at 1.5 m and 115.81% at 2 m.

14.1.3.2 Motion robustness of camera

In real life, random body movement will significantly interfere with vital signs detection based on Doppler radar. Unfortunately, this noise is difficult to remove due to its unpredictability [8]. Fig. 14.5(a) shows the calculated "random body movement (RBM)s" with 0-, 8-, and 16-mm/s velocities, respectively. The heartbeat motion is parasitic on the respiration motion. Fig. 14.5(b) shows the corresponding spectra. The insets of Fig. 14.5(b) show the enlarged regions close to the respiration and heartbeat frequencies. In the right inset, it is seen that when the velocity is increased to eight mm/s, the heartbeat spectrum disappears. Due to the possible phase destruction between the heartbeat spectrum and the local RBM spectra around the heartbeat frequency, the heartbeat spectrum can be even lower than the RBM noise floor. When the velocity is increased to 16 mm/s, the respiration spectrum is also concealed, as seen in the left inset. Such concealed spectra cannot be recovered by conventional high-pass filters.

On the contrary, the motion-induced noise could be eliminated properly in the camera-based approaches. It can be noted that the pulse component is a relatively micro signal in the camera-based approach. As shown in Fig. 14.6, the pulse signal varies within a tiny range −0.25–+0.25, while the whole range of the observed signal is 0–255 in Green channel. For this example, even the subject's face is fixed by a holder to minimize the motion; the unconscious adjustment is still significant and could be observed as the trend (black line in Fig. 14.6). Obviously, the desired pulse rate will be dominated by the motion-induced noise when there is any body movement during the monitoring.

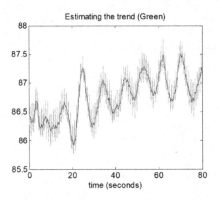

FIGURE 14.6 The observed G-channel signal of face in the static case.

Fortunately, the motion-induced noise could be properly eliminated by some prior knowledge as already mentioned. If the motion noise is aperiodic, it could be reduced by signal processing techniques such as detrending [17]. If the motion noise is periodic, e.g., in fitness, it will dominate the frequency spectrum and makes it even more difficult to extract the pulse signal. In this case, the separate motion signal of the face could be extracted from the video itself as side information. For example, as shown in Fig. 14.7, the frequency traces due to motion are mixed together with the trace due to pulse in the time-frequency map of the observed signal. By modeling the x- and y-coordinate of face region in video, we could estimate the motion signal from the video separately and notched the frequency traces due to motion from the time-frequency map [29]. Similarly, in [25], both the observed RGB signal and motion signal are decomposed by Singular Spectrum Analysis. The overlapping subbands are removed, and the remaining components are reconstructed to form the motion eliminated RGB signal for pulse extraction. As shown in Fig. 14.7 and Fig. 14.8, the estimated heart rate is quite close to the ground truth measured by a Polar H7 chest belt in a long-duration fitness session on a treadmill. In this way, the side in-

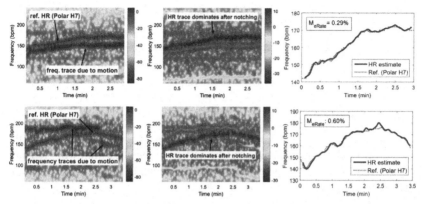

FIGURE 14.7 Comparison of spectrograms before (column 1) and after (column 2), notching the frequencies of fitness motions. Column 3: Heart-rate estimates when compared to the ECG-based reference measurement using a Polar H7 chest belt. Top row: from elliptical machine, URL: http://goo.gl/e0WCnc; bottom row: from treadmill, URL: http://goo.gl/8GLoLB [29].

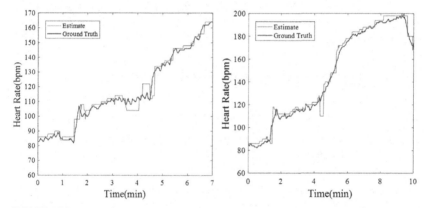

FIGURE 14.8 Heart-rate monitoring of subjects in long-duration strenuous exercises on treadmill. The ground truth is measured by the ECG-based Polar H7 chest belt [25].

formation (motion signal extracted from the video itself) helps to increase the motion robustness of camera-based approach.

14.2 A hybrid radar–camera sensing system with phase compensation for random-body movement cancellation (RBMC) Doppler vitals sign detection

In accordance with Sect. 14.1, we know that a reliable wireless vital signs detection in the presence of random human body movements remains a technical challenge. Since the radar operation is based on micro-Doppler effects through the sensing of tiny physiological movements in the range of mm or cm, RBM, which presents a displacement comparable tom or larger than the chest wall displacement due to breathing and heartbeat, is a substantial noise source that can submerge and destroy the signal of interest and significantly degrade the accuracy of detection and measurement. To achieve reliable and accurate non-contact vital signs detection, the noises caused by RBM should be effectively reduced.

14.2.1 Theory

In non-contact vital sign detection, the radar transmits a single-tone signal to the subject. Both the RBM and the periodic motion of the chest wall will modulate the single-tone signal in the phase. The reflected signal from the chest wall is received.

$$R(t) = A \cdot \cos\left(\omega t + \frac{4\pi x_V(t)}{\lambda} + \frac{4\pi x_B(t)}{\lambda} + \Delta\theta\right), \qquad (14.1)$$

where A is the amplitude, ω is the carrier frequency, λ is the wavelength, $X_v(t)$ is the vital sign signal, $X_b(t)$ is the RBM, $\Delta\theta$, and is the residual phase. From (14.1), we can see that all we need to do is remove the bulk momentum component from the resulting echo signal.

14.2.2 Three-phase compensation strategies for RBMC

Take the example of a homodyne architecture radar, as shown in Fig. 14.9. There are three kinds of method to perform random body-movement cancellation:

1. Phase Compensation at RF Front-End with a phase shifter.
2. Phase Compensation for Complex Signals.
3. Phase Compensation After Demodulation.

There are advantages and disadvantages associated with each approach, and we will now analyze each in detail.

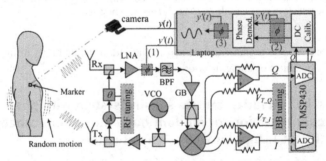

FIGURE 14.9 Block diagram of the proposed 2.4 GHz radar sensor system with RBMC. The camera measures the RBM that is fed to the radar system for motion compensation: phase compensation at RF (1) relieves the system's linearity burden, while baseband compensations (2)–(3) allow accurate fine-tuning [8,9].

14.2.2.1 Phase compensation at RF front-end

The first RBMC strategy is to compensate the phase information that is caused by RBM at the RF front-end, while retaining the phase information of the vital sign signals. This is realized by using a phase shifter at the RF front-end. It is seen from (14.1) that the phase modulation is due to RBM, which can be reduced if a phase shift is added to the carrier signal. The time-variant phase information can be derived to compensate for the radar-measured signal through the phase shifter to reduce the RBM, as shown in Fig. 14.9. Therefore, the signal that appears at the input of the mixer is

$$R'(t) = A_R \cdot \text{Re} \left\{ \exp j \left(\omega t + \frac{4\pi x_V(t)}{\lambda} + \frac{4\pi x_B(t)}{\lambda} + \Delta\theta \right) \right. $$
$$\left. \cdot \exp j \left(-\frac{4\pi x_S(t)}{\lambda} \right) \right\} \tag{14.2}$$
$$= A_R \cdot \cos \left(\omega t + \frac{4\pi x_V(t)}{\lambda} + \Delta x(t) + \Delta\theta \right),$$

where A_R is the amplitude considering the total receiver gain, x_S is Body movement components measured by other sensors, and $\Delta x(t)$ is the residual error remaining between the actual RBM and the sensor-measured RBM since the sensor-based body-movement detection has respective detection errors. Though that approach cannot completely remove the RBM, it significantly reduces the RBM and helps relieve the RF front-end from abrupt fluctuations that are caused by the RBM, which may easily saturate the receiver chain. Advanced signal processing in the baseband can be used to further improve the SNR.

14.2.2.2 Phase compensation for complex signals

If the RBM is not large enough to saturate the baseband amplifier, the phase compensation for RBMC can be realized at baseband in the digital domain. After quadrature down-conversion, the baseband signals are

$$B(t)_I = A_I \cos \left[\frac{4\pi x_V(t)}{\lambda} + \frac{4\pi x_B(t)}{\lambda} + \Delta\theta \right] + DC_I$$
$$B(t)_Q = A_Q \sin \left[\frac{4\pi x_V(t)}{\lambda} + \frac{4\pi x_B(t)}{\lambda} + \Delta\theta \right] + DC_Q, \tag{14.3}$$

where A_I, A_Q are the amplitudes, $\Delta\theta$ is the residual phase noise considering the initial position of the subject and the noise from the circuit, and DC_I, DC_Q are the DC offsets of the channels, respectively. After DC calibration removing the DC offset but retaining the DC information, the two channels can be combined to form a complex signal

$$C(t) = I + j \cdot Q = A_C \cdot \exp j \cdot \left(\frac{4\pi x_V(t)}{\lambda} + \frac{4\pi x_B(t)}{\lambda} + \Delta\theta \right), \tag{14.4}$$

where A_C is the amplitude. A similar phase compensation process as (14.2) can be employed at the baseband output to reduce the RBM. The large RBM means large phase modulation, which leads to long trajectory that may cover more than one quadrant of the constellation, resulting in phase discontinuity in the demodulation. The reduction of the large RBM before the demodulation leads to shorter trajectory, which may eliminate the necessity for phase unwrapping in the demodulation.

14.2.2.3 Phase compensation after demodulation

The third RBMC strategy is realized after the phase demodulation. It is the most straightforward approach that does not need any change to the framework of the existing radar systems. It compensates the phase information by directly adding the opposite phase information of RBM, as measured by a camera to the demodulated signal, which is

$$R_D(t) = \tan^{-1}\left(\frac{I}{Q}\right) = \frac{4\pi x_V(t)}{\lambda} + \frac{4\pi x_B(t)}{\lambda} + \Delta\theta. \qquad (14.5)$$

It is easy to implement without any change to the radar architecture or demodulation approaches. However, the downside of this strategy is that the large trajectory, which is caused by large RBM, may likely result in phase discontinuity in the demodulation. In this case, either the phase unwrapping or the DACM demodulation should be applied to compensate for the phase discontinuity.

14.2.3 Camera-based body-movement detection method

Small-amplitude motion in different ROIs captured by a video camera may also contain useful information about the body movement. Many researchers have analyzed video and image sequences to detect these motions with a dedicated illumination source or ambient light.

It should be noted that the actual amplitude of the RBM does not have to be known. The RBMC is realized by linearly scaling the camera-measured motion pattern in a range of values to find the optimum point that leads to the maximum SNR of respiration. Though the proposed algorithm may not completely extract the body movement, it reduces the computational load and provides a rough motion-pattern estimation, which helps to reduce the RBM to the extent that the vital sign signals can be identified.

14.2.3.1 Marker-tracking method

The simplest detection method is to measure by tracking a white point on a small piece of black paper attached to the shoulder of the subject, as shown in Fig. 14.10. The pixel difference of the point indicates the motion pattern of RBM. This method is simple and has good performance. However, it requires tagging the subject, which limits the environment in which this method can be used [9].

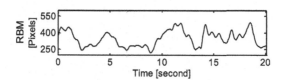

FIGURE 14.10 The marker tracking result [9].

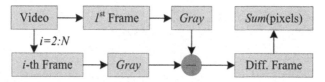

FIGURE 14.11 Flow chart of the simple algorithm for motion extraction [8].

14.2.3.2 Pixel-difference method

The flow chart of the algorithm is shown in Fig. 14.11. A camera-measured video has a resolution of 1900 × 1080 pixels and a frame rate of 30 frames/sec. If the subject is moving back and forth in front of the camera, the number of pixels occupied by the subject in the video is also changing according to the body motion. At the beginning of motion extraction, the first frame of the video is extracted, converted to gray image, and saved as a reference. The subsequent frames are also converted into gray images and then subtracted from the reference frame to generate the difference frames. The stationary objects in the video result in a black background (pixel value of zero) in the difference frame, while the moving subject leads to pixel values larger than zero. The summation of the pixels' values reflects the pattern of the subject movement.

This method removes the limitations of needing to tag a marker to the subject. But due to the limited resolution of the proposed algorithm and the intention of the subject to perform less RBM. And advanced video processing algorithms will extract more movement information, and thus will allow more complicated RBM to be compensated in radar vital sign detection [8].

14.2.3.3 Optical flow method

The optical flow method is another method that can extract the motion of an object, and it has the following advantages over the previous methods: It is a markerless motion-extract method, and it can approximate the playground that cannot be directly obtained from the sequence image and judge the motion of objects according to the size and direction of the optical flow (see Figs. 14.12 and 14.13) [27].

Optical flow can be regarded as the instantaneous velocity field generated by the motion of pixel points with a gray level in the image plane. It is a 2-D vector field formed by the motion of the target between two consecutive frames of images.

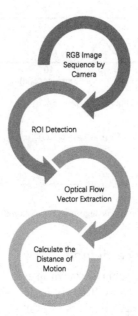

FIGURE 14.12 The flowchart of radar body-motion compensation algorithm based on the optical flow method [27].

The establishment of optical flow method has several significant assumptions: The brightness of the target pixel in the image does not change in two consecutive frames; the time between adjacent frames is short enough; adjacent pixels have similar motions.

According to these assumptions, the brightness mode can be expressed as:

$$I(x, y, t) = i(x + \Delta x, y + \Delta y, t + \Delta t). \tag{14.6}$$

The right side of the equation is expanded by first-order Taylor series. Let I_x, I_y, I_t represent the brightness component of the space-time image, $u = \frac{dx}{dt}$, $v = \frac{dy}{dt}$, which is the component of the light flow at this point in the xy direction, and we can get:

$$I_x u + I_y v = -I_t. \tag{14.7}$$

According to the last hypothesis and (14.7), we solve the optical vector equation of u and v. Then, we can establish the formula as follows:

$$\begin{aligned} I_x(p_1) u + I_y(p_1) v &= -I_t(p_1), \\ I_x(p_2) u + I_y(p_2) v &= -I_t(p_2), \\ I_x(p_n) u + I_y(p_n) v &= -I_t(p_n), \end{aligned} \tag{14.8}$$

p_n is any pixel in the region of interested (ROI). And a Lucas–Kanada Method was used to solve this equation. These equations can be written in matrix form

$AV = b$, where

$$A = \begin{bmatrix} I_x(p_1) & I_x(p_1) \\ \vdots & \vdots \\ I_x(p_n) & I_x(p_n) \end{bmatrix}, \quad V = \begin{bmatrix} u \\ v \end{bmatrix}, \quad b = \begin{bmatrix} -I_t(p_1) \\ \vdots \\ -I_t(p_n) \end{bmatrix}. \quad (14.9)$$

The Lucas–Kanade method obtains a compromise solution by the least squares principle. Namely, it solves the 2×2 system:

$$V = \left(A^T A \right)^{-1} A^T b \quad (14.10)$$

where A^T is the transpose of matrix A

$$\begin{matrix} u \\ v \end{matrix} = \begin{bmatrix} \sum I_x(p_i)^2 & \sum I_x(p_i)I_y(p_i) \\ \sum I_y(p_i)I_x(p_i) & \sum I_y(p_i)^2 \end{bmatrix}^{-1} \begin{bmatrix} \sum I_x(p_i)I_y(p_i) \\ -\sum I_x(p_i)I_y(p_i) \end{bmatrix}. \quad (14.11)$$

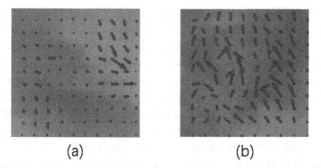

(a) (b)

FIGURE 14.13 The optical flow graphs of (a) non-motion (b) with body motion.

From Fig. 14.13 we can find that, as the object moves, the vector of optical flow changes. To simplify the calculations, we chose to use four corner pixels to calculate the Euclidean distance between them to represent the change in the distance between the human body and the camera. The equation is as follows:

$$dis = \sum_{i=1}^{3} \sum_{j>i}^{4} \sqrt{(P_i - P_j)^2}. \quad (14.12)$$

Fig. 14.14 shows the result containing 17 seconds of body movement detected by the camera. With the optical flow method, we can accurately detect body-movement and label it within the dashed lines.

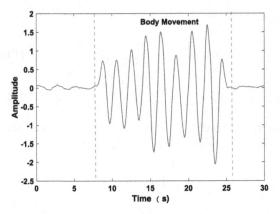

FIGURE 14.14 The optical flow based body-movement detection.

14.2.4 Results

Fig. 14.15 show the 20 sec time-domain signals with body movement and the result of the signal after high-pass filter and with RMBC. A chest belt and a pulse sensor were used as the reference signals for respiration and heartbeat. Since most body movements are slow, a straightforward method to remove RBM may use a high-pass filter. To compare with the proposed RBMC technique, a fourth-order Butterworth high-pass filter with a corner frequency of 0.25 Hz is used. From this picture, strong near-DC spectral components are observed as shown in Fig. 14.15(a). The heartbeat is invisible in the spectrum, and the time-domain signal has fluctuations due to the RBM. After applying the high-pass filter, it can be seen in Fig. 14.15(b) that the near-DC spectral components are significantly suppressed, and the heartbeat signal can be identified. Fig. 14.15(c) shows the results with RBMC. It is apparent that not only the near-DC interferences are suppressed but also the second harmonic of respiration becomes visible, while it is hard to be identified in Fig. 14.15(b). This is because the measured RBM contains more spectral information, and it can help to remove more spectral interferences. It is also seen that the time-domain signal using the proposed RBMC technique has a more uniform pattern than that using the high-pass filter. In both Figs. 14.15(b) and (c), the radar measured respiration and heartbeat match well with the reference signals from the pulse sensor and the chest belt due to the limited resolution of the proposed algorithm and the intention of the subject to perform less RBM in this work. Due to the limited resolution of the proposed algorithm and the intention of the subject to perform less RBM in this work. It should be noted that more advanced video processing algorithms will extract more movement information, and thus will enable more complicated RBMs to be compensated for in radar vital sign detection.

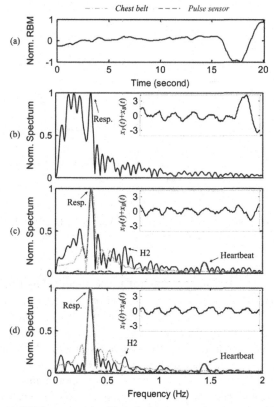

FIGURE 14.15 (a) RBM measured by iPhone5; (b) the raw signal measured by radar without RBMC; (c) signal after high-pass filtering; (d) signal with RBMC. The insets show the corresponding time-domain signals with duration of 20 sec [8].

14.3 Non-contact dual-modality emotion recognition system by CW radar and RGB camera

14.3.1 Introduction

Emotion is the comprehensive state of a person's opinion about whether objective things meet his or her needs. As an advanced function of the human brain, emotions are also a key factor in personality traits and psychopathology and play an important role in everyday life. Emotions are the psychological and physiological states that accompany cognitive and conscious processes and play a very important role in human communication. The study of emotions has a long history, and in recent years, with the development of perception technology and human-computer interaction technology, affective computing (AC) has gradually become an emerging field of emotion research.

Emotions are not only internal experiences withexternal behavioral manifestations, but they are also accompanied by complex neurological processes and

FIGURE 14.16 A graphical representation of the circumplex model of affect with the horizontal axis representing the valence dimension and the vertical axis representing the arousal or activation dimension [19].

physiological changes. Emotions can usually be identified by facial expressions [26], voice tones [26], body posture [11] and other outward features, but these features are easily falsified or concealed, and it is hard to know the true inner emotional state. However, the physiological responses accompanying emotions are governed by the neurological and endocrine systems, which are spontaneous and not easily controlled by subjective thoughts. Fig. 14.16 is Russell's circumplex model of emotion, whose subdivision of emotion into two dimensions, valence and arousal, has provided the theoretical basis for emotion recognition using physiological signals.

Fig. 14.17 shows examples of the physiological signals measured for a subject who is intentionally expressing anger and grief. There are four kinds of physiological signal: electromyogram (EEG), blood volume pressure (BVP), skin conductivity (SC), and respiration, each shown here for two emotions with visible differences. Therefore, it is evident that emotion recognition, based on the corresponding physiological signals, can obtain objective and realistic results and is also more suitable for practical applications.

Traditional physiological signals such as respiration and heartbeat are acquired by contact, which can cause discomfort to the subject and increase the measurement error, while non-contact monitoring can overcome these limitations. Non-contact vital signs monitoring includes infrared [1], video [22], radar [15], among others. Of these, radar-based and video-based sensors are now the two main ways of acquiring non-contact vital signs. Bio-Radar uses electromagnetic waves to monitor vital signs. Continuous-wave radar has the advantages of being less susceptible to environmental factors and having high penetrating power. By applying bio-radar technology, the physiological activities such as heartbeat and respiration can be monitored in a non-contact manner through clothing, bedding, and other objects. The video-based non-contact sensing is

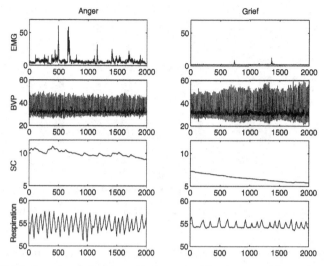

FIGURE 14.17 Examples of physiology signals measured from a user under anger and grief [18].

another popular non-contact detection solution which is based on the principle of the photoelectric volumetric pulse wave-tracing method. It requires only a device that can shoot video without direct contact with human skin; it does so through detection and tracking of the face, waveform detrending, independent component analysis, heart-rate extraction, and other algorithms to obtain the heart rate.

14.3.2 Physiological signal-optimization algorithms based on radar and video sensors

The two sensors, however, possess their own limitations in practical use. First, traditional video-based studies of emotion recognition rely on the quality of the video and are susceptible to the influence of various environmental factors. Secondly, the accuracy of the physiological signal extracted from the video will be affected by factors such as the lighting conditions, and in poor lighting conditions, the accuracy of the physiological signal measured by the video is low; at the same time, when conducting emotion-recognition experiments, the experimenter may unconsciously produce body shaking normally due to a subject's emotional fluctuations, and such body movements will have a great impact on the radar waveform.

Based on this, we can use the content of Sect. 14.2.3 to use video signals to correct the radar signal in the body-motion segment, and to automatically switch between radar and video as a source of physiological signals by determining the light intensity to improve the robustness and accuracy of physiological signals.

In a laboratory environment, four volunteers were invited to demonstrate the feasibility of the optimization algorithm and the general adaptation of the

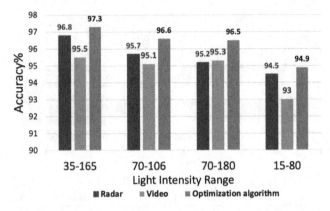

FIGURE 14.18 Results of heartbeat optimization algorithm based on light intensity.

whole system by comparing the heart-rate measurement error before and after the application of the optimization algorithm by multiple people under different light conditions. As shown in Fig. 14.18, the heart-rate measurement error rate is reduced and the accuracy of the measurement is improved after the optimization algorithm treatment in different light-intensity ranges. Thus, the dynamic feedback heartbeat-signal optimization algorithm based on light intensity can effectively improve the accuracy of heart-rate measurement and eliminate the influence of light intensity on heartbeat signal detection. This also fully reflects the advantages of the system, which is more accurate than the single-mode system, can adapt to different environments, and has stronger generalizability and robustness.

14.3.3 Multi-sensor data fusion

Multi-sensor data fusion deals with information that has more complex data types and processes than a single sensor to take advantage of the complementary nature of the various models to improve classification results. In general, approaches for modality fusion can be classified into two broad categories, namely, feature fusion (or early integration) and decision fusion (or late integration) [13].

14.3.3.1 Feature-level fusion

The common feature-level fusion block diagram is shown below as Fig. 14.19, where the signals from multiple sensors are extracted separately, and the features of each target are fused and sent to the classifier to obtain the final classification result. Feature-level fusion is an intermediate level of fusion. The feature information is extracted from the raw data from each sensor, the feature information is formed into feature vectors, and the grouped information is fused before the target is classified. Feature-level fusion compresses the raw data information to reduce the impact of large data volumes. It improves the real-time data processing and achieves a high level of accuracy. However, feature-level fusion requires that the feature vector must be divided into several different combinations before fusion. Feature-level fusion is mainly used for multi-sensor target tracking and identification, estimation of relevant parameters, etc. The common feature-level fusion block diagram is shown below.

Four types of classifiers can be used: Decision Trees, Support Vector Machines, K-nearest Neighbor Classifiers, and Ensemble Classifiers.

1. Decision Trees.
   A decision tree is a decision-support tool and is one of the most common machine learning methods. The purpose of a decision tree is to build a predictive model, which uses a tree structure or decision model. Consisting of a root node, multiple branch nodes, and multiple leaf nodes, the root node of the decision tree is the starting point of the decision tree and also the particular branch node, the most informative feature attribute. The branch node corresponds to the feature attributes of the extracted sample, which determines which branch the input present data goes to next. The leaf node corresponds to the final output classification result.

2. Support Vector Machines (SVM).
   A support vector machine is a supervised learning model whose basic concept is based on the training set. The set is used to find a "maximum separation" in the sample space of the hyperplane's division, and so separate the two types of samples.
   If there is a hyperplane that can be completely separated from the training samples, then the two types of samples are said to be linearly separable. Then one needs to find a sample local perturbations and noise "tolerance" of the best hyperplane, and this time the classifier is the most robust. In some cases, however, a hyperplane that accurately separates the two types of training samples completely may not exist. If the original space is finite dimensional, there must be a high-dimensional feature space to make the samples separable. Support vector machines use kernel methods in solving nonlinear samples for efficient classification. Commonly used kernel functions are: linear, polynomial, Gaussian, Laplace, and Sigmoid.

3. K-Nearest Neighbor Classifiers (KNN).
   The basic principle of the algorithm is: Given a test sample, based on some distance measure, find the k nearest training samples in the feature space, and the category to which most of these k samples belong is the category to which this sample belongs. The classifier makes predictions based on the information of the k neighbors and selects the category markers that appear most frequently in these k samples as the classification result. Assigning the average of the attribute attributes of these neighboring samples to the current sample yields the attribute for that sample.

4. Ensemble Classifiers.
   Ensemble classifiers, also known as multi-classifier systems, combine multiple learners in machine learning to perform learning tasks and can achieve

better predictive performance than a single learner. While a machine learning collection contains only a limited set of optional models, the structure of the combinations among these models is flexible and variable. Ensemble learners are typically built with a structure that generates multiple individual learners and then combines them by algorithmic strategies to meet specific performance requirements. Homogeneous integration means that there are only a few individual learners of the same type in an integration, which is called a "base learner". Heterogeneous integration means that there are different types of individual learners in the integration, where the individual learners are called "component learners".

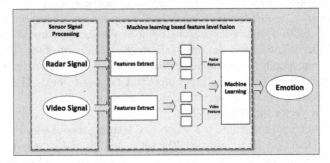

FIGURE 14.19 The structure of feature-level fusion.

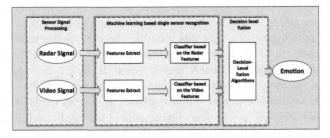

FIGURE 14.20 The structure of decision-level fusion.

14.3.3.2 Decision-level fusion

As Fig. 14.20 shows, decision-level fusion functions so that, before any fusion, each sensor processes the target data, obtains independent target-identification estimates, and realizes the detection and classification of the target. The classification results of each sensor are fused and processed according to certain decision criteria, and finally the classification results of the whole system are obtained. A decision fusion-based system can be constructed by using existing unimodal classification systems and enables us to model asynchronous characteristics of the modalities flexibly.

Methods commonly used for decision-level fusion include Bayesian estimation, artificial neural network methods, fuzzy set theory, expert systems, reliability theory, and D-S evidence inference.

The artificial neural network method is based on the structure of the human brain's neural processing of information, and builds a corresponding model from the perspective of the correlation of information processing to form multiple networks according to the different connections between nodes. The D-S evidence-inference method uses a fusion classification algorithm based on statistical methods, and, with the help of the combination rules of D-S evidence inference, the basic credibility of different evidence bodies is determined according to the correlation between sensors. Then the redundancy is merged to ultimately produce an overall plausibility distribution. Bayesian estimation is a statistical fusion algorithm based on Bayes' theorem that solves some of the unsolvable problems of classical probabilistic inference and differs from classical inference in that it can be used in the case of multiple hypotheses. The basic principle is that a measurement is added after a previous likelihood estimate is given, i.e., the likelihood function can be transformed to update the a priori density of a given hypothesis to the true a posteriori density. The plain Bayesian algorithm is the simplest of a class of algorithms applied for Bayesian estimation and is based on Bayes' theorem and feature condition independence. It has the advantages of being a simpler algorithm, requiring fewer parameters, allowing for missing data, and ensuring stable classification accuracy and efficiency.

14.3.4 Features extraction

Feature extraction is an important step in the emotion recognition algorithm; to recognize emotions based on physiological signals, representative features need to be selected that can reflect the differences in physiological signals in different emotional states. Therefore, corresponding features need to be extracted based on the characteristics of different physiological signals.

The human body's breathing and heartbeat will change accordingly under different emotional states, resulting in different breathing depths, respiratory rates, and heart rates. It is found that the respiratory depth and heart rate are greater in the happy state than in the calm state, and the respiratory rate and heart rate are faster in the fear state than in the calm state. So, we will study the radar echo characteristics of calm, happy, sad, and fear states to extract effective and representative physiological signal characteristics.

14.3.4.1 Respiratory features extraction

Respiratory features were employed to extract the characteristics. Some of them reflect some of the characteristics of the waveform in the time domain and the literature [18], such as the mean, standard deviation, mean of the absolute values of the first difference, and mean of the absolute values of the second difference of raw signal. In the frequency domain, we used respiratory rate and the power spectral density to represent the respiratory frequency information. As shown in [18], we studied four features that represent the average energy in each of the first four 0.1-Hz bands of the power spectral density range 0.0–0.4 Hz. Some features are nonlinear one including approximate entropy, detrended fluctuation analysis (DFA), and Poincaré plots. These features refer to the estimation and characterization of the phase space (or state space) of the cardiovascular system.

14.3.4.2   Heartbeat features extraction

There is a large literature on extracting emotion-dependent features from human heartbeats [28]. In addition to the features used in the breath signal previously addressed, which we also can extract from the heartbeat raw signal, we extracted some features such as Skewness, Kurtiosis, RMSSD, pNN50, and other common features from heart-rate variability (HRV) which was measured by the variation in the time intervals between heartbeats.

Among the features, 24 of them are extracted from the respiratory signal, and 39 of them are from the heartbeat signal.

14.3.4.3   Experiments and results

A total of nine healthy volunteers participated in the emotion-elicitation experiment, including six male and two female volunteers, all aged 24 ± 3 years. A total of 626 packets of valid data were obtained, and each packet included the physiological signals from three channels: radar breathing, radar heartbeat, and video heartbeat; the corresponding features were extracted from the breathing and heartbeat signals of each packet, and the corresponding artificial category labels 1–4 were added, corresponding to happy, relax, sad and fear four emotional states. To exclude the effects of individual differences and time, we normalized the characteristics of the other emotional states by subtracting the mean of each individual's characteristics when they were neutral. The volunteers sat in a chair facing the digital IF Doppler radar (2.4 GHz, 50 µW transmitting power) and an iPhone 6s (Sony Exmor RS IMX315 sensor, 12MP, 1.22 µm pixel size, f/2.2 aperture), with the videos recoded in a H.264 MOV format (1920 × 1080 pixels, 8-bit depth, 30 fps). The radar was set up one m away with the antenna facing the chest of volunteers. The iPhone was set up 1.3 mr away. The iPhone and the desktop computer to process the radar signals were synchronized to the same internet time to align the time stamps of data measured by the different modalities. In addition, a laptop was placed between the radar and camera to play the videos intended to stir up the subject's emotion.

Fig. 14.21 shows the comparison of single-sensor emotion recognition and dual-sensor optimized result. The results of feature-level fusion are improved relative to both radar and video sensors, with an increase of 9.1% relative to a single-radar sensor and a larger increase of 16.8% relative to the accuracy of a single-video sensor, making the fusion effect obvious. The accuracy of both feature-level fusion and decision-level fusion have been improved relative to the

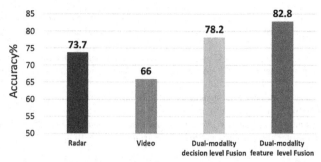

FIGURE 14.21 A comparison of single-sensor emotion recognition and dual-sensor optimized results.

accuracy of individual sensors, which fully illustrates the effectiveness of fusion and further demonstrates the advantages of the dual-sensor system.

References

[1] Abbas K. Abbas, et al., Neonatal non-contact respiratory monitoring based on real-time infrared thermography, Biomedical Engineering Online 10 (1) (2011) 93.

[2] J. Allen, Photoplethysmography and its application in clinical physiological measurement, Physiological Measurement 28 (3) (2007) R1–R39.

[3] Kun-Mu Chen, et al., An X-band microwave life-detection system, IEEE Transactions on Biomedical Engineering 7 (1986) 697–701.

[4] Harikesh Dalal, Ananjan Basu, Mahesh P. Abegaonkar, Remote sensing of vital sign of human body with radio frequency, CSI Transactions on ICT 5 (2) (2017) 161–166.

[5] G. de Haan, V. Jeanne, Robust pulse rate from chrominance-based rPPG, IEEE Transactions on Biomedical Engineering 60 (10) (2013) 2878–2886.

[6] W. Einthoven, Galvanometrische registratie van het menschilijk electrocardiogram, in: Herinneringsbundel Professor SS Rosenstein, 1902, pp. 101–107.

[7] Ramzie A. Fathy, Haofei Wang, Lingyun Ren, Comparison of uwb Doppler radar and camera based photoplethysmography in non-contact multiple heartbeats detection, in: 2016 IEEE Topical Conference on Biomedical Wireless Technologies, Networks, and Sensing Systems (BioWireleSS), IEEE, 2016, pp. 25–28.

[8] Changzhan Gu, et al., A hybrid radar-camera sensing system with phase compensation for random body movement cancellation in Doppler vital sign detection, IEEE Transactions on Microwave Theory and Techniques 61 (12) (2013) 4678–4688.

[9] Changzhan Gu, et al., Doppler radar vital sign detection with random body movement cancellation based on adaptive phase compensation, in: 2013 IEEE MTT-S International Microwave Symposium Digest (MTT), IEEE, 2013, pp. 1–3.

[10] A.B. Hertzman, The blood supply of various skin areas as estimated by the photoelectric plethysmograph, American Journal of Physiology 124 (2) (1938) 328–340.

[11] Yuxin Hou, et al., Soul dancer: emotion-based human action generation, ACM Transactions on Multimedia Computing, Communications, and Applications (TOMM) 15 (3s) (2020) 1–19.

[12] Jure Kranjec, et al., Non-contact heart rate and heart rate variability measurements: a review, Biomedical Signal Processing and Control 13 (2014) 102–112.

[13] Jong-Seok Lee, Cheol Hoon Park, Robust audio-visual speech recognition based on late integration, IEEE Transactions on Multimedia 10 (5) (2008) 767–779.

[14] Mario Leib, et al., Vital signs monitoring with a UWB radar based on a correlation receiver, in: Proceedings of the Fourth European Conference on Antennas and Propagation, IEEE, 2010, pp. 1–5.

[15] Changzhi Li, et al., A review on recent advances in Doppler radar sensors for noncontact healthcare monitoring, IEEE Transactions on Microwave Theory and Techniques 61 (5) (2013) 2046–2060.

[16] Qinyi Lv, et al., Doppler vital signs detection in the presence of large-scale random body movements, IEEE Transactions on Microwave Theory and Techniques 66 (9) (2018) 4261–4270.

[17] D.J. McDuff, M.-Z. Poh, R.W. Picard, Advancements in noncontact, multiparameter physiological measurements using a webcam, IEEE Transactions on Biomedical Engineering 58 (1) (2011) 7–11.

[18] Rosalind W. Picard, Elias Vyzas, Jennifer Healey, Toward machine emotional intelligence: analysis of affective physiological state, IEEE Transactions on Pattern Analysis and Machine Intelligence 23 (10) (2001) 1175–1191.

[19] Jonathan Posner, James A. Russell, Bradley S. Peterson, The circum-plex model of affect: an integrative approach to affective neuroscience, cognitive development, and psychopathology, Development and Psychopathology 17 (3) (2005) 715.

[20] Lingyun Ren, et al., Comparison study of noncontact vital signs detection using a Doppler stepped-frequency continuous-wave radar and camera-based imaging photoplethysmography, IEEE Transactions on Microwave Theory and Techniques 65 (9) (2017) 3519–3529.

[21] Lorenzo Scalise, et al., Non contact monitoring of the respiration activity by electromagnetic sensing, in: 2011 IEEE International Symposium on Medical Measurements and Applications, IEEE, 2011, pp. 418–422.

[22] Gaetano Scebba, Giulia Da Poian, Walter Karlen, Multispectral video fusion for non-contact monitoring of respiratory rate and apnea, IEEE Transactions on Biomedical Engineering (2020).

[23] L.O. Svaasand W. Verkruysse, J.S. Nelson, Remote plethysmographic imaging using ambient light, Optics Express 16 (26) (2008) 21434–21445.

[24] W. Wang, et al., Algorithmic principles of remote PPG, IEEE Transactions on Biomedical Engineering 64 (7) (2017) 1479–1491.

[25] K. Xie, et al., Non-contact heart rate monitoring for intensive exercise based on singular spectrum analysis, in: 2019 IEEE Conference on Multimedia InformationProcessing and Retrieval (MIPR), 2019, pp. 228–233.

[26] Zhihong Zeng, et al., A survey of affect recognition methods: audio, visual, and spontaneous expressions, IEEE Transactions on Pattern Analysis and Machine Intelligence 31 (1) (2008) 39–58.

[27] Hongyu Zhang, et al., Body movement cancellation based on hybrid radar-webcam sensing system, in: 2019 IEEE MTT-S International Microwave Biomedical Conference (IMBioC), vol. 1, IEEE, 2019, p. 13.

[28] Mingmin Zhao, Fadel Adib, Dina Katabi, Emotion recognition using wireless signals, in: Proceedings of the 22nd Annual International Conference on Mobile Computing and Networking, 2016, pp. 95–108.

[29] Q. Zhu, et al., Fitness heart rate measurement using face videos, in: 2017 IEEE International Conference on Image Processing (ICIP), 2017, pp. 2000–2004.

Index

Printed in the United States
by Baker & Taylor Publisher Services